THE UNIVERSAL HISTORY OF
NUMBERS

THE COMPUTER AND THE
INFORMATION REVOLUTION

GEORGES IFRAH

*Translated from the French, and with Notes, by
E. F. Harding*

*assisted by Sophie Wood, Ian Monk,
Elizabeth Clegg and Guido Waldman*

THE HARVILL PRESS
LONDON

First published in France with the title *Histoire universelle des chiffres*
by Editions Robert Laffont, Paris, in 1994

Part One of the original edition first published in Great Britain in 1998
and reissued as a two-volume set in 2000
Part Two of the original edition first published
in Great Britain in 2000 in this volume by
The Harvill Press
2 Aztec Row
Berners Road
London N1 0PW

www.harvill.com

1 3 5 7 9 8 6 4 2

© 1981, 1994 by Editions Robert Laffont S.A., Paris
Translation © 2000 by The Harvill Press Ltd

Georges Ifrah asserts the moral right to be
identified as the author of this work

A CIP catalogue record for this book
is available from the British Library

ISBN 1 86046 738 5

Designed and typeset in Stone Print at
Libanus Press, Marlborough, Wiltshire

Printed and bound in Italy

E670   510.

SUMMARY TABLE OF CONTENTS

Part One

Part Two

Part Three

References in this volume to Chapters 1 to 27 relate to the chapters in the earlier volume(s) of this work unless otherwise indicated.

The Translator's notes, whether inserted into the text or printed as footnotes, are identified by the abbreviation *Transl.*

PART ONE

CHAPTER 1

HISTORICAL SUMMARY OF ARITHMETIC, NUMERICAL NOTATION, AND WRITING SYSTEMS

The writing of words and the writing of numbers show many parallels in their histories.

In the first place, human life was profoundly changed by each system, which allowed spoken language on the one hand, and number on the other, to be recorded in lasting form.

Further, each system answered marvellously to the universal need, felt by every member of every advanced society, for a visual medium to embalm human thought – which otherwise would inevitably dissolve into dust.

Again, everyone became empowered to create a persistent record of what he had expressed or communicated: of words which were otherwise long silent, or of calculations long since completed.

Finally, and most importantly, each system granted direct access to the world of ideas and thoughts across space and time. By encapsulating thought, and by inspiring it in others, the writing down of thought imposed on it both discipline and organisation.

Number and letter have often worn the same clothes, especially at times when letters were used to stand for numerals. But this is superficial: at a much deeper level there is still a close correspondence between the alphabet and the positional number-system. Using an alphabet with a fixed number of letters, every word of a language can be written down. Using our ten digits 1, 2, 3, 4, 5, 6, 7, 8, 9, and 0, any whole number whatever can be written down.

So we perceive a perfect analogy between these two great discoveries, the final stage in the development of writing and the final stage in the development of numerical notation. They are among the most powerful intellectual attributes of the modern human race.

The analogy is not limited to this, however. Throughout its history, each written number-system evolved in a very similar way to the verbal writing system that it grew up with. This can be seen in the way that both reflected the spoken language or the cultural traditions to which the language was adapted; or, again; in the mannerisms of local scribes, and in the influence of the very materials used for writing.

The main purpose of the chapter is to present a recapitulation of the history of numerical notation and arithmetical calculation. But so close is the parallel with writing just noted, that we shall incorporate notes on the history of writing at relevant points; for clarity, these notes will be flagged with [W].

* * *

For all that, the writing of language and the notation of numbers differ radically in one respect. For something to be called writing, its signs must be related to a spoken language; it must reflect a conscious effort to represent speech: "Writing is a system of human communication which uses conventional signs, which are well defined and which represent a language, which can be sent out and received, which can be equally well understood by sender and receiver, and which are related to the words of a spoken language." [J. G. Février, *Histoire de l'écriture* (1959)]

By contrast, numerical notation needs no correspondence with spoken numbers. The mental process of counting is not linked to any particular act of speech; we can count to any number without speaking or even without thinking a single word. We need only create a "sign language for numbers", and in fact humankind devised many "number languages" before inventing even the *word* "number" and going on to use the human voice itself to measure concrete or abstract quantities.

While written characters correspond to the articulations of a spoken language, the signs in a numerical notation reflect components of thought, of a method of thinking much more structured than the sounds of speech. This method of thinking is itself a language (the language of numbers, no less), but to acquire this language we need first of all to have a concept of distinct units and the capacity to aggregate them. The language of number organises numerical concepts into a fixed order according to an idea which, on reflexion, we recognise as a general principle of *recurrence* or *recursion* [a principle according to which the evaluation of a complex entity is resolved by evaluating its component entities of a lower degree of complexity, which in turn ... , until the simplest level is reached, at which each entity can be evaluated immediately. *Transl.*] It also makes use of a scale of magnitude (or *base*) according to which numbers can be distributed over successive levels called first-order, second-order, ... units.

For a system of signs to constitute a written number-system, therefore, they must in the first place have a structure which its user can conceive, in his mind, as a hierarchical system of units nested each within the next. Then there must be a predetermined fixed number which gives the number of units on one level which must be aggregated together so as to constitute a single unit at the next level; this number is called the *base* of the number-system.

In summary: a notation for numbers is a very special human communication system using conventional signs called *figures* which have a well defined meaning, which can be sent and received, which are equally well understood by both communicating parties [in other words, a *code. Transl.*], and which are attached to the natural whole numbers according to a mentally conceived structure which obeys both the *recurrence principle* and the *principle of the base*.

Over the five thousand years which have elapsed since the emergence of the earliest number-system, of course people have not merely devised one single number-system. Nor have there been an indefinite quantity of them – as can be seen from our *Classification of the Written Number-systems of History* in Chapter 23, which brings together systems so separated in space and time as to be effectively isolated from each other.

At the end of this chapter, we shall present the main conclusions of this Classification in a series of comparative systematic tableaux which will exhibit the mathematical characteristics of each number-system.

The order of succession of the number-systems in this series of tables will not be purely chronological, but will follow a path which traces their evolution in logic, as well as in time, from the most primitive to the most advanced.

The number-systems which are found throughout history fall into three main types, each divided into several kinds (Fig. 1.40):

A. the *additive* type of number-system. They are based on the additive principle and each of their figures has a particular value which is always the same regardless of its position in the representation of the number (Fig. 1.14 to 16). Basically, they are simply written versions of more ancient methods of counting with objects (Fig. 1.1 to 13);

B. the *hybrid* type of number-system. These use a kind of mixed multiplicative and additive principle (Fig. 1.28 to 32), and are essentially transcriptions of oral number-systems of varying degrees of organisation;

C. the *positional* type of number-system. These are based on the principle that the value of a particular figure depends on its position in the representation of a number (Fig. 1.33 to 36), and therefore need a zero (Fig. 1.37). Number-systems of this type exhibit the greatest degree of abstraction, and therefore represent the final stage in the development of numerical notation (Fig. 1.38 and 39).

In the tables at the end of this chapter, the letters A, B and C will therefore indicate the above types of additive, hybrid, and positional number-systems respectively. The kinds within these types will be indicated by numbers attached to these letters, such as A1 for an additive number-system of the first kind, and so on.

By classifying them in this way we shall be able to perceive clearly the true nature of our modern system of numerical notation, and therefore to understand why no essential improvement of it has been found neces-sary – or indeed possible – since the time it was invented in all perfection fifteen hundred years ago in India. Its birth, then and there, came about through the remote chance that three great ideas came together, namely: well-conceived figures representing the base digits (1, . . . , 9); their use according to a principle of position; and their completion by a sign 0 for zero which not only served to mark the absence of a base digit in a given position but also – and above all – served to denote the null number.

When we talk of "our modern number-system", by the way, we do not only mean the way we now write numbers worldwide, but also any of the other number-systems also used in the Near East, in Central Asia, in India and in Southeast Asia, which have identical structure and therefore identical possibilities. See "Indian Written Numeral Systems (The mathematical classification of)" in the Dictionary of Indian Numerical Symbols (Chapter 24, Part II).

This "temporal logic" alone could demonstrate the deep unity of all human culture; but it is not merely a question of order and system revealed solely on lines of time – this would be to ignore both transmission within each culture and also, above all, transmission from one culture to another. It would also gloss over the true chronology of events in which, in our story, we see cultures both overlapping each other, and leapfrogging past each other. At certain times, some cultures have been far in advance of others. Some other peoples have clung to inadequate number-systems throughout their history, whether by failure to break out of the prison of an inadequate system, or through a conservatism which attached them to a poor tradition.

For these reasons, we now embark on a systematic chronological résumé which will trace out this logic of time. Fig. 1.41, to be found below, will show how from the different civilisations emerged our different classifications of number-systems (defined in Fig. 1.14 to 16, 1.28 to 32, and 1.38 to 39); and Fig. 1.41 is the inverse of Fig. 1.42, showing how the different number-systems emerged in order of time, according to civilisations.

Our chronological résumé traces the history of a graphical notation for numbers, whose prime function was to represent numbers obtained in the course of calculations or counts previously carried out, in order to compensate for the defects of human memory.

We shall also follow the principal stages of development of arithmetical calculation, which evolved in parallel with the writing of numbers. It began with counting on the fingers, and with pebbles; continued through many-coloured strings, the abacus, checkerboards and abacuses traced in wax or dust or sand, and finally made contact with written numerical notation when our modern positional numerical notation, and the zero on which it depends, were discovered.

The dates of "first appearance" given in the following Figures are the results of archaeological, epigraphical and palaeographical research, to mention a few of the domains of study from which the information has been gleaned. They do not correspond to the definite date of an invention or a discovery. A date which we give below merely means the established date, according to these researches, of the earliest known documentary evidence for the system or concept in question. They are therefore only approximate.

Nevertheless we must maintain a distinction between the date of invention or discovery of something, and the date by which it came into common use; and the latter must in turn also be distinguished from the dates of the earliest instances of which we are currently aware.

Quite possibly, a discovery occurred many generations prior to its popularisation, and there may well be a delay between this and the date of the earliest evidence we possess today. There may be many reasons for this. Often, a discovery long remained the esoteric property of a closed sect, or of a specialised elite who jealously guarded their monopoly of an arcane art. In many cases, no doubt, documents which had existed prior to those of which we know have perished; or perhaps they have yet to be discovered. Archaeological or documentary discoveries which remain to be made may yet cause changes in the conclusions which we present below.

Chronological summary

At undetermined dates, beginning in prehistory, the following sequence of developments occurred. Entries flagged with [W] refer specifically to developments in writing, rather than numbers.

? The human race was, in the earliest stages of its evolution, at the most primitive stage of the notion of number, which was confined to such number (up to four or five) as could be assimilated at a glance. This nevertheless awoke in the human mind a realisation of the concrete aspects of objects which it directly perceived.[1]

1 The developments marked with ? are plausible reconstructions based on the results of ethnographical and psychological investigations into the behaviour of contemporary "primitive" peoples, and of very young children (see Chapters 1 and 2 of previous volume).

? By force of necessity, aided by native intelligence and by the capacity for thought, human beings little by little learned to solve an increasing range of problems. For numerical magnitudes greater than four, people devised procedures based on the manipulation of concrete objects which enabled them to achieve, up to a point, results which met their needs of the moment. These were simply based on a principle of counting by one-to-one correspondence, and amongst them may be found especially the methods of using the fingers or other parts of the body; thus they had simple methods which were always to hand. These methods came to be expressed in articulated speech, accompanied by corresponding gestures.

? By force of habit, counting according to these parts of the body (once adopted according to an invariable routine) slowly became part abstract, part concrete, thereby suggesting less and less the specific part of the body and more and more the concept of a certain corresponding number which increasingly tended to become detached from the notion of the body and to become applicable to objects of any kind. (For a detailed account of the above, see Chapter 1.)

? The resulting necessity to make a distinction between the numerical symbol itself and the name of the concrete object or image led people finally to make a clean break between the two, and the relationship between them disappeared from their minds. Thenceforth, people progressively learned to count and to conceive of numbers in an abstract sense, not related to specific concrete counting tokens. In particular, as they learned to employ speech sounds for the purpose, the sounds themselves took over the role of the objects for which they had been created. Day by day, the notion of *successor* became established in the human mind, and what had been a motley collection of concrete objects became a structured abstract system, at first based on gesture before assuming verbal or written forms. It became a spoken system when the names of the numbers were invented as abstractions out of custom, usage and memory. Much later, in a similar way written systems came about when all kinds of graphic symbols were brought into use – scratched or drawn or painted lines, marks hollowed out of clay or carved in stone, various figurative symbols, and so on (see Chapter 2).

? This proliferation of representations created problems, which were solved by the invention of the principle of the *base* of a number-system (the base 10 being the most commonly used throughout history). Using every kind of object and device (the fingers, pebbles, strings of pearls, little rods, . . .) people gradually arrived at the abstractions embodied in the procedures of calculation and in the operations of arithmetic (see Chapter 2).

From this point in the history, we are able to assign approximate dates to the successive stages of development.

35000–20000 BCE. The earliest *notched bones* of prehistory are the most ancient known archaeological objects which had been used for numerical ends. They are in fact graphical representations of numbers, though we do not know what precise purposes they served (see Chapter 4).

20000 BCE [W]. The first pictures on rock appeared in Europe: these, among the earliest known visual representations of human thought, were made by engraved or painted lines.

9th–6th millennia BCE [W]. We see the simultaneous appearance in Anatolia (Beldibi), in Mesopotamia (Tepe Asiab), in Iran (Ganj Dareh Tepe), in Sudan (Khartoum), in Palestine (Jericho) and in Syria of the little *clay tokens* of various sizes and shapes (cones, discs, spheres, small balls, little rods, tetrahedra etc.). Some of them bear parallel lines, some crosses and other motifs, while others are decorated with carved figurines representing every kind of object (jugs, animal heads, etc.). These relief drawings (which surely had significance for their creators and users) are clear evidence of the development of symbolic thought. We do not know, however, whether these are the elements of some system nor whether this corresponds to one of the intermediate stages between a systematic purely symbolic expression of human thought and its formal expression in a spoken language (see Chapter 10).

9th–2nd millennium BCE. The peoples of the Middle East (from Anatolia and Palestine to Iran and Mesopotamia, from Syria to Sudan) made their calculations using cones, spheres, rods and other clay objects which stood for the different unit magnitudes of a number-system. Such systems can be found, from the fourth millennium BCE onwards, in Elam and in Sumer, in a much elaborated form which will give rise, not only to written counts, but also to some extent to the graphical forms of the Sumerian and proto-Elamite figures (see Chapters 10 and 12).

6th–5th millennia BCE [W]. The earliest *ceramic* artefacts, on which motifs have been painted, engraved, cut out or impressed on the raw clay, or engraved after firing, appear in the Middle East. These are evidently graphical representations emanating from some symbolic system, but we do not know their meaning or purpose.

6th–5th millennia BCE [W]. At the same time, in Asia Minor (Çatal Hüyük) and later in Mesopotamia, there appear the earliest *seals* (carved objects which can be used to impress a relief design on soft material such as clay).

4th millennium BCE (?). The people of Sumer have an oral number-system, to base 60 (see Chapters 8 and 9). This base 60 has come down to us via the Babylonians, the Greeks and the Arabs, and we use it yet for the minutes and seconds of time, and for the measure of angle in minutes and seconds of a degree.

3500 BCE [W]. The first *cylindrical seals* appear in Elam and in Mesopotamia. They are small cylinders of stone, precious or semi-precious, bearing an engraved symbolic design. Every man of a certain standing had one of these: it represented the very person of its bearer and therefore was associated with all economic or judicial aspects of his life. By rolling the cylinder onto any object of clay, the proprietor of the seal thereby impressed his "signature", or his right of property, upon it. The different designs did not constitute "writing" in the strict sense of the word; rather, they had a symbolic significance subject to every kind of interpretation.

3300–3200 BCE. The figures of the Sumerian number-system and the figures of the proto-Elamite number-system make a simultaneous appearance at this time. These are the most ancient written number-systems at present known (see Chapters 8 and 10).

3200–3100 BCE [W]. The writing signs of Sumer, the most ancient writing system known, make their appearance. These are pictograms which represent every kind of object, and they are found on clay tablets which seem to have been used for some economic purpose. However, this is still not a true writing system, since the signs are symbolic of objects rather than directly related to a spoken language. This latter step will occur only at the beginning of the third millennium BCE, at which time the Sumerian system will have become phonetic, will represent the various parts of speech, and will have become linked to spoken language, which is the most highly developed way of analysing and communicating reality (see Chapter 8).

3000 BCE [W]. In ancient Persia, the proto-Elamite writing signs appear (see Chapter 10).

3000–2900 BCE [W]. The signs of Egyptian hieroglyphic writing appear (see Chapter 14).

3000–2900 BCE. The figures of the Egyptian number-system appear (see Chapter 14).

2700 BCE [W]. The cuneiform characters (in the form of angles and wedges) of the Sumerian writing system appear on their clay tablets (see Chapter 8).

2700 BCE. The cuneiform figures of the Sumerian number-system appear (see Chapter 8).

2700–2300 BCE. For doing arithmetic, the people of Sumer now abandon their old *calculi* and invent their *abacus*, a kind of table of successive columns, ruled beforehand, which delimit the successive orders of magnitude of their sexagesimal number-system. By clever manipulation of small balls or rods on the abacus, they are able to perform all sorts of calculations (see Chapter 12).

2600–2500 BCE [W]. Egyptian hieratic writing appears, a cursive abbreviation of hieroglyphic writing and used alongside the latter for the sake of rapid writing on manuscripts (see Chapter 14).

2500 BCE. The Egyptian hieratic figures appear (see Chapter 14).

2350 BCE [W]. The Semites of Mesopotamia borrow the cuneiform characters of Sumer to write down their own speech. This is the beginning of the *Akkadian script* from which will emerge the Babylonian and Assyrian writing systems.

2350 BCE [W]. Appearance of the writing of Ebla (the capital of the Semite kingdom situated at Tell Mardikh, to the South of Aleppo in Syria), a cuneiform script cut into clay tablets, for their Western Semitic dialect which was close to Ugaritic, Phoenician and Hebrew.

2300 BCE [W]. Proto-Indian writing appears in the Indus valley at Mohenjo-daro and Harappâ (in what is now Pakistan). This writing of the ancient Indus civilisation (25th–18th centuries BCE) is separated by a hiatus of over two thousand years from the earliest written texts in any true Indian language and in true Indian writing. It is not known how to bridge this gap, nor, indeed, if it ever was bridged.

End of 3rd millennium BCE. The Semites of Mesopotamia are now slowly adopting a cuneiform decimal notation which has come down to them from their predecessors. In everyday use, this system will come to supplant the Sumerian sexagesimal system (of which, however, the base 60 will survive in the positional notation of the Babylonian scholars). At the same time, the ancient Sumerian abacus undergoes a radical transformation: instead of using beads or rods, they trace their cuneiform figures inside the ruled columns of a large clay tablet; in the course of calculation, these figures are successively erased according as the successive partial results are obtained (see Chapter 13).

2000–1660 BCE [W]. The hieroglyphic writing of the Minoan civilisation appears in Crete, found at Knossos and Mallia on bars and tablets of clay which appear to have been accountancy documents (see Chapter 15).

2000–1660 BCE. At the same time appear the hieroglyphic figures which they used for numbers (see Chapter 15).

1900 BCE [W]. The "Linear A" script of the Minoan civilisation appears in Crete, found at Haghia Triada, Mallia, Phaestos and Knossos on clay tablets which were undoubtedly inventories of some kind. Somewhat casual in style, this script occurs not only in administrative quarters but also in sanctuaries and probably in private houses too (see Chapter 15).

1900 BCE. At the same time appear the "Linear A" figures which they used for numbers (see Chapter 15).

1900–1600 BCE [W]. The cuneiform script of the Semites of Mesopotamia gradually supplants the Sumerian script and spreads across the Near East, where it will even become the official script of the chancelleries.

1900–1200 BCE. The decimal cuneiform number-system of the Semites of Mesopotamia spreads across the Near East.

1900–1800 BCE. The oldest known positional number-system comes on the scene: this is the cuneiform sexagesimal system of the Babylonian scholars, but it is not yet in possession of a zero (see Chapter 13).

17th century BCE [W]. The first known venture into an alphabetic script – the Semites who were in the service of the Egyptians in the Sinai made use of simple phonetic symbols derived from Egyptian hieroglyphics (the so-called "proto-Sinaitic inscriptions" of Serabit al Khadim).

17th century BCE. Notwithstanding the very rudimentary nature of their hieroglyphic and hieratic numerals, the Egyptians are able to make use of them for arithmetical calculations (see Chapter 14). These methods relieve the burden on the memory (since it is sufficient simply to know how to multiply and divide by 2), but they are not unified and they lack flexibility; they are time-consuming, and are very complicated in comparison with the procedures of our own day.

16th century BCE. By now, the Egyptian hieratic number-system has come to the end of its graphical evolution (see Chapter 14).

15th century BCE [W]. Desiring abbreviation, and keen to break away from the complicated Egyptian and Assyro-Babylonian writing systems then in use in the Near East, the Semites of the Northwest who were settled along the Syrian and Palestinian coasts develop the very first purely alphabetical writing system in history, thereby inventing the *alphabet*. This superior method of transcribing words, capable of being adapted to any spoken language, henceforth allows all the words of any language to be written by means of a small number of simple phonetic symbols called *letters* (see Chapter 17).

15th century BCE [W]. The hieroglyphic writing of the Hittite civilisation appears. This script will not only be used for religious and dedicatory purposes, but also – and above all – for secular purposes (see Chapter 15).

15th century BCE. At the same time, the Hittite hieroglyphic number-system appears (see Chapter 15).

1350–1200 BCE [W]. The Creto-Mycenaean script called "Linear B" appears in Crete (found at Knossos) and in Greece (Pylos, Mycenae, etc.); it is a modification of "Linear A", used to write an archaic Greek dialect (see Chapter 15).

1350–1200 BCE. At the same time, the "Linear B" numerals appear (see Chapter 15).

14th century BCE [W]. The oldest known entirely alphabetic script appears, found on tablets from Ugarit (Ras Shamra, near Aleppo in Syria). This is a cuneiform script whose alphabet has only thirty letters; it was used to write a Semitic language related to Phoenician and Hebrew (see Chapter 17).

End of the 14th century BCE [W]. One of the oldest known specimens of archaic Chinese writing appears at Xiao. It is found on inscriptions made on bones or on tortoise shells, and its main purpose was to enable communication between the world of the living and the world of the spirits by means of various divinatory and religious practices (see Chapter 21).

End of the 14th century BCE. At the same time, we find the oldest known Chinese numerals (archaic Chinese number-system, in inscriptions on the bones and tortoise shells at Xiao dun; see Chapter 21).

End of the 12th century BCE [W]. The earliest known specimens of the Western Semitic "linear" alphabet, a precursor of all modern alphabets (see Chapter 17), are used by the Phoenicians. Since they had dealings with a great variety of peoples, these notable merchants and bold navigators diffused their alphabet far and wide. In the East, they will pass it first to their immediate neighbours (Moabites, Edomites, Ammonites, Hebrews, etc.), including the Aramaeans who in turn will spread it from Syria to Egypt and to Arabia, and from Mesopotamia to the Indian sub-continent. From the ninth century BCE, it will spread also round the whole Mediterranean seaboard and be progressively adopted by the Western peoples who will adapt it to their own languages, and modify it by the addition of some further symbols.

9th century BCE [W]. Ancient Israelite inscriptions in palaeo-Hebraic characters, derived directly from the 22 Phoenician characters, appear (see Chapter 17).

9th century BCE. The Hebrews adopt the Egyptian hieratic numerals, which they especially make use of in correspondence (see Chapter 18).

9th–8th centuries BCE [W]. The alphabetic script of Phoenician origin spreads across the near East and the Eastern Mediterranean (Aramaeans, Hebrews, Greeks, etc.).

End of 9th century BCE [W]. The Greeks perfect the principle of the modern alphabetical system by adding symbols for the vowels to the consonants of the original Phoenician alphabet. This is the first alphabet to have a strict and integrated notation for the vowels (see Chapter 17). In turn, this alphabet will inspire the Italic alphabets (Oscan, Umbrian, Etruscan, etc.) and then Latin, later giving rise to the Coptic, Gothic, Armenian, Georgian and Cyrillic alphabets.

8th century BCE [W]. Egyptian demotic script appears, a cursive script, arising from a local branch of Egyptian hieratic script but more abbreviated, which it will later supplant in everyday use (see Chapter 14).

8th century BCE [W]. Appearance of the Italic alphabets, in particular the Etruscan (see Chapter 17).

8th century BCE. The earliest clearly differentiated forms of Egyptian demotic numerals appear (see Chapter 14).

8th century BCE. Italic numerals (Oscan, Umbrian, and especially Etruscan) appear (see Chapter 16).

8th century BCE. This is the period of the earliest known Western Semitic numerals; Aramaean numerals appear (see Chapter 18).

7th century BCE [W]. Archaic Latin writing appears.

6th century BCE. Archaic Latin numerals appear (see Chapter 16).

6th century BCE. Greek acrophonic numerals appear in Attica (see Chapter 16).

End of 6th century BCE. The earliest known Phoenician numerals (see Chapter 18).

5th century BCE [W]. The earliest known specimens of Zapotec writing in pre-Columbian Central America.

5th century BCE. The Zapotecs use an additive number-system with base 20, which can be found in use amongst all the pre-Columbian peoples of Central America (Mayas, Mixtecs, Aztecs, etc.) to within minor graphical variations (see Chapter 22). For purposes of calculation, these peoples certainly do not make use of this ill-adapted number-system; on the contrary, they make use of calculating instruments. Although Central-American archaeology has yielded up nothing relevant to this subject, we may nevertheless get an idea of it by appealing to ethnology and to history which provide us with numerous analogies. We can therefore suppose that, on the example of certain African societies, they made use of rods, each corresponding to an order of magnitude, along which they slid pierced pebbles. They possibly proceeded in the same way as the Apache, the Maidu and the Havasupai of North America who threaded pearls and shells onto coloured threads; or, perhaps more plausibly, like the Incas of South America who distributed pebbles, beans or grains of maize onto the squares of a checkerboard on a kind of tray made of stone, pottery or wood, or drawn on the floor (see Chapters 12 and 22).

5th century BCE [W]. Aramaean script becomes generally adopted for international correspondence in the Middle East, henceforth supplanting the Assyro-Babylonian script for this purpose.

5th century BCE. The Aramaean numerals, which have already reached their final form, spread over the Middle East (Mesopotamia, Syria, Palestine, Egypt, Northern Arabia, etc.).

5th century BCE. Earliest archaeological evidence for the use of the Greek abacus: tables made out of wood or marble, pre-set with small counters in wood or metal suitable for the mathematical calculations (see Chapter 16). The Persians in the time of Darius were to use this type of abacus, and after them the Etruscans and Romans. The Western Christian world was to inherit the use of this abacus, which they were to continue until the French Revolution (see Chapters 16, 25 and 26).

5th century BCE. The Greek acrophonic number-system spreads across the Hellenic world (see Chapter 16).

5th century BCE. The acrophonic numerals of Southern Arabia appear in the inscriptions of the kingdom of Sheba (see Chapter 16).

End of 4th century BCE. The earliest known records of the Greek alphabetic number-system appear in Egypt, showing that the Greek letter numerals are in general use by this time (see Chapters 17 and 18).

3rd century BCE. The first known case of the use of zero comes on the scene, as used by the Babylonian scholars. This was a cuneiform character. Although used in the positional Babylonian number-system to signify the absence of a sexagesimal unit, it is nevertheless still not perceived as a number in its own right (see Chapter 13).

3rd century BCE. The Greek alphabetical number-system spreads across the Middle East and the Eastern Mediterranean (see Chapter 17).

3rd century BCE [W]. The Aramaeo-Indian Kharoshthî writing appears in the edicts of the Emperor Aśoka. This is a cursive script derived from Aramaean writing, and used in the Northwest of India, as well as in the territories which are now Pakistan and Afghanistan (see Chapter 24).

3rd century BCE. At the same time, Kharoshthî numerals appear in Northwest India and in the various countries now subsumed in Pakistan and Afghanistan.

3rd century BCE [W]. Brâhmî script appears in the edicts of the Emperor Aśoka. (see Chapter 24). This is derived from the ancient alphabetic scripts of the Western Semitic world, no doubt with an intermediate Aramaean form of which no specimens have been found. This will become the earliest truly Indian script, and will be the origin of all of the alphabetic scripts of the Indian sub-continent and of Southeast Asia. Over the centuries, it will undergo many changes which culminate in many distinctly different types of writing, such as *Gupta, Bhattiprolu* and *Pâli*. Gupta in turn will split into *Nâgarî, Siddham,* and *Śhâradâ* from which will descend all the current scripts to be found in Central and Northern India, in Nepal, in Tibet, and in Chinese Turkestan. Bhattiprolu will give rise to the scripts of Southern India and Ceylon, while Pâli will give rise to the scripts of Southeast Asia. The apparently considerable differences between all these scripts are due either to the natures of the languages and traditions to which they have been adapted, or to regional differences between scribes and differences between writing materials. (See "Indian Styles Of Writing" and "Indian Styles of Writing (The materials of)" in the Dictionary of Indian Numerical Symbols, Chapter 24, Part II.)

3rd century BCE. Brâhmî numerals appear in the edicts of the Emperor Aśoka, which can be found throughout the Maurya Empire. These, the earliest truly Indian numerals, occur more and more frequently in later inscriptions (Shunga, Ândhra, Śhaka, Kshatrapa etc.); and they are the prototypes of all the numerical notations which flourished later in India, Central Asia, and Southeast Asia. Although the number-system did not at the time follow a principle of position, the figures which correspond to the first nine digits are clear precursors of the digits 1 to 9 in our own number-system and in the modern Arabic number-system (see Chapter 24).

3rd century BCE to 4th century CE [W]. Greek manuscript writing splits into three types: "book" script, official script, and the script used for private documents.

2nd century BCE. In Plutarch, we find mention of the sand abacus alongside the abacus with tokens. A board, with a raised border, is filled with fine sand on which lines are drawn to mark out the columns, and the numbers are written in, using an iron stylus. The same type of abacus is later found amongst the Christian population of the mediaeval West, and they carry out their calculations using either Roman numerals or the Greek alphabetic numerals (see Chapter 16).

2nd century BCE [W]. This epoch sees the Chinese invention of paper. (According to some, it was invented by Cai-Lun who achieved it by boiling up unravelled tissues and old fishing nets.) They also invent *xylography*: the text to be reproduced is written on the polished surface of a wooden board, and then the wood surrounding the writing is cut away so as to leave it in relief; ink is applied, and a sheet of paper is pressed on. (See further information about paper in the course of the entry "Pâtîganita" in the Dictionary of Indian Numerical Symbols, Chapter 24, Part II .)

2nd century BCE. The earliest known documents which refer to use of the Chinese abacus, and to "calculation with rods" (*suan zî*) in which small bamboo sticks are placed in successive squares of a checkerboard (see Chapter 21).

2nd century BCE. The earliest known documents which affirm the use of a positional decimal notation by the Chinese. This system does not however have a zero (see Chapter 21), and in fact is simply a written counterpart of the method of "calculation with rods" (see above).

2nd century BCE [W]. The earliest documentary reference to "square Hebrew": Hebrew writing in its modern form, but whose squat and massive letters are derived from the cursive Aramaean script (see Chapter 17).

2nd century BCE. The earliest documents in which we can see the use of modern Hebrew alphabetic numerals (see Chapter 17).

2nd century BCE to 3rd century CE. Indian arithmeticians perform their calculations by tracing their nine digits in Brâhmî notation on the floor, within consecutive columns already delineated, with a pointed rod. A similar procedure will later be used by the Arabs, especially those of the Maghreb and of Andalusia. Boards covered with fine sand, with flour or with some other powder are also used, with a stylus whose point is used to trace the figures and its flat end to erase them. This board might be placed on the floor, on a stool or on a table, or might be furnished with legs like those used much later in the Arab, Turkish and Persian administrations. The board might be made in a small version which could be kept in a case. (See Chapters 24 and 25, and also the entry "Dhûlîkarma" in the Dictionary of Indian Numerical Symbols, Chapter 24, Part II.)

2nd century BCE to 2nd century CE [W]. The reform of Chinese writing, and the emergence of the *lì shū* graphics which will evolve towards the modern system of Chinese characters (see Chapter 21).

1st century BCE. Horace notes the use of the wax abacus, as well as the abacus with rods, by the Romans: a real "portable calculator" which could be hung over the shoulder, this consists of a board made of bone or wood, covered with a thin layer of black wax, on which the lines for the columns, and the figures, are drawn with an iron stylus (see Chapter 16).

Start of the Common Era [W]. A cursive branch of the ancient Aramaean evolves to give rise to Arabic script.

1st century CE. At this period we find the earliest archaeological evidence of the Roman "pocket abacus". This is a small metal plate with parallel slots along which mobile beads can be slid; each is associated with a numerical order of magnitude. It is therefore very similar to the bead abacus which in modern times still holds an important place in the Far East and in certain Eastern countries (see Chapters 16 and 21).

2nd–3rd centuries CE [W]. The Roman script undergoes a change which will give rise to two new forms of Latin script: the New Common Writing and the Uncial.

End of 3rd century CE [W]. The oldest known specimens of Maya writing in pre-Columbian Central America (see Chapter 22).

End of 3rd century CE. The earliest examples of use of the "Long Count" for dates by Mayan astronomers (see Chapter 22).

3rd–4th centuries CE [W]. The earliest known cases of runic script used by Germanic peoples (*Futhark alphabet*).

Beginning of 4th century CE [W]. Pharnavaz, first king of the country which lay between Armenia and the Caucasus, is inspired by Greek to invent the *Mkhedrouli* alphabet, ancestor of the Georgian alphabet (see Chapter 17).

4th century CE. Appearance of Ethiopian numerals in the inscriptions from Aksum in the kingdom of Abyssinia (see Chapter 19).

4th century CE [W]. The first appearance of the Chinese *kǎi shū* writing, a form of modern Chinese writing (see Chapter 21).

4th century CE [W]. Bishop Wulfila draws on the Greek alphabet to invent the Gothic alphabet, for the purpose of recording the Germanic language of the Goths (see Chapter 17).

4th–5th centuries CE [W]. The earliest forms of the Indian Gupta alphabet appear, from which all the alphabetical scripts of central India, Nepal, Tibet and Chinese Turkestan will be derived.

4th–5th centuries CE. The earliest forms of the Indian Gupta numerals appear, from which all the numerical notations of central India, Nepal, Tibet and Chinese Turkestan will be derived (see Chapter 24, and also the entry "Gupta Numerals" in the Dictionary of Indian Numerical Symbols, Chapter 24, Part II).

4th–5th centuries CE. The first nine digits of the Indian system, derived from the old Brâhmî notation, acquire a positional significance in a decimal base, and they are completed by an additional symbol in the form of a small circle or a dot which represents zero; this therefore is the birth of the Indian positional decimal notation, which was the ancestor of our modern numerical notation (see Chapter 24).

4th–6th centuries CE. During this period, the Indian arithmeticians radically transform their traditional methods of calculation. They do away with the columns of their ancient sand abacus, and attribute values to written digits according to their decimal position. This, therefore, is the beginning of the modern number-system. It is also, however, the beginning of modern written arithmetic.

To begin with, their techniques, albeit liberated from the columns of the abacus, were but written imitations of the abacus procedures. As formerly practised, on a medium as inconvenient as the sand abacus, with intermediate results noted after erasing the previous ones, these constrained the role of human memory and made it difficult if not impossible to check the calculation and correct errors made along the way. The Indian and Arab scholars subsequently developed procedures which did not involve erasure, but involved writing intermediate results above the working. While certainly advantageous for checking purposes, since every intermediate error remained to be seen, nevertheless this resulted in a cluttered worksheet from which it was difficult to get a clear view of the progression of the calculation.

Because of this kind of complication, even using the nine digits and the zero, the new methods long remained beyond the grasp of ordinary mortals. Writing their calculations on a board with chalk without worrying about how many figures there were or, better still, rubbing them out successively with a cloth: such was the convenient and relaxed method which, even before the advent of pen and paper, allowed the Indian arithmeticians and their Arab and European successors to work in their own way and with unfettered imagination to arrive at simplifications of the rules and methods and ultimately to create the techniques which would give rise to our modern methods of written arithmetic. (See Chapters 24 and 25, and also the entry "Indian Methods of Calculation" in the Dictionary of Indian Numerical Symbols, Chapter 24, Part II.)

4th–6th centuries CE [W]. The earliest forms of the Bhattiprolu and Pâlî scripts appear: from these will be derived, respectively, all the alphabetic writing systems of South India, and those of Southeast Asia (see Chapter 24).

4th–6th centuries CE. The earliest forms of the Bhattiprolu and Pâlî numerals appear: from these will be derived, respectively, all the number-systems of South India, and those of Southeast Asia (see Chapter 24).

4th–9th centuries CE. This is probably the period during which the positional notation, with base 20 and a zero, of the Mayan astronomer-priests emerged. However, as a result of its forced conformity to the peculiarities of the Mayan calendar, this number-system exhibited an irregular use of the base 20 beyond the third digit position which robbed it, along with its zero, of practical operational value (see Chapter 22).

5th century CE [W]. The earliest known specimens of Arab writing found in pre-Islamic inscriptions. This script was cursive in style, and in time diversified to give rise to the Kufic script and the Naskhî script during the early centuries of Islam (see Chapters 19 and 25).

5th century CE [W]. The priest Mesrop Machtots draws inspiration from Greek to invent the Armenian alphabet (see Chapter 17).

5th–7th centuries CE [W]. Emergence of the Ogham script in Celtic inscriptions in Ireland and Wales.

510. The Indian astronomer Âryabhata invents a special numerical notation for which it is necessary to have a full awareness of the concepts of zero and the principle of position. He further makes use of a remarkable method for calculating square and cube roots which it is impossible to perform unless the numbers are written down using the principle of position, the nine digits, and a tenth sign which plays the role of zero. (See Chapter 24, and also the entries "Âryabhata," "Âryabhata's Number-System," "Indian Mathematics, The history of" and "Square Roots, How Âryabhata calculated his" in the Dictionary of Indian Numerical Symbols, Chapter 24, Part II.)

628. The Indian mathematician and astronomer Brahmagupta publishes *Brahmasphutasiddhânta*, which displays total mastery of positional decimal notation, using the nine digits and a zero. (See Chapters 24 and 25, and also the entries "Brahmagupta" and "Indian Mathematics, The history of" in the Dictionary of Indian Numerical Symbols, Chapter 24, Part II.)

629. The mathematician Bhâskara publishes a Commentary on the *Âryabhatîya*. This work not only reveals complete mastery of the use of zero and of the positional decimal number-system: it also shows that the author is quite at ease with the Rule of Three and with arithmetical fractions, which he writes in a way very similar to ours, though lacking the horizontal bar which will not be introduced until several centuries later, by Arab mathematicians. (See Chapters 24 and 27, and also the entry "Bhâskara" in the Dictionary of Indian Numerical Symbols, Chapter 24, Part II.)

7th century CE [W]. The earliest distinct forms of the Indian *Nâgarî* writing appear, from which the scripts of North and Central India will be derived (Bengâlî, Gujarâtî, Oriyâ, Kaîthî, Maithilî, Manipurî, Marâthî, Mârwarî, etc. (See Chapter 24, and also the entries "Indian Styles of Writing" and "Indian Styles of Writing, The materials of," in the Dictionary of Indian Numerical Symbols, Chapter 24, Part II.)

7th century CE. The earliest distinct forms of the Indian *Nâgarî* numerals appear, from which the numerals of North and Central India will be derived (Bengâlî, Gujarâtî, Oriyâ, Kaîthî, Maithilî, Manipurî, Marâthî, Mârwarî, etc., see Chapter 24).

7th century CE [W]. The earliest distinct forms of the stylised scripts of Southeast Asia appear (Khmer, Malaysian, Shan, Kawi, etc., see Chapter 24).

7th century CE. The earliest distinct forms of the Indian numerals which will give rise to the stylised numerals of Southeast Asia appear (Khmer, Malaysian, Shan, Kawi, etc., see Chapter 24).

7th–8th centuries CE. During this period, the Indian decimal notation, with the zero, spreads to the Indianised civilisations of Southeast Asia (Cambodia, Shan, Java, Malaysia, Bali, Borneo, etc., see Chapter 24).

7th–8th centuries CE [W]. The era of the oldest known manuscripts in which we find Latin writing of "Visigoth" and "Luxeuil" type.

7th–10th centuries CE [W]. The earliest distinct forms which will give rise to the stylised scripts of South India (Tamil, Malayâlam, Tulu, Telugu, Kannara, etc., see Chapter 24).

8th century CE. Under the influence of Indian Buddhist monks, the zero, of Indian origin, takes its place in the Chinese positional decimal number-system of "bar numbers" (the *suan zí* system, see Chapter 24).

8th century CE [W]. The first appearance of the so-called "minuscule" Greek writing (which will replace the older system in books from the ninth century onwards).

End of 8th century CE. At this time, the positional decimal notation with a zero enters the world of Islam. In the hands of the Arab scribes, the figures will undergo changes of form, in some cases far enough from the original forms that they appear to be new (see Chapter 25).

8th–11th centuries CE [W]. Runic inscriptions from Viking times (from Uppland province in Sweden).

8th–11th centuries CE [W]. The earliest distinct forms of Carolingian writing (the Corbie studio, manuscripts of the Bible written under the direction of the monk Maurdramnus, the dedication of the Gospels by Charlemagne, etc.).

820–850. The period of the great Muslim astronomer and mathematician Al Khuwārizmī,[2] whose works contributed greatly to the knowledge and dissemination of the numerals and arithmetical methods which originated in India (see Chapter 25).

9th century CE. The Ghubār numerals of the Maghreb and Andalusian Arabs now appear (they are of Indian origin, and their form anticipates that of the European numerals of the Middle Ages and the Renaissance, before giving rise to our modern numerals. See Chapter 25.)

9th century CE [W]. The earliest distinct forms of the Indian Shârâdâ script appear (a southern variant of the Gupta script), which will give rise to the scripts of Northwest India (Dogrî, Tâkarî, Multânî, Sindhî, Punjabî, Gurûmukhî, etc. (See Chapter 24.)

9th century CE. The earliest forms of the Indian Shârâdâ numerals appear, which will give rise to the numerals of Northwest India (Dogrî, Tâkarî, Multânî, Sindhî, Punjabî, Gurûmukhî, etc., see Chapter 24).

9th century CE [W]. With the aim of converting the Bulgars, the bishop Cyril draws inspiration from Greek to invent the Glagolitic alphabet.

9th century CE [W]. The appearance of Japanese writing, properly speaking.

10th century CE [W]. In order to record the sounds of the Slavic languages, the bishop Clement draws inspiration from Greek to invent the Cyrillic alphabet. The first simplification of this writing which will later give rise to the modern Russian alphabet will be brought about by Peter the Great in the eighteenth century.

972–982. In the course of a voyage to Spain, the monk Gerbert d'Aurillac

---

2 From whose name the word "algorithm" is derived. See also the entry for the 12th century CE below. [*Transl.*]

from Auvergne (later to become Pope Sylvester II, in 999) learns the "Arab" numerals and introduces them to Western Europe (see Chapter 26).

976–992. Two manuscripts from non-Muslim Spain illustrate the forms of nine figures which are very similar to numerals of the Ghubār type. These are the oldest known evidence of the presence of "Arab" numerals in Western Europe (see Chapter 26).

10th–12th centuries CE. Europeans are carrying out arithmetic operations using the abacus with columns, of Roman origins and perfected by Gerbert d'Aurillac and his pupils. They use counters made of horn (called *apices*) marked with the Arab numerals from 1 to 9, or with the Greek alphabetic numerals from α to θ, or with the Roman numerals from I to IX (see Chapter 26).

11th century CE [W]. The master calligrapher Lanfranc, in the Carolingian tradition, creates the script which will become the most beautiful *pontifical writing* of the twelfth century.

Middle of 11th century CE [W]. Printing is invented by the Chinese. They make use of separate characters made of baked clay, which later will be made of lead and then in copper. This invention is related by Qin Guo in 1056; he attributes it to Bi Xing and dates it at 1041.

12th century CE. The Indian sign for zero is introduced to Europe. The European arithmeticians henceforth do their calculations with the zero and the nine "Indo-Arabic" digits. Also, the rules of arithmetic, of Indian origin, are now called *algorisms*.[3]

12th–13th centuries CE [W]. Gothic script gradually replaces Carolingian script.

12th–13th centuries CE [W]. Aztec writing emerges (see Chapter 22).

12th–16th centuries CE. A ferocious dispute takes place between the *Abacists* (adherents of methods of calculating by counters on the abacus, and prisoners of a system seamed with ancient number-systems such as the Roman numerals and the Greek alphabetic numerals) and the *Algorists*, proponents of methods of written arithmetic using the Indian numerals and the zero (see Chapter 26).

12th–15th centuries CE. The period where the forms of the "Arabic" numerals become established in Europe, where they will eventually evolve into their modern forms (see Chapter 26).

1202. Following his travels in North Africa and the Middle East, the Italian

---

3 Nowadays usually *algorithms*, though *algorism* is in fact correct. [*Transl.*]

mathematician Leonard of Pisa, better known as Fibonacci, publishes *Liber Abaci* ("A Treatise on the Abacus"). Over the ensuing three centuries, this book will prove to be a most fruitful source of development of arithmetic and algebra in Western Europe (see Chapter 26).

13th century CE. In this period we find the earliest documents which illustrate the use of the Chinese abacus (*suan pan*). This is a rectangular frame of wood traversed by a certain number of rods along which slide seven wooden balls. A longitudinal wooden slat divides the interior into two parts, on one side (the lower) of which there are five balls on each rod, and on the other side (the upper) two. Each rod corresponds to a power of 10, increasing to the left.

Of all the ancient calculating instruments, the Chinese abacus is the only one to provide a simple means to carry out all the operations of arithmetic; Western observers are usually astonished at the speed and dexterity with which even the most complicated arithmetic can be done. The same kind of instrument is still employed in modern times, and not only in Japan: it may be found also in Russia (the *stchoty*), in Iran, in Afghanistan (the *choreb*), in Armenia and in Turkey, but in these cases it has a different structure, and is of more basic design than the *suan pan*. The Japanese *soroban*, on the other hand, will later benefit from a considerable refinement. In the nineteenth century, it lost one of the upper pair of balls; and during the Second World War it lost one of the lower five. These balls are in fact superfluous to the strict needs of the Chinese instrument, so it might seem that the Japanese instrument, having been reduced to its necessary essentials, represents the ultimate perfection of design. However, a skilled operator can use the extra balls to represent an intermediate result whose value exceeds nine and thereby gain speed and facility; learning to use the Japanese abacus well requires a longer and more difficult training, and the acquisition of a more elaborate and precise finger technique (see Chapter 21).

14th–15th centuries CE [W]. The Italians develop the Humanist script, a scholarly style based on the Carolingian script of the ninth, tenth, and eleventh centuries.

c.1440 [W]. In Holland, the first attempts at typographic printing are made. The printer Laurens Janszoon first printed playing cards from woodcuts, then whole pages of text, and was led on to make single characters out of wood which he then used to print a small eight-page book called *Horarium*.

c.1540 [W]. Printing is reinvented, this time in the West by Johannes Gensfleisch, known as Gutenberg, in Mainz. Recognising the inconvenience of mobile characters made of wood, and that they are ill suited to a good impression, he develops metallic characters, completely regular and adjustable, together with the requisite typographic techniques.

This achievement will later have at least two important consequences. One will be that the rapid spread of typographic procedures will replace the ancient Gothic and Humanistic calligraphies, and will accelerate the evolution of handwriting, which it will cause to settle into a more and more standardised form; the second, the more important, will be that it becomes possible to produce as many copies as one wants of any literary, scientific, or philosophical work, which will lead to a wider and wider dissemination of knowledge within Western Europe. This will lead to a radical transformation of society and the inauguration of a new era in Europe.

15th–16th centuries CE. After undergoing various apparently major changes, which are however simply due to the natural tendencies of handwriting, the "Arab" numerals take on a fixed form once and for all, thanks to the upsurge of printing in Europe (see Chapter 26).

15th–16th centuries CE. A progressive generalisation of calculation methods occurs, due to the use of "Arabic" numerals and the zero. The *Algorists* have triumphed and the *Abacists* are in retreat. Calculation on the abacus will continue to be done by tradesmen, financiers and other businessmen, and only with the French Revolution will these archaic methods disappear (see Chapter 26).

1478. The publication of the Treviso *Arithmetic*, a manual of practical arithmetic by an anonymous author, evinces the diffusion of "Arabic" numerals and the increasing favour which the new methods are finding in Western Europe (see Chapter 25).

1654. The French mathematician Blaise Pascal gives the first general definition of a number-system to base m where m is an arbitrary integer greater than or equal to 2 (see Fig. 1.39).

Recapitulation

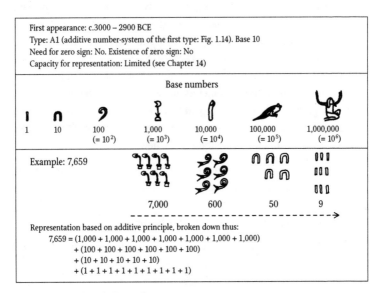

First appearance: c.3000 – 2900 BCE
Type: A1 (additive number-system of the first type: Fig. 1.14). Base 10
Need for zero sign: No. Existence of zero sign: No
Capacity for representation: Limited (see Chapter 14)

Base numbers

| 1 | 10 | 100 $(= 10^2)$ | 1,000 $(= 10^3)$ | 10,000 $(= 10^4)$ | 100,000 $(= 10^5)$ | 1,000,000 $(= 10^6)$ |

Example: 7,659

7,000   600   50   9

Representation based on additive principle, broken down thus:
7,659 = (1,000 + 1,000 + 1,000 + 1,000 + 1,000 + 1,000 + 1,000)
    + (100 + 100 + 100 + 100 + 100 + 100)
    + (10 + 10 + 10 + 10 + 10)
    + (1 + 1 + 1 + 1 + 1 + 1 + 1 + 1 + 1)

FIG. 1.1. *Egyptian hieroglyphic number-system*

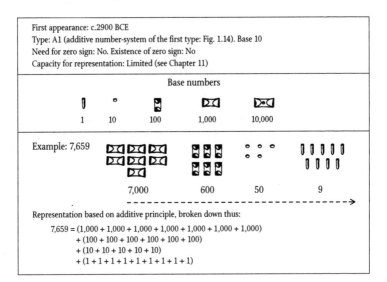

First appearance: c.2900 BCE
Type: A1 (additive number-system of the first type: Fig. 1.14). Base 10
Need for zero sign: No. Existence of zero sign: No
Capacity for representation: Limited (see Chapter 11)

Base numbers

| 1 | 10 | 100 | 1,000 | 10,000 |

Example: 7,659

7,000   600   50   9

Representation based on additive principle, broken down thus:
7,659 = (1,000 + 1,000 + 1,000 + 1,000 + 1,000 + 1,000 + 1,000)
    + (100 + 100 + 100 + 100 + 100 + 100)
    + (10 + 10 + 10 + 10 + 10)
    + (1 + 1 + 1 + 1 + 1 + 1 + 1 + 1 + 1)

FIG. 1.2. *Proto-Elamite number-system*

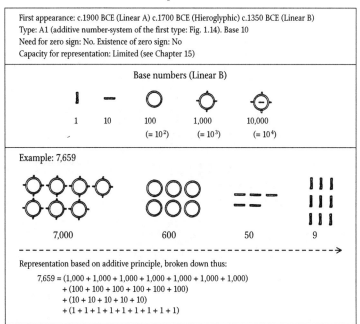

First appearance: c.1900 BCE (Linear A) c.1700 BCE (Hieroglyphic) c.1350 BCE (Linear B)
Type: A1 (additive number-system of the first type: Fig. 1.14). Base 10
Need for zero sign: No. Existence of zero sign: No
Capacity for representation: Limited (see Chapter 15)

Base numbers (Linear B)

| 1 | 10 | 100 ($= 10^2$) | 1,000 ($= 10^3$) | 10,000 ($= 10^4$) |

Example: 7,659

| 7,000 | 600 | 50 | 9 |

Representation based on additive principle, broken down thus:

$7,659 = (1,000 + 1,000 + 1,000 + 1,000 + 1,000 + 1,000 + 1,000)$
$+ (100 + 100 + 100 + 100 + 100 + 100)$
$+ (10 + 10 + 10 + 10 + 10)$
$+ (1 + 1 + 1 + 1 + 1 + 1 + 1 + 1 + 1)$

FIG. 1.3. *Cretan number-system (hieroglyphic and Linear A and B)*

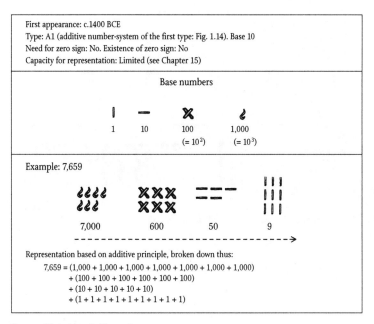

First appearance: c.1400 BCE
Type: A1 (additive number-system of the first type: Fig. 1.14). Base 10
Need for zero sign: No. Existence of zero sign: No
Capacity for representation: Limited (see Chapter 15)

Base numbers

| 1 | 10 | 100 ($= 10^2$) | 1,000 ($= 10^3$) |

Example: 7,659

| 7,000 | 600 | 50 | 9 |

Representation based on additive principle, broken down thus:

$7,659 = (1,000 + 1,000 + 1,000 + 1,000 + 1,000 + 1,000 + 1,000)$
$+ (100 + 100 + 100 + 100 + 100 + 100)$
$+ (10 + 10 + 10 + 10 + 10)$
$+ (1 + 1 + 1 + 1 + 1 + 1 + 1 + 1 + 1)$

FIG. 1.4. *Hittite hieroglyphic number-system*

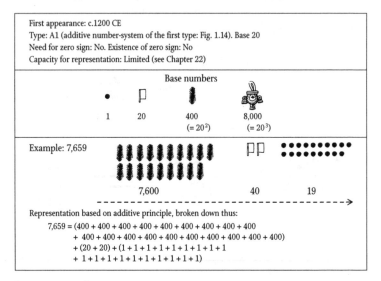

FIG. 1.5. *Aztec number-system*

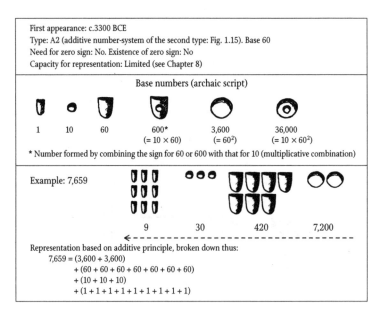

FIG. 1.6. *Sumerian number-system*

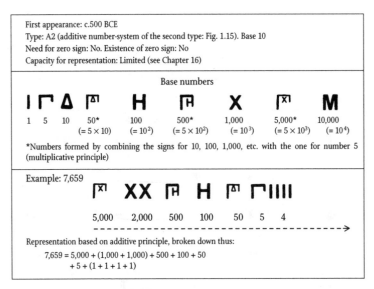

First appearance: c.500 BCE
Type: A2 (additive number-system of the second type: Fig. 1.15). Base 10
Need for zero sign: No. Existence of zero sign: No
Capacity for representation: Limited (see Chapter 16)

Base numbers

1 5 10 50* 100 500* 1,000 5,000* 10,000
(= 5 × 10) (= 10²) (= 5 × 10²) (= 10³) (= 5 × 10³) (= 10⁴)

*Numbers formed by combining the signs for 10, 100, 1,000, etc. with the one for number 5 (multiplicative principle)

Example: 7,659

5,000 2,000 500 100 50 5 4

Representation based on additive principle, broken down thus:

$7,659 = 5,000 + (1,000 + 1,000) + 500 + 100 + 50$
$+ 5 + (1 + 1 + 1 + 1)$

FIG. 1.7. *Greek acrophonic number-system*

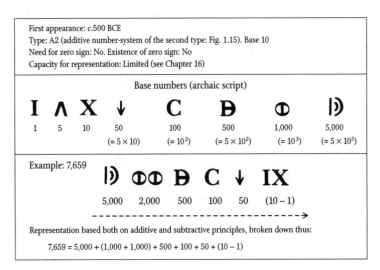

First appearance: c.500 BCE
Type: A2 (additive number-system of the second type: Fig. 1.15). Base 10
Need for zero sign: No. Existence of zero sign: No
Capacity for representation: Limited (see Chapter 16)

Base numbers (archaic script)

1 5 10 50 100 500 1,000 5,000
(= 5 × 10) (= 10²) (= 5 × 10²) (= 10³) (= 5 × 10³)

Example: 7,659

5,000 2,000 500 100 50 (10 − 1)

Representation based both on additive and subtractive principles, broken down thus:

$7,659 = 5,000 + (1,000 + 1,000) + 500 + 100 + 50 + (10 − 1)$

FIG. 1.8. *Roman number-system*

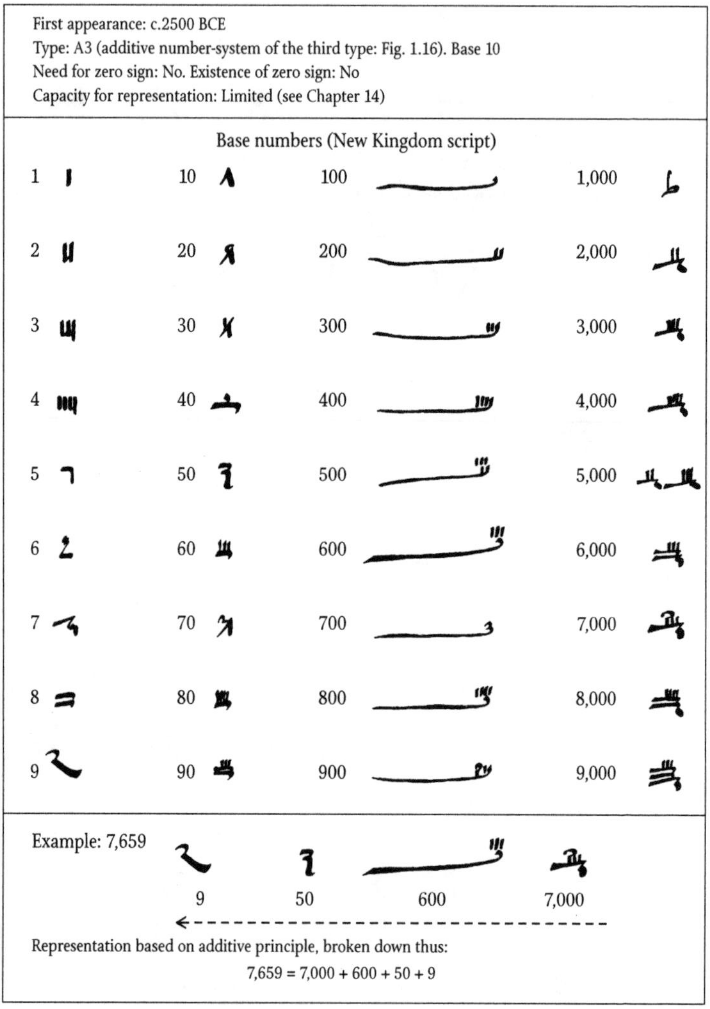

First appearance: c.2500 BCE
Type: A3 (additive number-system of the third type: Fig. 1.16). Base 10
Need for zero sign: No. Existence of zero sign: No
Capacity for representation: Limited (see Chapter 14)
Base numbers (New Kingdom script)
1
2
3
4
5
6
7
8
9
10
20
30
40
50
60
70
80
90
100
200
300
400
500
600
700
800
900
1,000
2,000
3,000
4,000
5,000
6,000
7,000
8,000
9,000
Example: 7,659
9
50
600
7,000
Representation based on additive principle, broken down thus:
7,659 = 7,000 + 600 + 50 + 9

FIG. 1.9. *Egyptian hieratic number-system*

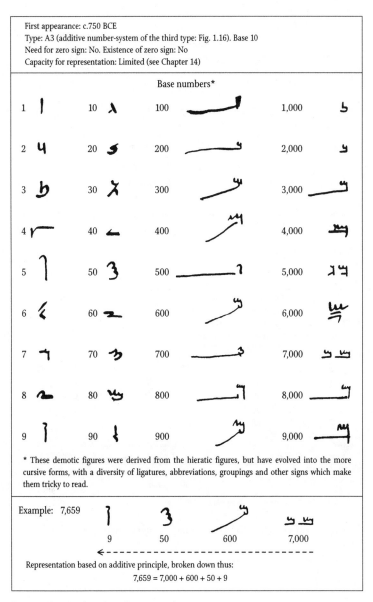

First appearance: c.750 BCE
Type: A3 (additive number-system of the third type: Fig. 1.16). Base 10
Need for zero sign: No. Existence of zero sign: No
Capacity for representation: Limited (see Chapter 14)

Base numbers*

1	10	100	1,000
2	20	200	2,000
3	30	300	3,000
4	40	400	4,000
5	50	500	5,000
6	60	600	6,000
7	70	700	7,000
8	80	800	8,000
9	90	900	9,000

* These demotic figures were derived from the hieratic figures, but have evolved into the more cursive forms, with a diversity of ligatures, abbreviations, groupings and other signs which make them tricky to read.

Example: 7,659

9   50   600   7,000

Representation based on additive principle, broken down thus:
$$7,659 = 7,000 + 600 + 50 + 9$$

Fɪɢ. 1.10. *Egyptian demotic number-system*

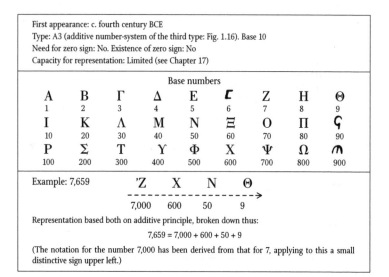

First appearance: c. fourth century BCE
Type: A3 (additive number-system of the third type: Fig. 1.16). Base 10
Need for zero sign: No. Existence of zero sign: No
Capacity for representation: Limited (see Chapter 17)

				Base numbers				
A	B	Γ	Δ	E	Ϛ	Z	H	Θ
1	2	3	4	5	6	7	8	9
I	K	Λ	M	N	Ξ	O	Π	Ϟ
10	20	30	40	50	60	70	80	90
P	Σ	T	Υ	Φ	X	Ψ	Ω	ϡ
100	200	300	400	500	600	700	800	900

Example: 7,659

'Z   X   N   Θ

7,000   600   50   9

Representation based both on additive principle, broken down thus:

$$7,659 = 7,000 + 600 + 50 + 9$$

(The notation for the number 7,000 has been derived from that for 7, applying to this a small distinctive sign upper left.)

FIG. 1.11. *Greek alphabetic number-system*

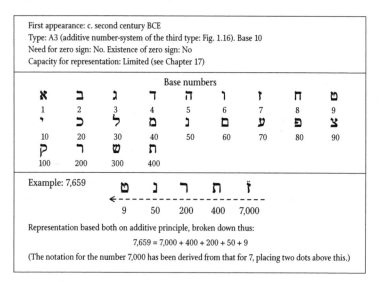

First appearance: c. second century BCE
Type: A3 (additive number-system of the third type: Fig. 1.16). Base 10
Need for zero sign: No. Existence of zero sign: No
Capacity for representation: Limited (see Chapter 17)

				Base numbers				
א	ב	ג	ד	ה	ו	ז	ח	ט
1	2	3	4	5	6	7	8	9
י	כ	ל	מ	נ	ס	ע	פ	צ
10	20	30	40	50	60	70	80	90
ק	ר	ש	ת					
100	200	300	400					

Example: 7,659

ט   נ   ר   ת   ז̈

9   50   200   400   7,000

Representation based both on additive principle, broken down thus:

$$7,659 = 7,000 + 400 + 200 + 50 + 9$$

(The notation for the number 7,000 has been derived from that for 7, placing two dots above this.)

FIG. 1.12. *Hebraic alphabetic number-system*

First appearance: c.400 CE
Type A3 (additive number-system of the third type: Fig. 1.16). Base 10
Need for zero sign: No. Existence of zero sign: No
Capacity for representation: Limited (see Chapter 17)

Base numbers
(Line 1, lower case; line 2, upper case)

ա	բ	գ	դ	ե	զ	է	ը	թ
Ա	Բ	Գ	Դ	Ե	Զ	Է	Ը	Թ
1	2	3	4	5	6	7	8	9
ժ	ի	լ	խ	ծ	կ	հ	ձ	ղ
Ժ	Ի	Լ	Խ	Ծ	Կ	Հ	Ձ	Ղ
10	20	30	40	50	60	70	80	90
ճ	մ	յ	ն	շ	ո	չ	պ	ջ
Ճ	Մ	Յ	Ն	Շ	Ո	Չ	Պ	Ջ
100	200	300	400	500	600	700	800	900
ռ	ս	վ	տ	ր	ց	ւ	փ	ք
Ռ	Ս	Վ	Տ	Ր	Ց	Ւ	Փ	Ք
1,000	2,000	3,000	4,000	5,000	6,000	7,000	8,000	9,000

Example: 7,659

Lower case:	ւ	ո	ծ	թ
Upper case:	Ւ	Ո	Ծ	Թ
	7,000	600	50	9

Representation based on additive principle, broken down thus:

7,659 = 7,000 + 600 + 50 + 9

FIG. 1.13. *Armenian alphabetic number-system*

Number-systems of this type fall into three kinds whose mathematical characteristics are summed up in Fig. 1.14 to 1.16: they require the adoption of a new writing convention based on a certain order of magnitude in order to note down high numbers (see Chapter 23).

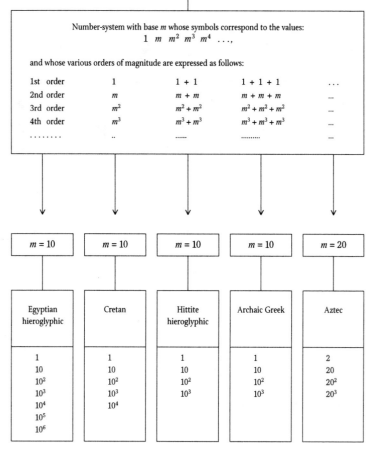

Systems of Type A1
(additive number-systems of the first type)

Number-system with base m whose symbols correspond to the values:
$$1 \quad m \quad m^2 \quad m^3 \quad m^4 \quad \ldots,$$

and whose various orders of magnitude are expressed as follows:

1st order	1	$1 + 1$	$1 + 1 + 1$	...
2nd order	m	$m + m$	$m + m + m$	...
3rd order	m^2	$m^2 + m^2$	$m^2 + m^2 + m^2$	...
4th order	m^3	$m^3 + m^3$	$m^3 + m^3 + m^3$	...
....	..	......	....	...

$m = 10$	$m = 10$	$m = 10$	$m = 10$	$m = 20$
Egyptian hieroglyphic	Cretan	Hittite hieroglyphic	Archaic Greek	Aztec
1	1	1	1	2
10	10	10	10	20
10^2	10^2	10^2	10^2	20^2
10^3	10^3	10^3	10^3	20^3
10^4	10^4			
10^5				
10^6				

FIG. 1.14. *Classification of additive number-systems (Type A1)*

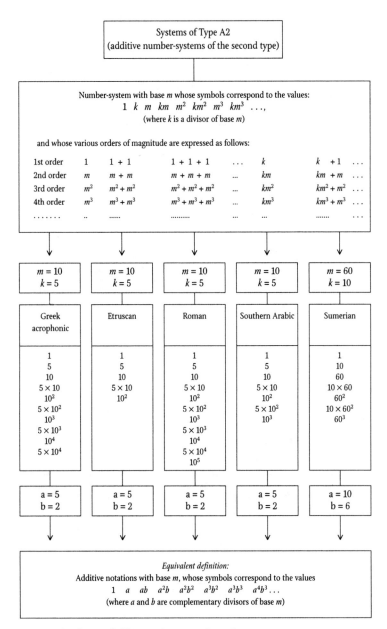

FIG. 1.15. *Classification of additive number-systems (type A2)*

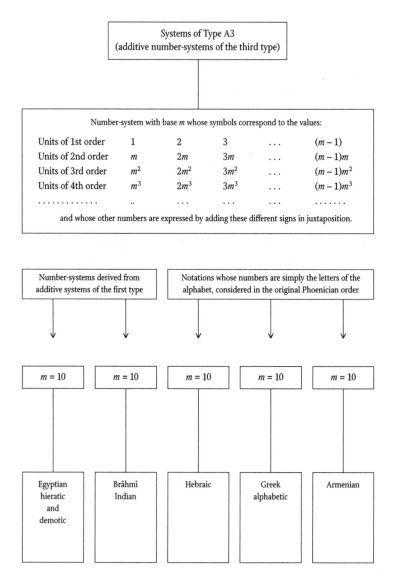

FIG. 1.16. *Classification of additive number-systems (Type A3)*

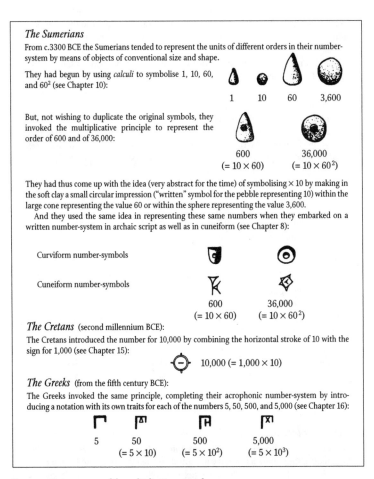

The Sumerians

From c.3300 BCE the Sumerians tended to represent the units of different orders in their number-system by means of objects of conventional size and shape.

They had begun by using *calculi* to symbolise 1, 10, 60, and 60^2 (see Chapter 10):

1   10   60   3,600

But, not wishing to duplicate the original symbols, they invoked the multiplicative principle to represent the order of 600 and of 36,000:

600   36,000
$(= 10 \times 60)$ $(= 10 \times 60^2)$

They had thus come up with the idea (very abstract for the time) of symbolising × 10 by making in the soft clay a small circular impression ("written" symbol for the pebble representing 10) within the large cone representing the value 60 or within the sphere representing the value 3,600.

And they used the same idea in representing these same numbers when they embarked on a written number-system in archaic script as well as in cuneiform (see Chapter 8):

Curviform number-symbols

Cuneiform number-symbols

600   36,000
$(= 10 \times 60)$ $(= 10 \times 60^2)$

The Cretans (second millennium BCE):

The Cretans introduced the number for 10,000 by combining the horizontal stroke of 10 with the sign for 1,000 (see Chapter 15):

10,000 $(= 1,000 \times 10)$

The Greeks (from the fifth century BCE):

The Greeks invoked the same principle, completing their acrophonic number-system by introducing a notation with its own traits for each of the numbers 5, 50, 500, and 5,000 (see Chapter 16):

5   50   500   5,000
$(= 5 \times 10)$ $(= 5 \times 10^2)$ $(= 5 \times 10^3)$

FIG. 1.17. *First emergence of the multiplicative principle*

Thus from the beginning of history, people have sometimes introduced the multiplication rule into systems essentially based on the additive principle. But during this first stage, the habit was confined to certain particular cases and the rule served only to form a few new symbols.

But in the subsequent stage, it gradually became clear that the rule could be applied to avoid not only the awkward repetition of identical signs, but also the unbridled introduction of new symbols (which always ends up requiring considerable efforts of memory).

And that is how certain notations that were rudimentary to begin with were often found to be extensible to large numbers.

The Greeks

This idea was exploited by ancient Greek mathematicians whose "instrument" was their alphabetic number-system: in order to set down numbers superior to 10,000, they invoked the multiplicative rule, placing a sign over the letter M (initial of the Greek word for 10,000, μυριοι) to indicate the number of 10,000s (see Chapter 17):

α	β	γ	ιβ
M	M	M	M
10,000	20,000	30,000	120,000
(= 1 × 10,000)	(= 2 × 10,000)	(= 3 × 10,000)	(= 12 × 10,000)

The Arabs

Using the twenty-eight letters of their number-alphabet the Arabs proceeded likewise, but on a smaller scale: to note down the numbers beyond 1,000, all they had to do was to place beside the letter *ghayin* (worth 1,000 and corresponding to the largest base number in their system) the one representing the corresponding number of units, tens or hundreds (see Chapter 19):

بغ	جغ	يغ	نغ
2,000	3,000	10,000	50,000
(= 2 × 1,000)	(= 3 × 1,000)	(= 10 × 1,000)	(= 50 × 1,000)

The ancient Indians

The same idea was invoked by the Indians from the time of Emperor Aśoka until the beginning of the Common Era in the numerical notation that related to Brâhmî script (see Chapter 24). To write down multiples of 100, they used the multiplicative principle, placing to the right of the sign for 100 the sign for the corresponding units. For numbers beyond 1,000 they wrote to the right of the sign for 1,000 the sign for the corresponding units or tens:

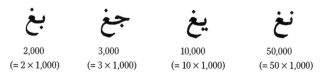

400	4,000	6,000	10,000
(= 100 × 4)	(= 1,000 × 4)	(= 1,000 × 6)	(= 1,000 × 10)

FIG. 1.18A. *First extension of the multiplicative principle*

The Egyptian hieroglyphic system (late period)

In Egyptian monumental inscriptions we find (at least from the beginning of the New Kingdom) a remarkable diversion from the "classical" system: when a tadpole (hieroglyphic sign for 100,000) was placed over a lower number-sign, it behaved as a multiplicator. In other words, by placing a tadpole over the sign for 18, for instance, the number 100,018 (= 100,000 + 18) was no longer being expressed, but rather the number 100,000 × 18 = 1,800,000 (a number which in the classical system would have been expressed by setting eight tadpoles adjacent to the hieroglyphic for 1,000,000).

Example: 27,000,000
Expressed in the form:
100,000 × 270

100,000

270

Taken from a Ptolemaic hieroglyphic inscription (third – first century BCE)

The Egyptian hieratic system

But the preceding irregularity was actually the result of the way the hieroglyphic system was influenced by hieratic notation: this used a more systematic method to note down numbers above 10,000 according to the rule in question. (See Chapter 14)

	Early Kingdom	Middle Kingdom	New Kingdom
70,000	ЩЩ	𓏲	𓏲
90,000		𓏲	
200,000	𝒆𝒆	𓏲	
700,000	𝒆𝒆𝒆 𝒆𝒆	𓏲	𓏲
1,000,000	𓏲	𓏲	𓏲
2,000,000	𓏲𓏲	𓏲𓏲	𓏲
10,000,000			𓏲

Example: The number 494,800

(From the Great Harris Papyrus; 73, line 3. New Kingdom)

$$800 + 4{,}000 + \overset{10{,}000}{\underset{9}{\times}} + \overset{100{,}000}{\underset{4}{\times}}$$

Fig. 1.18b.

LARGE ROMAN NUMBERS

To note down large numbers the Romans and the Latin peoples of the Middle Ages developed various conventions. Here are the principal ones (see Chapter 16):

1. *Overline rule*
This consisted in multiplying by 1,000 every number surmounted by a horizontal bar:

$\overline{X} = 10,000$ $\overline{C} = 100,000$ $\overline{CXXVII} = 127 \times 1,000 = 127,000$

2. *Framing rule*
This consisted in multiplying by 100,000 every number enclosed in a sort of open rectangle:

$\lceil\ X\ \rceil = 1,000,000$ $\lceil CCLXIV \rceil = 264 \times 100,000 = 26,400,000$

3. *Rule for multiplicative combinations*
The rule is occasionally found in Latin manuscripts in the early centuries CE, but most often in European mediaeval accounting documents. To indicate multiples of 100 and 1,000, first the number of hundreds and thousands to be entered are noted down, then the appropriate letter (C or M) is placed as a coefficient or superscript indication:

100	C			1,000	M	
200	II.C	or	II^c	2,000	II.M	or	II^m
300	III.C	or	III^c	3,000	III.M	or	III^m
. . . .				. . . .		
900	VIIII.C	or	VIIII^c	9,000	VIIII.M	or	VIIII^m

Examples taken from Pliny the Elder's *Natural History*, first century CE (VI, 26; XXXIII, 3).

LXXXIII.M   for 83,000
CX.M   for 110,000

The same system is to be found in the Middle Ages, notably in King Philip le Bel's Treasury Rolls, one of the oldest surviving Treasury registers. In this book, dated 1299, we find what is reproduced here below, drawn up in Latin (from Registre du Trésor de Philippe le Bel, BN, Paris, Ms. lat. 9783, fo. 3v, col.1, line 22):

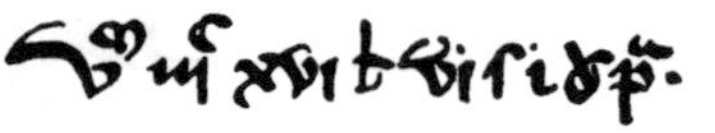

V^m. IIIe.XVI.l(ibras). VI.s(olidos)   "5,316 livres, 6 sols &
I. d(enarios). p(arisiensium)   1 denier parisis"

FIG. 1.19. *Latin notation of large numbers (late period)*

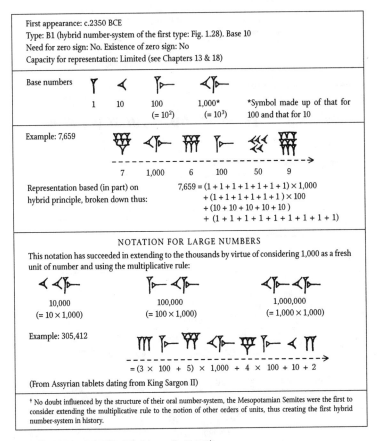

First appearance: c.2350 BCE
Type: B1 (hybrid number-system of the first type: Fig. 1.28). Base 10
Need for zero sign: No. Existence of zero sign: No
Capacity for representation: Limited (see Chapters 13 & 18)

Base numbers

1   10   100   1,000*   *Symbol made up of that for
            $(= 10^2)$ $(= 10^3)$ 100 and that for 10

Example: 7,659

7   1,000   6   100   50   9

Representation based (in part) on
hybrid principle, broken down thus:

$$7,659 = (1 + 1 + 1 + 1 + 1 + 1 + 1) \times 1,000$$
$$+ (1 + 1 + 1 + 1 + 1 + 1) \times 100$$
$$+ (10 + 10 + 10 + 10 + 10)$$
$$+ (1 + 1 + 1 + 1 + 1 + 1 + 1 + 1 + 1)$$

NOTATION FOR LARGE NUMBERS

This notation has succeeded in extending to the thousands by virtue of considering 1,000 as a fresh unit of number and using the multiplicative rule:

10,000   100,000   1,000,000
$(= 10 \times 1,000)$ $(= 100 \times 1,000)$ $(= 1,000 \times 1,000)$

Example: 305,412

$$= (3 \times 100 + 5) \times 1,000 + 4 \times 100 + 10 + 2$$

(From Assyrian tablets dating from King Sargon II)

† No doubt influenced by the structure of their oral number-system, the Mesopotamian Semites were the first to consider extending the multiplicative rule to the notion of other orders of units, thus creating the first hybrid number-system in history.

FIG. 1.20. *Common Assyro-Babylonian number-system*†

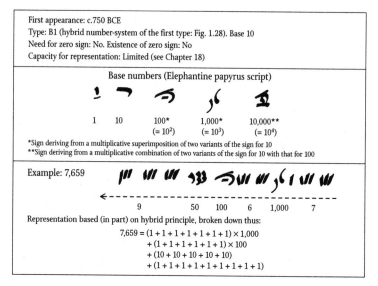

First appearance: c.750 BCE
Type: B1 (hybrid number-system of the first type: Fig. 1.28). Base 10
Need for zero sign: No. Existence of zero sign: No
Capacity for representation: Limited (see Chapter 18)

Base numbers (Elephantine papyrus script)

| 1 | 10 | 100* ($= 10^2$) | 1,000* ($= 10^3$) | 10,000** ($= 10^4$) |

*Sign deriving from a multiplicative superimposition of two variants of the sign for 10
**Sign deriving from a multiplicative combination of two variants of the sign for 10 with that for 100

Example: 7,659

9   50   100   6   1,000   7

Representation based (in part) on hybrid principle, broken down thus:

$$7,659 = (1 + 1 + 1 + 1 + 1 + 1 + 1) \times 1,000$$
$$+ (1 + 1 + 1 + 1 + 1 + 1) \times 100$$
$$+ (10 + 10 + 10 + 10 + 10)$$
$$+ (1 + 1 + 1 + 1 + 1 + 1 + 1 + 1 + 1)$$

FIG. 1.21. *Aramaean number-system*

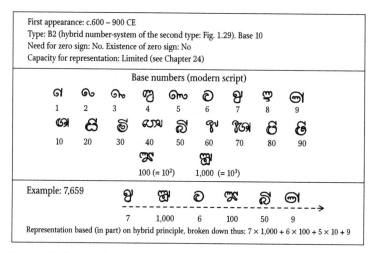

First appearance: c.600 – 900 CE
Type: B2 (hybrid number-system of the second type: Fig. 1.29). Base 10
Need for zero sign: No. Existence of zero sign: No
Capacity for representation: Limited (see Chapter 24)

Base numbers (modern script)

| 1 | 2 | 3 | 4 | 5 | 6 | 7 | 8 | 9 |

| 10 | 20 | 30 | 40 | 50 | 60 | 70 | 80 | 90 |

100 ($= 10^2$)   1,000 ($= 10^3$)

Example: 7,659

7   1,000   6   100   50   9

Representation based (in part) on hybrid principle, broken down thus: $7 \times 1,000 + 6 \times 100 + 5 \times 10 + 9$

FIG. 1.22. *Singhalese number-system*

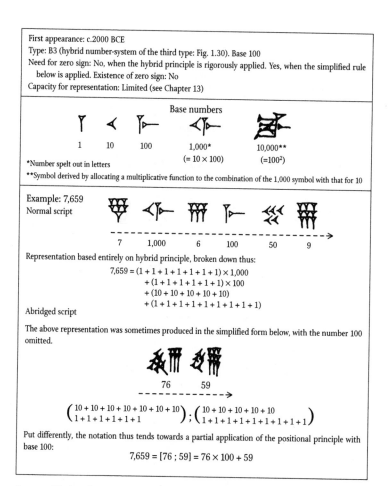

First appearance: c.2000 BCE

Type: B3 (hybrid number-system of the third type: Fig. 1.30). Base 100

Need for zero sign: No, when the hybrid principle is rigorously applied. Yes, when the simplified rule below is applied. Existence of zero sign: No

Capacity for representation: Limited (see Chapter 13)

Base numbers

1	10	100	1,000*	10,000**
			(= 10 × 100)	(=100²)

*Number spelt out in letters

**Symbol derived by allocating a multiplicative function to the combination of the 1,000 symbol with that for 10

Example: 7,659
Normal script

7   1,000   6   100   50   9

Representation based entirely on hybrid principle, broken down thus:

$$7{,}659 = (1 + 1 + 1 + 1 + 1 + 1 + 1) \times 1{,}000$$
$$+ (1 + 1 + 1 + 1 + 1 + 1) \times 100$$
$$+ (10 + 10 + 10 + 10 + 10)$$
$$+ (1 + 1 + 1 + 1 + 1 + 1 + 1 + 1 + 1)$$

Abridged script

The above representation was sometimes produced in the simplified form below, with the number 100 omitted.

76   59

$$\left(\begin{array}{c}10 + 10 + 10 + 10 + 10 + 10 + 10\\1 + 1 + 1 + 1 + 1 + 1\end{array}\right) ; \left(\begin{array}{c}10 + 10 + 10 + 10\\1 + 1 + 1 + 1 + 1 + 1 + 1 + 1 + 1\end{array}\right)$$

Put differently, the notation thus tends towards a partial application of the positional principle with base 100:

$$7{,}659 = [76 ; 59] = 76 \times 100 + 59$$

FIG. 1.23. *Mari number-system*

First appearance: c.350 CE
Type: B4 (hybrid number-system of the fourth type: Fig. 1.31). Base 100
Need for zero sign: No. Existence of zero sign: No
Capacity for representation: Limited (see Chapter 19)

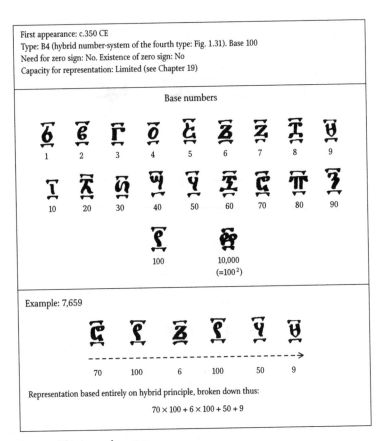

Base numbers

1	2	3	4	5	6	7	8	9
10	20	30	40	50	60	70	80	90

100

10,000
$(=100^2)$

Example: 7,659

70	100	6	100	50	9

Representation based entirely on hybrid principle, broken down thus:

$$70 \times 100 + 6 \times 100 + 50 + 9$$

FIG. 1.24. *Ethiopian number-system*

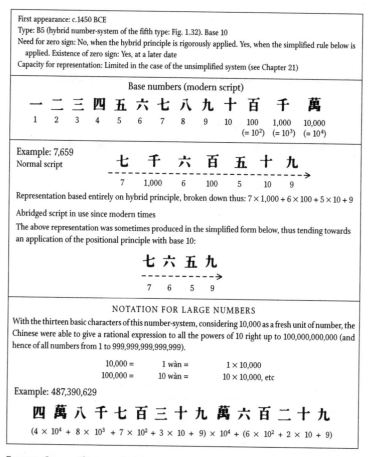

First appearance: c.1450 BCE

Type: B5 (hybrid number-system of the fifth type: Fig. 1.32). Base 10

Need for zero sign: No, when the hybrid principle is rigorously applied. Yes, when the simplified rule below is applied. Existence of zero sign: Yes, at a later date

Capacity for representation: Limited in the case of the unsimplified system (see Chapter 21)

Base numbers (modern script)

一	二	三	四	五	六	七	八	九	十	百	千	萬
1	2	3	4	5	6	7	8	9	10	100 $(= 10^2)$	1,000 $(= 10^3)$	10,000 $(= 10^4)$

Example: 7,659
Normal script

七	千	六	百	五	十	九
7	1,000	6	100	5	10	9

Representation based entirely on hybrid principle, broken down thus: $7 \times 1,000 + 6 \times 100 + 5 \times 10 + 9$

Abridged script in use since modern times

The above representation was sometimes produced in the simplified form below, thus tending towards an application of the positional principle with base 10:

七	六	五	九
7	6	5	9

NOTATION FOR LARGE NUMBERS

With the thirteen basic characters of this number-system, considering 10,000 as a fresh unit of number, the Chinese were able to give a rational expression to all the powers of 10 right up to 100,000,000,000 (and hence of all numbers from 1 to 999,999,999,999,999).

| 10,000 = | 1 wàn = | $1 \times 10,000$ |
| 100,000 = | 10 wàn = | $10 \times 10,000$, etc |

Example: 487,390,629

四 萬 八 千 七 百 三 十 九 萬 六 百 二 十 九

$(4 \times 10^4 + 8 \times 10^3 + 7 \times 10^2 + 3 \times 10 + 9) \times 10^4 + (6 \times 10^2 + 2 \times 10 + 9)$

FIG. 1.25. *Common Chinese number-system*

First appearance: c.600 – 900 CE

Type: B5 (hybrid number-system of the fifth type: Fig. 1.32). Base 10

Need for zero sign: No, when the hybrid principle is rigorously applied. Yes, when the simplified rule below is applied.

Existence of zero sign: Not before the modern era

Capacity for representation: Limited in the case of the unsimplified system (see Chapter 24)

System used among the Tamils (southern India)

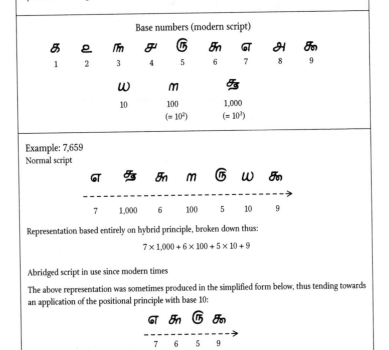

Base numbers (modern script)

க	உ	௩	௪	௫	௬	எ	அ	௯
1	2	3	4	5	6	7	8	9

ω	m	௲
10	100 ($= 10^2$)	1,000 ($= 10^3$)

Example: 7,659

Normal script

எ ௲ ௬ m ௫ ω ௯

- →

7   1,000   6   100   5   10   9

Representation based entirely on hybrid principle, broken down thus:

$$7 \times 1,000 + 6 \times 100 + 5 \times 10 + 9$$

Abridged script in use since modern times

The above representation was sometimes produced in the simplified form below, thus tending towards an application of the positional principle with base 10:

எ ௬ ௫ ௯

- - - - - - - - - - - - →

7   6   5   9

FIG. 1.26. *Tamil number-system*

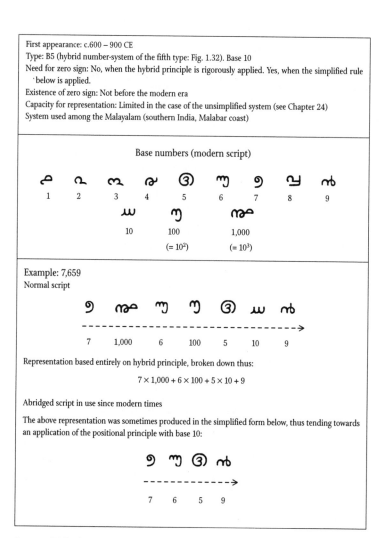

First appearance: c.600 – 900 CE

Type: B5 (hybrid number-system of the fifth type: Fig. 1.32). Base 10

Need for zero sign: No, when the hybrid principle is rigorously applied. Yes, when the simplified rule below is applied.

Existence of zero sign: Not before the modern era

Capacity for representation: Limited in the case of the unsimplified system (see Chapter 24)

System used among the Malayalam (southern India, Malabar coast)

Base numbers (modern script)

| 1 | 2 | 3 | 4 | 5 | 6 | 7 | 8 | 9 |

| 10 | 100 ($= 10^2$) | 1,000 ($= 10^3$) |

Example: 7,659
Normal script

7   1,000   6   100   5   10   9

Representation based entirely on hybrid principle, broken down thus:

$$7 \times 1,000 + 6 \times 100 + 5 \times 10 + 9$$

Abridged script in use since modern times

The above representation was sometimes produced in the simplified form below, thus tending towards an application of the positional principle with base 10:

7   6   5   9

FIG. 1.27. *Malayalam number-system*

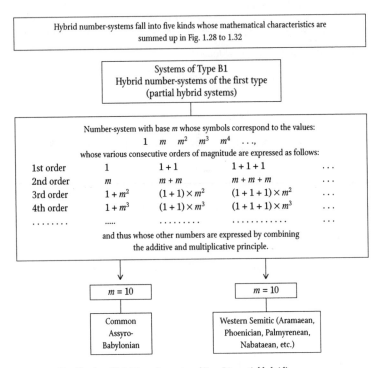

FIG. 1.28. *Classification of hybrid number-systems (Type B1, partial hybrid)*

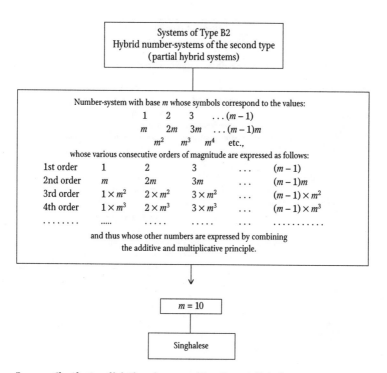

FIG. 1.29. *Classification of hybrid number-systems (Type B2, partial hybrid)*

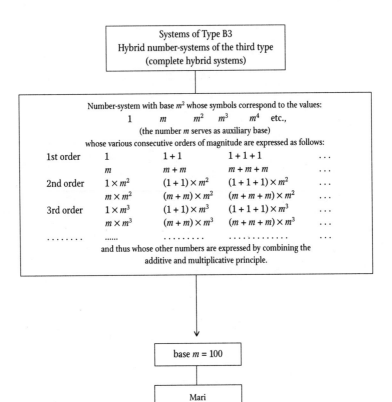

FIG. 1.30. *Classification of hybrid number-systems (Type B3, complete hybrid)*

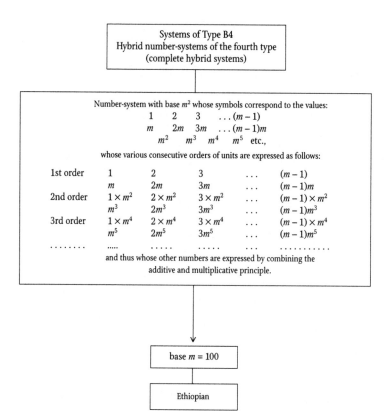

FIG. 1.31. *Classification of hybrid number-systems (Type B4, complete hybrid)*

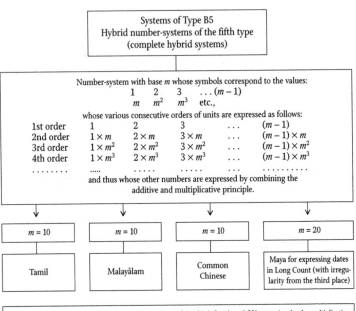

FIG. 1.32. *Classification of hybrid number-systems (Type B5, complete hybrid)*

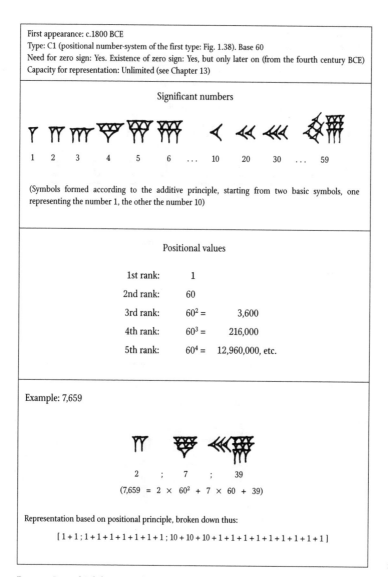

First appearance: c.1800 BCE
Type: C1 (positional number-system of the first type: Fig. 1.38). Base 60
Need for zero sign: Yes. Existence of zero sign: Yes, but only later on (from the fourth century BCE)
Capacity for representation: Unlimited (see Chapter 13)

Significant numbers

| 1 | 2 | 3 | 4 | 5 | 6 | ... | 10 | 20 | 30 | ... | 59 |

(Symbols formed according to the additive principle, starting from two basic symbols, one representing the number 1, the other the number 10)

Positional values

1st rank:	1
2nd rank:	60
3rd rank:	$60^2 =$	3,600
4th rank:	$60^3 =$	216,000
5th rank:	$60^4 =$	12,960,000, etc.

Example: 7,659

| 2 | ; | 7 | ; | 39 |

$(7,659 = 2 \times 60^2 + 7 \times 60 + 39)$

Representation based on positional principle, broken down thus:

$[1+1;1+1+1+1+1+1+1;10+10+10+1+1+1+1+1+1+1+1+1]$

FIG. 1.33. *Learned Babylonian number-system (the first positional number-system in history)*

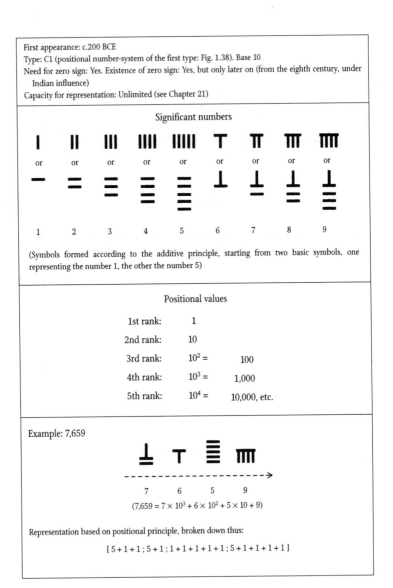

First appearance: c.200 BCE
Type: C1 (positional number-system of the first type: Fig. 1.38). Base 10
Need for zero sign: Yes. Existence of zero sign: Yes, but only later on (from the eighth century, under Indian influence)
Capacity for representation: Unlimited (see Chapter 21)

Significant numbers

| 1 | 2 | 3 | 4 | 5 | 6 | 7 | 8 | 9 |

(Symbols formed according to the additive principle, starting from two basic symbols, one representing the number 1, the other the number 5)

Positional values

1st rank:	1
2nd rank:	10
3rd rank:	$10^2 =$	100
4th rank:	$10^3 =$	1,000
5th rank:	$10^4 =$	10,000, etc.

Example: 7,659

7 6 5 9

$(7,659 = 7 \times 10^3 + 6 \times 10^2 + 5 \times 10 + 9)$

Representation based on positional principle, broken down thus:

$$[\,5+1+1\,;\,5+1\,;\,1+1+1+1+1\,;\,5+1+1+1+1\,]$$

FIG. 1.34. *Learned Chinese number-system*

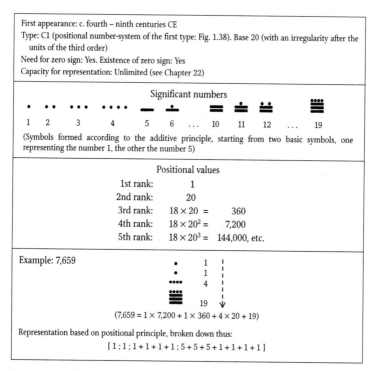

First appearance: c. fourth – ninth centuries CE
Type: C1 (positional number-system of the first type: Fig. 1.38). Base 20 (with an irregularity after the units of the third order)
Need for zero sign: Yes. Existence of zero sign: Yes
Capacity for representation: Unlimited (see Chapter 22)

Significant numbers

1   2   3   4   5   6   ...   10   11   12   ...   19

(Symbols formed according to the additive principle, starting from two basic symbols, one representing the number 1, the other the number 5)

Positional values

1st rank:	1
2nd rank:	20
3rd rank:	$18 \times 20 =$ 360
4th rank:	$18 \times 20^2 =$ 7,200
5th rank:	$18 \times 20^3 =$ 144,000, etc.

Example: 7,659

1
1
4
19

$(7,659 = 1 \times 7,200 + 1 \times 360 + 4 \times 20 + 19)$

Representation based on positional principle, broken down thus:

$[\,1\,;\,1\,;\,1+1+1+1\,;\,5+5+5+1+1+1+1\,]$

FIG. 1.35. *Learned Maya number-system*

First appearance: c. fourth century CE
Type: C2 (positional number-system of the second type: Fig. 1.39). Base 10
Need for zero sign: Yes. Existence of zero sign: Yes
Capacity for representation: Unlimited (see Chapter 24)

Base numbers (present-day script)

1   2   3   4   5   6   7   8   9

(Symbols devoid of all direct visual intuition)

Positional values

1st rank: 1	3rd rank: $10^2 =$ 100
2nd rank: 10	4th rank: $10^3 =$ 1,000, etc.

Example: 7,659

7   6   5   9

$(7,659 = 7 \times 10^3 + 6 \times 10^2 + 5 \times 10 + 9)$

FIG. 1.36. *Modern number-system*

	MAYA	BABYLONIAN	INDIAN	MODERN
ZEROS AND SYSTEMS	Base 20 with an irregularity from the 3rd order	Base 60	Base 10	Base 10
	Positional number rule			
	Basic significant figures according to the additive principle, from the symbols:		Basic significant figures devoid of any direct visual associations:	

This sign (which in the first instance is synonym for "empty") serves to mark the absence of units of a certain order in the representation of the numbers.

Attested: In median position	Attested: In median position	Attested: In median position	Used: In median position

$9 \times 7{,}200 + 0 \times 360 + 0 \times 20 + 7$	$9 \times 60^3 + 0 \times 60^2 + 0 \times 60 + 7$	$9 \times 10^3 + 0 \times 10^2 + 0 \times 10 + 7$	
In final position	In final position (only among Babylonian astronomers)	In final position	In final position

$6 \times 7{,}200 + 4 \times 360 + 9 \times 20 + 0$	$6 \times 60^3 + 4 \times 60^2 + 9 \times 60 + 0$	$6 \times 10^3 + 4 \times 10^2 + 9 \times 10 + 0$	

This zero is a mathematical operator: if it is added to the end of a number, the number's value is multiplied by the base.

Example:

		After a certain era this symbol was taken to mean "the number zero" having the meaning "nought".	Symbol representing "zero value" or "nought".
			This zero is at the root of all algebra and all present-day mathematics.

FIG. 1.37. *Classification of zeros*

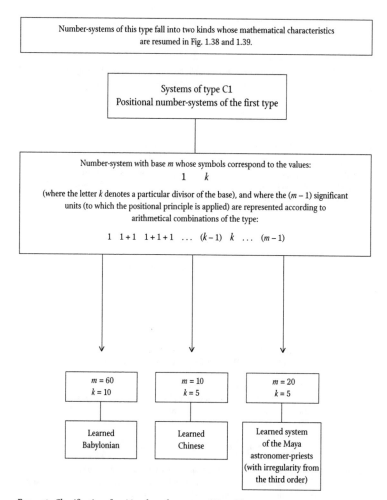

Number-systems of this type fall into two kinds whose mathematical characteristics are resumed in Fig. 1.38 and 1.39.

Systems of type C1
Positional number-systems of the first type

Number-system with base m whose symbols correspond to the values:
$$1 \quad k$$

(where the letter k denotes a particular divisor of the base), and where the $(m - 1)$ significant units (to which the positional principle is applied) are represented according to arithmetical combinations of the type:

$$1 \quad 1+1 \quad 1+1+1 \quad \ldots \quad (k-1) \quad k \quad \ldots \quad (m-1)$$

$m = 60$
$k = 10$

$m = 10$
$k = 5$

$m = 20$
$k = 5$

Learned Babylonian

Learned Chinese

Learned system of the Maya astronomer-priests (with irregularity from the third order)

FIG. 1.38. *Classification of positional number-systems (Type C1)*

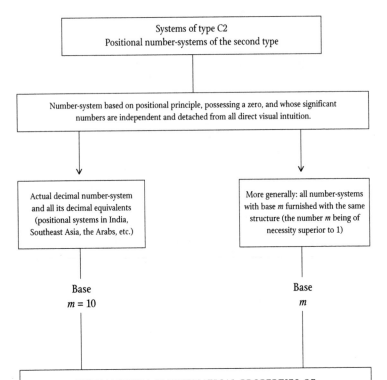

| Systems of type C2 |
| Positional number-systems of the second type |

Number-system based on positional principle, possessing a zero, and whose significant numbers are independent and detached from all direct visual intuition.

Actual decimal number-system and all its decimal equivalents (positional systems in India, Southeast Asia, the Arabs, etc.)

More generally: all number-systems with base m furnished with the same structure (the number m being of necessity superior to 1)

Base $m = 10$

Base m

FUNDAMENTAL MATHEMATICAL PROPERTIES OF POSITIONAL NUMBER-SYSTEMS WITH BASE m

1. The number of digits (including zero) is equal to m.

2. Every integer x may be represented uniquely as a polynomial of degree $k - 1$, with base m as variable, and with coefficients all less than m. In other words, any number x may be written uniquely in the form:

$$x = u_k\, m^{k-1} + u_{k-1}\, m^{k-2} + \cdots + u_4\, m^3 + u_3\, m^2 + u_2\, m + u_1$$

where the integers u_k, u_{k-1}, ... u_2, u_1, all less than m, are symbolised by digits of the system under consideration. One may adopt the convention of writing the number x as follows (where the horizontal bar avoids any confusion with the product $u_k\, u_{k-1} \ldots u_4\, u_3\, u_2\, u_1$):

$$x = \overline{u_k\, u_{k-1} \ldots u_4\, u_3\, u_2\, u_1}$$

3. The four fundamental arithmetical operations (addition, subtraction, multiplication and division) are easily carried out in such a system, according to simple rules entirely independent of the base m of the system.

4. This positional notation may be extended easily to fractions which have a power of the base as denominator, and thus to a simple and coherent notation for all numbers, rational and irrational, by introducing a decimal point, according to an expansion in positive and negative powers of m, by analogy with decimal numbers.

FIG. 1.39. *Classification of positional number-system (Type C2)*

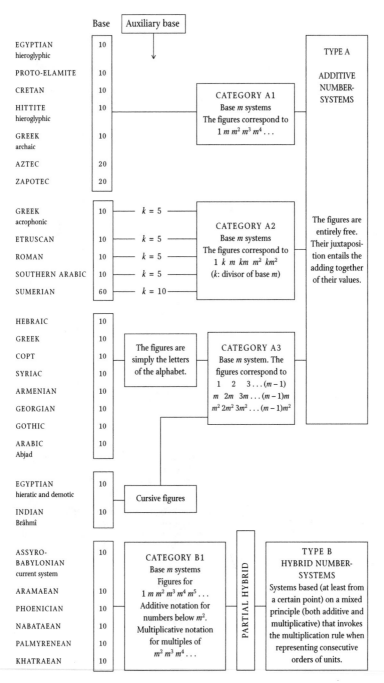

FIG. 1.40A. *Classification of written number-systems*

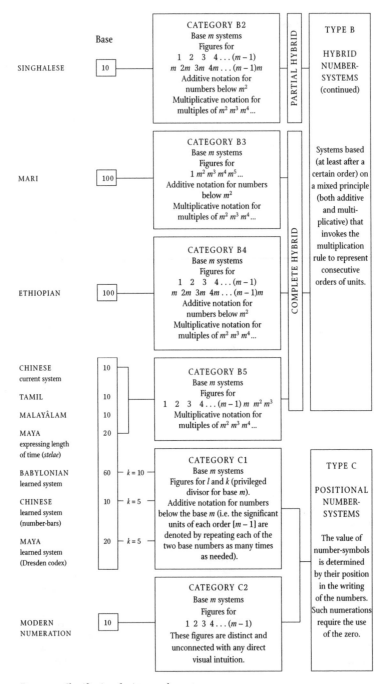

FIG. 1.40B. *Classification of written number-systems*

CONCORDANCE

From Classification to Chronology

TYPES	SYSTEMS	FIGURES	DATES
A-B-C	*Classification of written number-systems*	1.40	
A1	*Additive number-systems of the first type*	1.14	
	Egyptian hieroglyphic	1.1	−3000
	Proto-Elamite	1.2	−2900
	Cretan (Linear A)	1.3	−1900
	Cretan (hieroglyphic)	1.3	−1700
	Hittite (hieroglyphic)	1.4	−1400
	Cretan (Linear B)	1.3	−1350
	Zapotec	22.19	−450
	Aztec	1.5	+1200
A2	*Additive number-systems of the second type*	1.15	
	Sumerian	1.6	−3300
	Greek acrophonic	1.7	−500
	Southern Arabian (Minaean and Sheban)	16.19	−450
	Etruscan	16.35	−650
	Roman	1.8	−500
A3	*Additive number-systems of the third type*	1.16	
	Egyptian hieratic	1.9	−2500
	Egyptian demotic	1.10	−750
	Greek alphabetic	1.11	−350
	Brâhmî (India)	24.70	−260
	Hebraic alphabetic	1.12	−180
	Coptic alphabetic (Egypt)	17.39	+270
	Georgian alphabetic	17.41	+300
	Armenian alphabetic	1.13	+410
	Gothic alphabetic	17.42	+370
	Syriac alphabetic	19.1	+550
	Arabic alphabetic	19.4	+750

Fig. 1.41A.

CONCORDANCE

From Classification to Chronology

TYPES	SYSTEMS	FIGURES	DATES
B1	*Hybrid number-systems of the first type*	1.28	
	Common Assyro-Babylonian	1.20	−2350
	Aramaean	1.21	−750
	Phoenician	18.2	−550
	Kharoshthî (Northwest India)	24.54	−260
	Nabataean	18.2	−150
	Khatraean	18.2	+100
	Palmyrenean	18.2	+100
B2	*Hybrid number-systems of the second type*	1.29	
	Singhalese	1.22	+600/900
B3	*Hybrid number-systems of the third type*	1.30	
	Mari	1.23	−2000
B4	*Hybrid number-systems of the fourth type*	1.31	
	Ethiopian	1.24	+350
B5	*Hybrid number-systems of the fifth type*	1.32	
	Common Chinese	1.25	−1350
	Maya (long form of dates)		+292
	Tamil (South India)	1.26	+600/900
	Malayâlam	1.27	+600/900
C1	*Positional number-systems of the first type*	1.38	
	Learned Babylonian	1.33	−1850
	Learned Chinese	1.34	−200
	Learned Mayan	1.35	+400/900
C2	*Positional number-systems of the second type*	1.39	
	Modern (Indian origin)	1.36	+458
0	*Classification of zeros*	1.37	

FIG. 1.41B.

CONCORDANCE

From Chronology to Classification

DATE OF APPEARANCE	SYSTEM	CLASS	TYPE
–3300	Sumerian	A	2
–3000	Egyptian hieroglyphic	A	1
–2900	Proto-Elamite	A	1
–2500	Egyptian hieratic	A	3
–2350	Common Assyro-Babylonian	B	1
–2000	Mari	B	3
–1850	Learned Babylonian	C	1
–1800	Cretan (Linear A)	A	1
–1600	Cretan (hieroglyphic)	A	1
–1400	Hittite (hieroglyphic)	A	1
–1350	Common Chinese	B	5
–1350	Cretan (Linear B)	A	1
–750	Egyptian demotic	A	3
–750	Aramaean	B	1
–650	Etruscan	A	2
–550	Phoenician	B	1
–500	Greek acrophonic	A	2
–500	Southern Arabian (Minaean and Sheban)	A	2
–500	Roman	A	2
–450	Zapotec	A	1
–350	Greek alphabetic	A	3
–260	Kharoshthî (Northwest India)	B	1
–260	Brâhmî (India)	A	3
–200	Learned Chinese	C	1
–180	Hebraic alphabetic	A	3
–150	Nabataean	B	1
+100	Khatraean	B	1
+100	Palmyrenean	B	1
+270	Coptic alphabetic (Egypt)	A	3
+292	Maya (long form of dates)	B	5
+300	Georgian alphabetic	A	3
+350	Ethiopian	B	4
+370	Gothic alphabetic	A	3
+410	Armenian alphabetic	A	3
+458	Modern (Indian origin)	C	2
+400/900	Tamil (South India)	B	5
+550	Syriac alphabetic	A	3
+600/900	Learned Maya	C	1
+600/900	Singhalese	B	2
+600/900	Malayâlam	B	5
+750	Arabic alphabetic	A	3
+1200	Aztec	A	1

Fɪɢ. 1.42.

CHAPTER 2

FROM THE PARTICULAR TO THE GENERAL: ARITHMETIC LEADS TO ALGEBRA

Many believe that the Arabs invented algebra, but this is certainly a misapprehension resulting from loose language and from misunderstanding of the history of this science.

Algebra has, in fact, existed since Antiquity, as is shown by the fact that the Sumerians, the Babylonians, the Egyptians, the Greeks, and the Chinese knew how to make use – for example in their geometrical calculations – of various numerical relationships which imply, beyond any doubt, that they grasped some elementary algebraic concepts.

We also know that the Babylonians could solve equations of the first and second degrees (linear and quadratic equations), and that the Chinese were able to solve systems of equations in several unknowns (using numerical checkerboards with rods) by a method quite closely related to the modern approach based on matrices and determinants (see Chapter 21).

But none of these peoples grasped these algebraic ideas in the abstract sense, that is to say in a way which could lead on to mathematical generalisation, with the possible exception of the Chinese who had a positional decimal notation for numbers (though this lacked the zero which was only introduced much later).

It could hardly have been otherwise, in fact: this is yet another consequence of the inadequate notational systems of ancient times.

Even though the Chinese discovered many important results, the imperfections of their positional number-system stood in the way of further development of their methods, and prevented them from creating a science which could well have achieved the heights of contemporary mathematics (see above, Fig. 1.34 and 38).

The Indian thinkers were, in fact, the first to attain concepts which marked the beginnings of algebra in a properly abstract sense (see the references to Brahmagupta in Chapter 24, the various Dictionary entries for "Alphabet", and the Dictionary entry for "Indian mathematics"). They succeeded, in effect, in converting their arithmetical methods into algorithms (rules of procedure) which could then, in principle, be applied to numbers of any size independently of the quantities these represented: probably the first time, in all history, that such a step had been achieved.

These algorithms could be applied to abstract numbers where zero and negative quantities also stood as fully-fledged numbers and not as mere conventional indicators.

We may aptly say, indeed, that the mathematicians of the Indian sub-continent were the first to discern the general in the particular (cf. A. Marc, *Dialectique de l'affirmation*, p. 128).

Their number-system, with its zero and its positional notation, had already given them the possibility of developing a science of algebra, capable of supporting an abstract theory of numbers, and of advancing mathematical knowledge.

> With this kind of notation, numerical values are no longer represented by symbols, with a separate symbol for each value. All at once, they lose their specific and intrinsic characters; instead, having to a degree come to depend on the mutual positions of the digits in their representations, it is their relative characteristics which come to the surface. The cyclical regularity of their construction yields up a law, which permits the generation of the whole infinity of numbers by a procedure which never varies, and yet which never obscures exact perception of their relationships. (L. Massignon and R. Arnaldez)

However, the algebra of the Indians was incapable of attaining the level of our modern algebra, since it lacked a general notation.

The following summary will give us a clearer idea of how things developed. According to Tobias Dantzig, algebraic notation passed through three successive stages.

The first was *terminological algebra*, devoid of symbols except for the use of words themselves as symbols. So it is that, when we say that "the product of two numbers is independent of the order in which they are taken," we are using this same terminological algebra even though, in modern algebraic notation, the same thing is expressed by the formula $a \times b = b \times a$.

In the second stage, algebra became *abbreviated*: certain words of frequent use were replaced by shortened forms, and thus became "algebraic ideograms". As time passed, this abbreviation became more extreme until, finally, there remained no evident link between the symbols and the words originally used for the operations they represented. At this stage, the abbreviated word had become a true algebraic symbol.

We can illustrate this by looking at subtraction. In Western Europe, subtraction was originally denoted by the word *minus*; then this was abbreviated to $\bar{m}$; and finally the m disappeared, leaving only the bar, so that we arrive at our modern symbol '−' for subtraction from which even the m of the original *minus* has disappeared.

In the third and final phase we arrive at *modern symbolic algebra*, to which we shall turn shortly.

Meanwhile, the algebra of Âryabhata, Bhâskara, Brahmagupta, Mahâvîrâchârya and even of Bhâskarâchârya remained at the *abbreviated* stage. The symbols of the Indian mathematicians were simply the first syllables of the words for the corresponding operations. Moreover, their statements and solutions of mathematical problems were usually in verse, and therefore subject to the constraints of Indian verse metre (see "Bîjaganita" and "Poetry and the writing of numbers" in the Dictionary of Indian Numerical Symbols).

We can thus understand how Indian algebraic symbols remained wrapped in a kind of verbiage, subject to interpretation and with the interpretation itself liable to as many variations as there were interpreters; and that the symbols were locked into the same restrictions as secular usage had imposed on the words and letters of the Sanskrit language (see "Indian mathematics (history of)" in the Dictionary of Indian Numerical Symbols).

In this domain, therefore, the seeds of Indian civilisation were destined to germinate on foreign soil.

It was the Arab and Persian mathematicians Al Khuwārizmī, Abū Kāmil, Al Karajī, 'Umar al Khayyām, As Samaw'al al Maghribī, Nasīr ad dīn at Tusī, Ghiyat ad dīn Ghamshīd al Kāshī, Al Qalaṣādī and others who, much later, brought the science of algebra to its earliest spectacular developments, and turned it in the direction of the modern algebra which we know. Graced with a practical turn of mind, and with thought that tended to synthesis, these thinkers were better equipped than their Indian predecessors to convert arithmetic into a technique of algebraic operations; they finally, in the words of P. Boutroux (1922), created "a positive and constructive science" (see Chapter 25).

However, the immense flowering of algebra in later epochs is due principally to the essential work of European mathematicians, both of the Renaissance and of the Classical eras.

We may cite Fibonacci, Luca Pacioli, Nicolas Chuquet, Jean Bourrel, Christoph Rudolff, Michael Stifel, Scipione dal Ferro, Gerolamo Cardano, Raffaele Bombelli, Johannes Buteo, Robert Recorde, François Viète, Johann Widmann, Jacques Peletier du Mans, René Descartes, etc. among the principal personages whose fundamental contributions paved the road from *particular reasoning* to *global reasoning*.

As Condillac (*Histoire moderne*, XX, 6) says: "Algebra is to Arabic numerals as these are to Roman numerals." To paraphrase Jean Le Rond d'Alembert (*Explications générales des connaissances humaines*): Arithmetic comes in two kinds – one is arithmetic by numbers; the other is arithmetic in algebraic terms according to procedures of calculation which are applied to magnitudes in general.

And, because "algebra is the one well made language," modern mathematics "is a correctly treated science," since the language of mathematics is made up of the terms and expressions of algebra (Condillac, *La Langue des calculs*, Introduction, II, p. 420).

Dantzig has emphasised a striking analogy between the history of algebra and the history of arithmetic. We ourselves, in this very work, have seen how humankind has beaten its head for millennia against the problems of inadequate notation, through being deprived not only of a positional notation but also – and above all – of a symbol to represent a null quantity. "In algebra, the lack of a general notation reduced this subject, for a long time, to a collection of rules, established *ad hoc*, for solving particular numerical equations. Just as the discovery of zero created modern arithmetic, so the development of algebraic notation (i.e. the use of letters and other signs) launched a new era in the history of algebra."

This was a decisive step not only in the history of notation, but also in the history of symbolic thought, since algebraic notation reflects a move from the level of abstraction of the grammarian to that of the pure logician, for whom abbreviation becomes a mere symbol [see A. Koyré in Taton (1957–1964)].

Until this step was taken, in fact, mathematical propositions remained trapped, either in a purely terminological language vulnerable to random interpretations, or locked into a semi-concrete mode of thought like that of grammarians who use general rules, but apply them to concrete instances. Similarly, prior to algebraic notation, people dealt only with specific mathematical expressions, which could only be handled in terms of their specific characteristics.

On the other hand, the mere notion of using letters to denote variables or unknowns or undetermined constants liberated algebra from the yoke of the word. In contrast with the miscellaneous names and locutions employed until that time to express preconceived ideas such as "number", our modern 'x' is completely independent of the nature of what it is supposed to represent.

Algebraic notation is a bridge from the individual to the collective. An expression such as $ax^2 + bx + c$ is no longer a symbol which stands for a particular number, but is instead a characteristic form for a whole category of numbers.

We have come to see the equivalence between mathematical propositions expressed verbally, and the corresponding expressions formed solely from letters and symbols which represent arbitrary numbers. In so doing, we enabled ourselves to move from specific reasoning, bearing on particular objects, to general reasoning which bears on the properties which all instances of a particular kind have in common. The science of algebra has thus risen to a level far above a simple technique of transcription.

It is precisely this aspect which has not only brought about the general theory of functions (one of the foundations of modern applied mathematics), but has also facilitated the algebraicisation of analysis, and the emergence of analytical geometry. Modern algebra is a remarkable advance on simple arithmetic!

Chapter 3 describes the most important milestones of its development, while Chapter 5 demonstrates the extent to which this led to the abstraction which is characteristic of modern mathematics and science, and also how it led to the origins of the modern computer.

FROM CALCULATION
TO CALCULUS

The word *calculation*, and the more specifically mathematical word *calculus*, are derived from the Latin *calculus* whose primary meaning is "pebble". By extension, *calculus* in Latin also means "ball" or "counter". In connection with calculation, this etymology evokes not only the ancient techniques of manipulating the abacus, but also the even more ancient pile of pebbles from which our prehistoric ancestors acquired their initiation into the tricks of elementary arithmetic (see Chapters 1, 2, 11 and 12).

The ancient Romans used pebbles and counters to teach their children to do sums, whereby the word *calculus* came to refer to the basic operations of arithmetic (addition, subtraction, multiplication and division). The word *calculator* (the same as in Latin) has been used since the sixteenth century to refer to someone who calculates, or who is skilled in arithmetical procedures, or, again by extension, who is skilled in planning and scheming.

Such simple meanings, however, became very greatly extended as the science of mathematics developed over the centuries, and have diversified into many domains.[1]

Calculation originally, and over a long period, meant *numerical calculation* – the repertoire of arithmetical operations such as addition, subtraction, multiplication and division, and procedures such as extraction of square roots, which could be applied to arbitrary numbers: integers, fractions, and rational, irrational and transcendental numbers. Subsequent developments of such procedures into powerful methods for the solution of general mathematical problems in numerical terms came to be known, in English, as *numerical analysis* (though all the basic mechanisms are the traditional arithmetical operations).

[Quite early, however, in the history of modern mathematics the meaning of *calculation* took on a more general sense, well illustrated by the following quotation from Gottfried Wilhelm Leibniz (1646–1716) as cited by E. T. Bell:[2]

1 While an occurrence in English of the word *calculus* can almost always be directly translated into the French *calcul*, it is less often so in the reverse direction. Both *calculus* and *calcul* are generally used to refer to some unified body of mathematical theory and methods: examples will be cited below. The French use of the word *calcul* is more extended than the English use of *calculus*. Thus where the French may refer to *calcul algébrique*, the English would simply use *algebra*. The French *calcul numérique*, if used to refer to the study of arithmetical operations on arbitrary numbers, would be *arithmetic* in English; while *calcul numérique* in the sense of the study of procedures developed for the numerical solution of mathematical problems would be *numerical analysis* in English. [*Transl.*]

2 E. T. Bell, *Men of Mathematics*, p. 123.

Newton's miraculous year 1666 was also the great year for Leibniz. In what he called a "schoolboy's essay," *De arte combinatoria*, the young man of twenty aimed to create "a general method in which all truths of the reason would be reduced to a kind of calculation. At the same time this would be a sort of universal language or script, but infinitely different from all those projected hitherto; for the symbols and even the words in it would direct the reason; and errors, except those of fact, would be mere mistakes in calculation. It would be very difficult to form or invent this language or characteristic, but very easy to understand it without any dictionaries."

[In this vision of Leibniz, thought and reason, and the resolution of the theoretical problems of philosophy or the practical conundrums of the law, could be brought down to the manipulation of the symbols of his dream language, the "universal characteristic". It was not a wild vision: the symbolic logics developed in the nineteenth and twentieth centuries, their modern developments in computation and the theory of computability, and the practical fields of artificial intelligence and expert systems, are realisations of large parts of Leibniz's dream.

[Meanwhile, in the above quotation from Leibniz we see the word *calculation* used in the most general sense imaginable of the mechanistic manipulation of entities which symbolise objects of thought and the relations between them. *Transl.*]

In fact, any comprehensive set of mathematical methods applicable to a particular class of mathematical problems which are bound into an integrated whole by an appropriate mathematical theory, is known as a *calculus*, and nowadays there are many of them.

[Perhaps best known are the *differential calculus* and its inverse theory the *integral calculus*, invented (probably independently) by both Newton and Leibniz, whose elements are part of the more advanced mathematical teaching in secondary schools and are the stock-in-trade of every scientist and engineer. Between them, they embrace the mathematical theories needed for the quantitative treatment of continuously variable quantities. The area of mathematics commonly referred to as *calculus* is the two taken together; by virtue of the underlying technique of studying small variations which are allowed to be indefinitely small, it is also known as *Infinitesimal Calculus*.

[Especially in physics and engineering, *vector calculus* plays a vital role. This is the mathematics of *vectors*, quantities which not only have numerical magnitude but also direction, such as velocity (e.g. a velocity of 2,000 ft/sec Northwest at an elevation of 20 degrees above the horizontal; as it might be in gunnery) and force (e.g. an object is attracted by the Earth's gravitational field with a force whose magnitude is its weight, towards the centre of the Earth; as it might be, the shell fired in the previous example). *Transl.*]

An advanced extension of vector calculus is *tensor calculus*, the mathematics of *tensors*: quantities which have several simultaneous numerical magnitudes and also depend on several directions at once (compared with the single magnitude and direction of vectors). [Originally developed in the nineteenth century by the Italian mathematicians G. Ricci and T. Levi-Cività out of earlier work of the German B. Riemann which in turn generalised the work of Gauss on the mathematics of curved surfaces, this was the mathematics of curved geometrical spaces. Its most prestigious application was its adoption by Einstein as the mathematical language of his General Theory of Relativity (1915), but it is also much used in more down-to-earth subjects such as Elasticity and the theory of deformation of non-rigid objects, as well as in its original territory of *differential geometry*.

[As an example of a *tensor*, consider the deformation in the neighbourhood of a given point within a rubber ball which is squeezed. Prior to squeezing, all the points at a constant separation from the given point are on the surface of a sphere. After squeezing, these points lie at different separations from the given point, and the displacement of such a point (which is a vector, having magnitude and direction) depends in its original distance and direction from the original point (which is another vector). The full mathematical description of the deformation, namely the aggregate of the displacements of these points, depends on two vectors, and so it is a tensor.

[A more abstract example is the *predicate calculus*, a formal theory for the construction and manipulation of abstract logical expressions. This, which clearly descends from Leibniz's vision of a "universal characteristic", plays an important role not only in abstract mathematical logic but also in the construction of certain kinds of computer programme. *Transl.*]

A *calculus*, therefore (as can indeed be seen from the following chronology), may take on a wide range of meaning. In general, the word refers to a body of theory and methods applicable to a domain symbolising objects of a certain kind (numbers, vectors, tensors, logical propositions, etc.), and adapted to obtaining the solution of problems formulated in this domain by means of more or less mechanistic manipulation of the symbols of the domain according to the rules and methods of the calculus.

18th century BCE. The Mesopotamian scholars could solve equations of the second degree (quadratic equations), as well as some equations of the third (cubic) and fourth (quartic) degrees. They also knew how to solve certain systems of simultaneous equations in two unknowns, including some cases where one equation was of the first degree (linear) and one of the second (quadratic).

16th century BCE. The Egyptians were in possession of a primitive form of algebra at this period, which they were able to use for solving various problems concerning the division of goods and chattels which involved the solution of algebraic equations of the first degree (linear). They also had an elementary form of algebraic notation. The hieroglyph *hau*, meaning a heap, was used to denote the unknown in these equations. The pictogram of a leg was often used as a mathematical sign to represent addition or (in a symmetrical form) subtraction, or sometimes to denote raising to the second power (squaring).

5th century BCE. This was the time of the Greek mathematician and philosopher Pythagoras, forerunner of the great Greek mathematicians of Antiquity. Pythagoras and his school adhered, however, to a mystical and occult philosophy of number. According to tradition, Pythagoras discovered that it is impossible to find two whole numbers such that the square of one is twice the square of the other, or (in geometrical terms) that the diagonal of a square is *incommensurable* with the side of the square. This amounted to discovering that $\sqrt{2}$ is an *irrational number* (i.e. cannot be expressed as the ratio of two integers). In keeping with the mystical attitude of the Pythagoreans to number, this "imperfection" was long concealed from public view (see Chapter 27). This discovery exposed the existence of the first known of the class of irrational numbers, which are neither integers nor fractions. Other discoveries of the Pythagoreans (stripped of their mystical motivations) include studies of odd and even numbers, and of *figurate numbers*: integers such that such a number of balls or counters can be packed into a regular geometrical figure on a plane or in three dimensions, giving rise to square numbers, triangular numbers, cubic numbers, pyramidal numbers, and so on.

3rd century BCE. The time of the great Greek mathematician Archimedes. He extended the Greek number-system so as to enable the expression of very large numbers, and for instance used it to calculate the numbers of grains of sand that could be contained in the Universe (see Chapter 17). He also showed that the number π (the ratio of the circumference of a circle to its diameter) was between 3.14084 (= 3 +10/71) and 3.14285 (= 3 + 10/70):

by calculating the areas and circumferences of regular polygons of n and $2n$ sides inscribed to, and circumscribing, a circle, he was able to calculate lower and upper limits for the area and the circumference of the circle to a known degree of approximation, from which the limits given above for π were obtained. This method of calculation is still known as *Archimedes's algorithm*.

3rd century BCE. This was also the period of the famous Euclid, another of the great Greek mathematicians. In his *Elements*, he established the foundations of classical (now universally known as Euclidean) geometry. He devoted Book IX of the *Elements* to the study of odd and even numbers, and of the even perfect numbers (for the classical Greeks, the study of number was deeply linked to their study of Geometry). Book VIII was devoted to the integer powers of fractions (or rather, at that time, to whole numbers in geometric progression since fractions were only conceived of as ratios between whole numbers).

Beginning of the Common Era. At this time the *Jiu zhang suan shu* ("The art of calculation in nine chapters") was compiled, a major work which demonstrates that the Chinese already knew how to solve various sorts of equation and of systems of equations using their abacus with rods (*suan zi*) along with a conventional representation of positive and negative numbers. Much later (towards the eleventh century) they even managed to use the abacus to solve systems of algebraic equations by a method related to the modern theory of matrices.

2nd century CE. The neo-Pythagorean Greek mathematician Nicomacchus of Gerasa published *Introduction to Arithmetic*, which expounds the theory of figurate numbers at length.

3rd century CE. The Greek mathematician Diophantus of Alexandria published *Arithmetica*. Here, for the first time in the history of Greek mathematics, fractions are recognised as genuine numbers. He not only deals with equations of the first degree, but also those of higher degree. He introduces an elementary algebraic notation by using the first syllable of the Greek word *arithmos*, which means "*number*", to denote the unknown. His problems of indeterminate analysis laid the foundations for the modern theory of equations (and the study of integer solutions of so-called "Diophantine equations" remains an area of active research in modern Theory of Numbers). Despite the primitive character of his symbolic notation and the inelegance of his methods, he can be considered as one of the pioneers of modern algebra.

4th–5th centuries CE. At this time occurred the Indian discovery of zero and of the positional decimal number-system, out of which grew the generalisation of number which led to algebra and the growth of modern mathematics.

628. The Indian mathematician Brahmagupta published *Brahmasphut-asiddhânta*, showing complete mastery of positional decimal notation with zero. He described methods of calculation very similar to those we use today; his fundamental rules of algebra set out a system embracing positive and negative numbers, in which the zero appears as a true number (the null quantity). He also speaks of mathematical infinity which he clearly defines as the inverse of zero (see "Infinity" in the Dictionary of Indian Numerical Symbols).

629. The Indian mathematician Bhâskara published his Commentary on the *Âryabhatîya*, in which he gives valuable information about contemporary Indian mathematics and astronomy. The work shows that he is not only completely at ease with the zero and the positional decimal notation, but also is very skilful at the rule of three and with fractions, which he writes in a manner similar to our own, though lacking the horizontal bar (which was only introduced later, by Arab mathematicians).

820–850. The Arab astronomer Al Khuwārizmī worked at this time. He contributed greatly to the diffusion of the Indian number-system, and of the Indian methods of numerical and algebraic calculation, into the Arab and European worlds. He also made important contributions of his own to these methods, and the latinised form of his name – *Algorismi* or *Algorismus* – gave rise to the word *algorism* (later also *algorithm*).

An *algorism*, originally, was a method of written calculation using Indian numerals, as opposed to methods based on the use of an abacus. In due course, however, it acquired its more general and abstract (and modern) meaning of a systematic procedure of calculation (applicable to any numbers or indeed to any formal entities whatever).

The very word *algebra*, moreover, we also owe to him: it comes from the title of his book *Al jabr wa'l muqābala* ("transposition and comparison"), which describes the fundamental algebraic manipulative technique of transferring a negative term from one side of an equation to the other (where it becomes positive), and the reduction of several similar algebraic terms to one (*al muqābala*).

10th century CE. This is the period of three notable Arab mathematicians:

The great algebraist Abū Kāmil, who continued the work of Al Khuwā rizmī, and whose advances would later be put to use by the Italian mathematician Fibonacci.

Abū'l Wafā' al Bujzānī: he wrote commentaries on Euclid, Diophantus and Al Khuwārizmī. It is thanks to his work that the *Arithmetica* of

Diophantus and his studies of algebra and the theory of numbers came to the notice of European mathematicians.

Al Uqlīdisī, who published an important treatise on decimal fractions.

End of 10th century CE. The Arab mathematician Al Karajī carried out important studies of the arithmetic of fractions. Drawing on the work of Diophantus and of Abū Kāmil, he developed an algebra in which, as well as the familiar equations of the second degree (quadratic equations), he also discussed certain equations of degree $2n$. In his work on the manipulation of irrational numbers we can see an alliance between rigour of treatment and an approach more flexible than the purely geometric algebraic methods of the Greeks. Indeed, this is the beginning of a revolution in mathematical symbolism which will ultimately eliminate all geometrical methods of representation from Arab algebra and arithmetic.

11th century CE. The period of two important mathematicians:

The Arab Kushiyar ibn Labbān al Gilī, who worked on the Indian arithmetical methods and on the Babylonian sexagesimal arithmetic.

The famous Persian poet and mathematician 'Umar al Khayyām (Omar Khayyam), who made important studies of the theory of proportions and of equations of the third degree (cubic equations).

12th century CE. The period of As Samaw'al ibn Yahyā al Maghribī, a Jew converted to Islam, who carried on the work started by Al Karajī.

1202. By this time Leonard of Pisa, more familiarly known as Fibonacci, had travelled to North Africa and the Middle East and had acquainted himself with the Indian number-system based on nine digits. In 1202 he published *Liber Abaci* ("The Book of the Abacus"), destined to become a fertile source of later Western European developments in algebra and arithmetic.

13th century CE. The Arab mathematician and astronomer Nasīr ad dīn aṭ Ṭūsī, who did important work in arithmetic and algebra.

15th century CE. The period of:

Ulugh Bek, the enlightened ruler of Samarkand, noted for his Trigonometrical Tables, which are amongst the most accurate drawn up by Islamic scholars.

The Persian mathematician Ghiyat ad dīn Ghamshīd al Kāshī. He made advances in several areas, amongst which were: algebra; sexagesimal arithmetic; the binomial formula; calculation with decimal fractions; integer powers of numbers; nth roots; irrational numbers.

1478. The year of publication of the Treviso *Arithmetic*, an anonymous manual of practical arithmetic. This clearly shows that arabic numerals had become widespread, and that the Indian methods of written calculation were finding increased favour. This work also makes deliberate efforts to develop abbreviated notations for the operations of algebra and arithmetic, whose functions had previously been expressed in words.

1484. Nicolas Chuquet, a French mathematician, made skilled use of the zero and the written methods of calculation which had come from India. Especially, however, he employed negative numbers, which he was almost certainly the first to introduce into Western Europe. In order to devise a more tractable notation, he proposed a fully developed technical terminology to express powers of 10 above 1,000. He also introduced an exponential notation for positive and negative powers which is somewhat similar to the one we use today.

1487. Luca Pacioli (Luca di Borgo), an Italian, published an important book on arithmetic and algebra. This gives the principal Indian methods, and rules for solving numerous algebraic equations. He made use of a primary unknown which he called *cosa* (thing) to which he sometimes adjoined a secondary unknown called *quantità* (quantity).

[Pacioli's most famous work is the 1494 *Summa de Arithmetica, Geometria, Proportioni et Proportionalita*, the first printed work on mathematics, a very complete treatise on the mathematical knowledge of the time which included several chapters on accountancy methods. *Transl.*]

1489. The German mathematician Johann Widmann of Eger introduced the signs + and − instead of the letters *p* (for più, greater) and *m* (for minus, less) which had hitherto been used to represent addition and subtraction.

1525. The German mathematician Christoph Rudolff introduced the square root sign √, an abbreviated way of writing the letter *R* (for the Latin word *radix*, root).

1544. The German mathematician Michael Stifel published *Arithmetica integra*, a treatise on arithmetic and algebra. This mainly deals with rational and irrational numbers, and consciously strives for simplification of theories and notations.

1545–1560. The Italian mathematicians Gerolamo Cardano and Raffaele Bombelli introduced "imaginary numbers" (what would generally be called complex numbers nowadays) in order to obtain solutions to cubic equations.[3] This extension of the concept of number was destined to play a fundamental role in later developments of abstract mathematics, as well as in many branches of applied mathematics (especially, from the nineteenth

---

3 Mathematical completeness inspires the desire that every equation should have a solution; but there is no "real" number such that $x^2 + 1 = 0$, or $x^2 = -1$. Cardano and Bombelli said, in effect, "let there be one", so that this number is $x = \sqrt{-1}$. Since it is not "real", it is "imaginary". They were then able to obtain the

century, in electricity and magnetism, and in the study of wave motion and oscillations of all kinds).

1557. The English mathematician Robert Recorde introduced the symbol = to express equality.

1582. The Flemish mathematician Simon Stevin (Stevinus), born in Bruges, published *De Thiende* ("The Tenth"; variously dated also to 1585 and 1586). This was the first work published in Europe which was devoted to the general theory of decimal fractions (i.e. fractions expressed not as ratios but as series of digits). Such fractions had, of course, been considered at much earlier times (e.g. by Al Uqlīdīsī (952), Emmanuel Bonfils de Tarascon (1350), Regiomontanus (1463), Christoph Rudolff (1525), Elias Mizrachi (1535) and Franciscus Vieta (François Viète) (1579). With the possible exception of the work of Ghiyat ad dīn Ghamshīd al Kāshī (first half of fifteenth century) which was unknown in the West, no-one else apart from Stevin had had the idea of using such fractions in place of ordinary fractions; and he was the only one to develop a system of notation which brought the methods of calculation with such fractions into line with the methods used for calculating with integers.

1582. Simon Stevin made the decisive step in the direction of our modern notation for decimal numbers: where we would write 679.56 he wrote 679 (0) 5 (1) 6 (2), denoting 679 units, 5 tenths (decimal units of the first order), and 6 hundredths (decimal units of the second order).

1591. The French mathematician Franciscus Vieta (François Viète) introduced alphabetic notation into algebraic expressions. He represented the unknowns of an equation by vowels (A, E, ...) and undetermined constants by consonants (B, C, ...).

1592. The Swiss mathematician and watchmaker Jost Bürgi simplified Stevin's notation for decimal fractions by suppressing the redundant indications of fractional order, and simply placing a ring sign above the digit in the units position: 679̊ 56 (= 679.56).

1592. The Italian Magini brought decimal notation to its modern form, replacing the ring sign by a dot placed between the units digit and the first fractional digit, as in 679.56.

1608. The Dutch mathematician and scientist Willebrord Snellius of Leyden made the change from Magini's decimal point (dot) to the comma (679,56) commonly used in Continental Europe.

---

solutions of cubic equations in the form $a + b \sqrt{-1}$, a "complex number", in which a and b are ordinary "real" numbers. Gauss later (1799) obtained the essence of the proof that every algebraic equation of degree n has a solution of this form. There is, of course, nothing imaginary about an "imaginary" number: it is simply a more complicated mathematical construct than a "real" one. [*Transl.*]

1614. The Scottish mathematician John Napier of Merchiston had discovered (c.1594) the correspondence between the terms of an arithmetic progression (proceeding by successive additions of a constant quantity) and the terms of a geometric progression (proceeding by successive multiplication by a constant factor).[4] This was the basis for his invention of logarithms, by which multiplication can be achieved by using addition. He published the first table of logarithms, to the base e.[5] Jost Bürgi had independently made the same discovery in 1604.

1615. Henry Briggs (later Savilian professor of geometry at Oxford) found Napier's logarithms inconvenient for practical calculation (as a consequence of using base e). Having stayed a month with Napier in 1615, Briggs constructed a table of logarithms to base 10 (a change that Napier had already envisaged), for the first 1,000 integers, and with the entries given to 14 decimal places.

1631. Thomas Harriot, the English mathematician, introduced the symbols < and > to represent inequality. [He is also noted for his discovery (the most fundamental ever made in algebra) that an algebraic equation of the nth degree can be expressed as a product of n equations of the first degree (linear equations). *Transl.*]

1632. William Oughtred, another English mathematician, introduced the sign × for multiplication.

1637. The great French mathematician and philosopher René Descartes invented Analytical Geometry. [This consists of the use of numbers (coordinates) to represent the positions of points, and the use of algebraic equations to specify the sets of points which constitute lines, curves, and surfaces, thereby bringing geometrical questions under the power of the algebraic method. *Transl.*] He also created the modern algebraic notation, using the early letters of the alphabet ($a, b, c, \dots$) to represent constants and the later letters ($\dots, x, y, z$) to represent unknowns, and he introduced the modern exponential notation to represent positive powers of a quantity (a^2, a^3, etc.).

1640. Pierre de Fermat, another great French mathematician (and recently famous following the final solution of the problem of finding a proof to "Fermat's Last Theorem"), laid down the basis of combinatorial analysis in

4 For instance, $2 = 2^1$, $4 = 2 \times 2 = 2^{1+1} = 2^2$, $8 = 4 \times 2 = 2^2 \times 2^1 = 2^{2+1} = 2^3$ and so on. [*Transl.*]

5 The familiar logarithm tables are to base 10 ("common logarithms"). Any number can be expressed as a positive or negative power of any fixed number (the base) which is not equal to 1. Thus, $2 = 10^{0.3010\cdots} = 0.1^{-0.3010\cdots}$, for instance; the power is the logarithm of the number to the given base so that $\log_{10} 2 = 0.3010 \dots$. The base used by Napier is the number denoted by e whose value is equal to 2.7182818... (a non-terminating decimal). This apparently strange number is in fact fundamental in mathematics. Strictly speaking, Napier's logarithms were not to base e though very closely related. [*Transl.*]

his theory of magic squares (square arrays of numbers, all numbers different, such that all the rows and all the columns add up to the same number).

1654. In correspondence with Pierre de Fermat and the gambling Chevalier Georges de Méré, Blaise Pascal laid the foundations of the calculus of probabilities.

1656. The Dutch scientist Christiaan Huygens published *De ratiociniis in ludo aleae* ("Speculations on games of dice"), which was the first complete treatise on probabilities.

1656. The English mathematician John Wallis extended the exponential notation for powers of a quantity to negative and fractional powers, so that

$a^{-1} = 1/a$, $a^{-2} = 1/a^2$, $a^{-3} = 1/a^3$, $a^{1/2} = \sqrt{a}$, $a^{2/3} = \sqrt[3]{a^2}$, etc., and he introduced the symbol ∞ to represent infinity.

1670–1684. [Gottfried Wilhelm Leibniz and Isaac Newton invented the infinitesimal calculus: more or less independently, with Newton somewhat earlier in time, it would seem, though inevitably with some interaction over the several years of their work. Newton's method was called the *method of fluxions* (rates of change), Leibniz's the *differential method*. The two authors used different principles of notation. For Newton, the rate of change of a variable quantity x was called its *fluxion* and denoted by $\dot{x}$; for Leibniz, it was a ratio of *differentials* and denoted by dx/dt (nowadays the *derivative*); but these denoted the same thing. The methods of obtaining the fluxion (or derivative) of a variable given in terms of another were relatively mechanical; the inverse problem, of recovering the expression of the variable from the expression of its fluxion, was much harder, but both workers succeeded in solving it for several cases: it amounts to finding an equation for the area under a curve, the equation for the curve itself being given; and the solutions that Newton and Leibniz respectively obtained depended on the powerful techniques for manipulating infinite series which each had developed. The notations used by Leibniz both for the derivative and for the inverse (integration) proved much more useful than Newton's notations, and ultimately came to dominate mathematical usage, though both mathematical theories were essentially equivalent. Leibniz's *differential calculus* is Newton's *method of fluxions*; Leibniz's *integral calculus* is Newton's *inverse method of fluxions. Transl.*]

Leibniz, in 1684, applied his calculus to create "a new method for determining maxima and minima".

1679–1690. Leibniz established the earliest foundations of symbolic logic.

1696. The Marquis de l'Hospital, a French mathematician, published *Analyse des infiniment petits pour l'intelligence des lignes courbes* ("Analysis of infinitely small quantities, for the investigation of curved lines"). This was the first of the classical treatises on differential calculus. [A supplement which extended the treatment to integral calculus was published in 1754–1756 by L. A. de Bougainville. *Transl.*]

1737. The great Swiss mathematician Leonhard Euler demonstrated that the number e (the base of natural logarithms) is irrational, as also is e^2; he brought logarithmic calculation to its modern form.

1748. Euler published a treatise on the general theory of functions, especially the exponential, logarithmic and trigonometric functions, in which the developments of these special functions as infinite series and infinite products are notable. The close relationships which exist between the trigonometric and exponential functions were established.

1750. The Swiss mathematician Gabriel Cramer undertook the earliest systematic study of systems of equations.

1760. The German mathematician Johann Heinrich Lambert demonstrated that the numbers e and π are irrational (cannot be expressed exactly as ratios of integers).

1770. The French mathematician Alexandre Vandermonde worked on the solutions of systems of equations and on determinants, and studied combinatorial theory.

1786. The French mathematician A. M. Legendre embarked on a deep study of elliptic integrals (functions which are the solutions of differential equations such as describe the motion of a pendulum).

1795. The Law of 18 Germinal Year III of the Republic (7 April 1795) was passed, which gives the first definition of the metre (as unit of length, one ten-millionth of the distance over the Earth's surface from the North Pole to the Equator) and fixed the current nomenclature for weights and measures of the metric system (see Chapter 2.)

1797. The French mathematician Louis de Lagrange published *Théorie des fonctions analytiques* ("Theory of analytic functions") [in which, along with his later work *Leçons sur le calcul des fonctions* ("Lessons on the calculus of functions") published in 1801, he undertook a revision of the foundations of differential and integral calculus so as to eliminate the difficulties of Newton's method of fluxions and of Leibniz's infinitesimal quantities. This laid the foundations for the later fundamental work of Cauchy and subsequent workers, which gave rise to the rigorous theories of modern mathematical analysis. *Transl.*]

1799. The German mathematician Karl Friedrich Gauss (one of the greatest of all time) gave a rigorous proof of the Fundamental Theorem of Algebra, that every algebraic equation has a root: more precisely, that for any polynomial of degree n with integer coefficients, there is a complex number for which the value of the polynomial is zero. This theorem had already been stated, but no proper proof had been given; [indeed, the proof is not purely algebraic but depends on the analytic structure of the complex numbers, on considerations of continuity, and on topological matters (namely, that a continuous closed curve drawn on a plane, which winds at least once round a given point, cannot be continously deformed and shrunk into a single point without at some stage passing through the given point). Indeed, the "fundamental theorem of algebra" is not so much a statement about algebra as a statement about the properties of the continuum of complex numbers.[6] *Transl.*]

1799. The first international conference to consider the question of the universal adoption of the metric system met in Paris. The proposition was, however, considered too radical for the different nations at that time.

1800. Gauss gave precise definitions of the concepts of upper and lower bound, and upper and lower limit, of an aggregate of real numbers. [These definitions lie deep in the foundations of modern mathematical analysis, since they are indispensable in discussing the limiting process of mathematics. *Transl.*]

1807–1822. The French mathematician Joseph Fourier created the theory of trigonometric series now known as Fourier series, [according to which any continuous function of a real variable can be expressed over a finite interval as the sum of an infinite series of trigonometric functions (sines and cosines). The techniques of Fourier series allowed the solution of whole fields of differential equations, especially of the kind that occur in Physics,[7] and indeed Fourier's fully developed theory was published, in 1822, in his immortal book *Théorie analytique de la chaleur* ("Analytical theory of heat"). *Transl.*]

1821. The French mathematician Augustin Cauchy published *Analyse algébrique* ("Algebraic analysis") which laid down a rigorous theory of the elementary functions (algebraic, exponential, logarithmic, trigonometric) and, for the first time, dealt clearly and systematically with the concept of continuity in mathematics.

6 See, for instance, G. Birkhoff and S. MacLane, *A Survey of Modern Algebra*, New York: Macmillan, 1977, Chapter 5, section 3.

7 The conditions under which the use of Fourier series is valid involves special subtleties. At a later period (c.1872–1874) Georg Cantor, who began studying the convergence of Fourier series, was led on to the first of his profound discoveries in the theory of sets (the result that there are exactly as many roots of algebraic equations of all degrees as there are integers, no more and no less: both numbers are infinite, but the two sets are in a one-to-one correspondence). This result was published in 1874; Cantor's method of proof became a paradigm which was used by Gödel in the proof of his famous Incompleteness Theorem, and by Turing in his abstract theory of computer programmes. [*Transl.*]

1821–1893. This extended period saw great developments in the theory of ordinary and partial differential equations. Cauchy demonstrated the existence of solutions of certain ordinary differential equations (ODEs) and partial differential equations (PDEs). The theory of the existence of solutions to ODEs was further developed by Lipschitz (1877) and decisively improved by Picard's method of successive approximations (1893). On the other hand, methods for obtaining the solutions of PDEs developed by stages towards a formal theory of PDEs by Pfaff (1814–1815), most remarkably by Hamilton (1834–1835), and then by Jacobi (1837 onwards) whose completed works were collected together in 1866.

1826. A Memoir, "on a general property of a very extensive class of transcendental functions", by the Norwegian mathematician Niels Henrik Abel was read to the Academy of Sciences of Paris by Hachette. [This created the theory of elliptic functions (the inverse of the elliptic integrals discussed by Legendre) which was the inspiration of a rich field of later research. The manuscript was then lost, found again, and finally published in 1841, after Abel's death. *Transl.*]

1826–1829. Abel, and the young French mathematician Évariste Galois, worked on the solutions of algebraic equations, which would later lead to the discovery of algebraic numbers (numbers which can be solutions of an algebraic equation with integer coefficients).

1828. William Rowan Hamilton, an Irish mathematician, gave the first strict definition of a complex number.

1829–1854. In 1829, Nicolai Lobachevski (a Russian mathematician) and, in 1833, Wolfgang Bolyai (a Hungarian) invented non-Euclidean geometry.[8] In 1854, the German mathematician Bernhard Riemann created a general theory of non-Euclidean geometry, later known as Riemannian geometry, [which gave rise to the tensor calculus of Ricci and Levi-Città and became the geometrical framework of Einstein's general theory of relativity. *Transl.*]

1829. The French mathematician Peter Gustave Lejeune-Dirichlet

---

8 Euclidean geometry is defined in terms of points and straight lines joining points; the theorem of Pythagoras holds for right-angled triangles; and the sum of the three angles of any triangle is 180 degrees. A non-Euclidean geometry is likewise defined in terms of points and by paths between points which share the shortest-distance property of straight lines. But these paths may be "curved"; Pythagoras's theorem does not hold for right-angled triangles; and the sum of the three angles of a triangle may differ from 180 degrees. A down-to-earth example is given by sailing on the shortest closed voyage between three points on the ocean. Each leg, the shortest path between two successive points, is a great-circle arc; if one of the angles between two successive legs is 90 degrees the sum of the squared lengths of these two legs is greater than the squared length of the third leg (so Pythagoras's theorem does not hold); and the sum of the angles between the legs at each of the three points is greater than 180 degrees. The geometry of great-circle arcs on the Earth's surface is therefore non-Euclidean, reflecting the fact that this is a curved space (as opposed to the space of Euclidean geometry, which is flat). [*Transl.*]

developed his method of summation of Fourier series, and applied it to the study of their convergence.

[**1830–1832.** Évariste Galois created the algebraic theory of groups. This grew out of his earlier work on the possibility of solving algebraic equations of higher than the fourth degree, in which an important role is played by certain aggregates of patterns of substitution and permutation: such aggregates have a common structure which is taken as the defining characteristic of a mathematical group.[9] *Transl.*]

1830–1859. During this period, Cauchy published more than six hundred articles on his fundamental researches into the convergence of infinite series.

1840. On 1 January 1840, the law came into force under which the adoption of the metric system became compulsory in French education.

1843. Hamilton created his theory of quaternions (four-dimensional entities more general than vectors, less general than tensors) which are the earliest instance of a non-commutative algebra [(i.e. in which "multiplication", denoted here by $\times$, depends on the order of the terms: $A \times B \neq B \times A$). *Transl.*]

1843–1844. Hamilton, and the German mathematician Hermann Grassmann, almost simultaneously published their work on the early developments of vector calculus.

1844. The French mathematician Joseph Liouville gives the first examples of transcendental numbers (numbers which cannot be the solution of any algebraic equation with integer or rational coefficients). [He gave a necessary condition for a number to be a root of such an algebraic equation; failure to satisfy this is a sufficient condition for a number to be transcendental. By devising constructions designed to break the condition, he was able to exhibit examples of transcendental numbers.[10] However, where the "common" numbers of mathematics are concerned, it was not until 1873 that Hermite proved the transcendence of e, and 1882 that Lindemann proved the transcendence of π. *Transl.*]

1847. The English mathematician and logician Augustus de Morgan published *Formal Logic*, thus laying claim to be one of the founders of mathematical logic.

---

9 Groups pervade mathematics, and have an abstract theory of their own. The positive and negative integers, with zero, form a group with respect to the operations of addition and subtraction; the positive real numbers form a group with respect to multiplication and division; the orientations of a body in space form a group with respect to the possible rotations of the body about a fixed point; theories of modern physics are founded on group structures which relate the physical quantities that occur in them. [*Transl.*]
10 The number defined by $\sum_{n=1}^{\infty} \frac{1}{2^{n!}}$ is an example constructed in this way, where $n!$ denotes the factorial of n. [*Transl.*]

1847–1854. George Boole, an English mathematician, published *The Mathematical Analysis of Logic* in 1847, and *An Investigation of the Laws of Thought* in 1854. These works defined an algebraic model for formal logic, and under the name of Boolean Algebra it is nowadays one of the essential tools for the logic of automatic computation.

1858. Arthur Cayley, an English lawyer and original mathematician, invented matrix algebra.

1866. Use of the metric system became legally permissible in the United States.

1872. Richard Dedekind, a German mathematician who was one of the founders of the rigorous development of number and mathematics, gave a precise definition of irrational numbers and of real numbers. "The straight line", he explains, "is infinitely richer in points than the set of algebraic numbers is."

1873–1882. The French mathematician Charles Hermite proved (1873) that e is transcendental; the German Ferdinand von Lindemann proved (1882) that π is transcendental.

1882. The French mathematician Paul DuBois-Reymond published *Théorie des fonctions* ("Theory of functions"). He distinguished between a special theory of functions, and a general theory in which he studies the properties of completely arbitrary functions (whether or not capable of analytic representation).

1882–1883. The German mathematician Georg Cantor, of Russian origin, discovered transfinite numbers and, by a generalisation of the principle of one-to-one correspondence, created the notion of the power (number of elements) of a set; he applied this to the denumerable (countable) set of integers, and to non-denumerable sets such as the continuum of points on a line. He subsequently laid the foundations of the modern theory of sets and of modern topology, especially in its applications to mathematical analysis.

1883. Cantor proved that the set of algebraic numbers is countable (there are exactly as many algebraic numbers as there are integers).

1893. Maurice d'Ocagne published the first edition of *Calcul simplifié par les procédés mécaniques et graphiques* ("The calculus made simple by mechanical and graphical methods").

1897. Use of the metric system became legally permissible in Great Britain. However, not until 1970 did the British monetary system become decimal.

1899. The German mathematician David Hilbert published *Grundlagen der Geometrie* ("Foundations of geometry") which develops the fundamental principles of elementary geometry from a certain number of axioms.

1900. The Italians Ricci and Levi-Cività published their first systematic account of tensor calculus.

1931. The Austrian mathematician and logician Kurt Gödel published his famous Incompleteness Theorem, which proves that any formal mathematical system which is powerful enough to contain arithmetic is either self-contradictory or contains propositions which cannot be decided within the system (i.e. there is no formal chain of reasoning generated by the rules of the system which could either prove the proposition or prove its contrary).

1934. The Russian mathematician Alexander Ossipovich Gelfond proved that any number a^b is transcendental, if a is an algebraic number other than 0 or 1, and b is an irrational algebraic number.

First half of the 20th century. This period saw fundamental contributions from Western mathematicians and logicians to the development of contemporary mathematics and logic, especially in the areas of algorithmic logic and symbolic calculus. Notable figures in this history are G. Frege, G. Peano, B. Russell, A. N. Whitehead, D. Hilbert, E. Zermelo, E. Steinitz, R. Carnap, E. Artin, K. Gödel, E. Post, S. C. Kleene, A. Turing, A. Church, J. von Neumann, A. A. Markov, P. S. Novikov, H. Cartan, C. Chevalley, J. Delsarte, J. Dieudonné, A. Weil; and of course many others.

TAILPIECE

We have now met the five fundamental numbers of mathematics: 0 (the zero of the number-system); 1 (the unit); $e = 2.7182818 \ldots$ (the base of natural logarithms); $\pi = 3.1415926 \ldots$ (the ratio of the circumference of a circle to its diameter); and finally the "imaginary" number, the square root of –1, denoted by $i = \sqrt{-1}$. Between these five numbers there holds a startling relationship:

$$e^{\pi i} + 1 = 0. \ [\textit{Transl.}].$$

CHAPTER 4

BINARY ARITHMETIC AND OTHER NON-DECIMAL SYSTEMS

[Decimal number-systems are those in which the successive unit orders of magnitude increase as powers of the *base* 10: 1, 10, 100, ... , 10^n, ... All others are non-decimal systems. As well as the relics of mixed systems, such as the 16 ounces to a pound, 14 pounds to a stone, 12 inches to a foot, 3 feet to a yard, 12 pence to a shilling and 20 shillings to a pound, which may still be found in traditional English systems of weights and measures, the world still uses elements of systems with bases other than 10: 60 seconds to a minute of angle, 60 minutes to a degree, 60 seconds to a minute of time, 60 minutes to an hour (but 24 hours to a day).

[Many mixed systems have well-known ancient origins, and so do some systems with bases other than 10. But the *binary* system, to base 2 and proceeding by powers of 2 (1, 2, 4, 8, 16, ...), now known to many at least by name because it is the system underlying the operation of modern computers, nevertheless apparently did not emerge earlier than the seventeenth century CE. *Transl.*]

This chapter will summarise the chronology of non-decimal and binary number-systems.

THE EARLIEST STAGES

In the beginning, people produced an effective solution to the problem of representing numbers by inventing the principle of the *base* for number-systems. By the end of the fourth millennium BCE, the Sumerians had a non-positional written system with base 60. By 1900 BCE, the people of Mari in Mesopotamia had a partially positional number-system with base 100, and in the course of the ensuing century a sexagesimal (base 60) system – which, however, had no symbol for zero prior to the third century BCE – emerged at the hands of learned Babylonians.

Much later, in the fifth century BCE and almost on the opposite side of the world, the Zapotecs of Central America had a non-positional system with base 20; similar systems would later be used by other pre-Columbian peoples such as the Maya, the Aztecs, the Mixtecs, etc.

YIN AND YANG

Also at about this time, by the fourth century BCE, the Chinese had developed their philosophy of *yin* and *yang*. Despite being an essentially bipolar view of the world, this did not represent the birth of binary arithmetic even though many may think that it did.

The *yin-yang* system is based on the opposition, alternation, complementarity or interaction of two energies.

One of these, the *yin*, is "female" and is represented by a broken line: – –. The other, the *yang*, is "male" and is represented by a continuous line: —.

According to this doctrine, *yin* and *yang* are the bipolar manifestation of the Supreme Ultimate whose material forms were supposed to be the sky and the earth.

Combinations of these representative broken or continuous lines gave rise to a system corresponding to elements of nature, for example as trigrams with three lines such as ☷ or ☰, or as hexagrams with six such lines. Three continuous lines, the trigram ☰, represented the perfection of activity: the sky, the South, Summer, productive energy, the male, the Sun; the trigram ☷, three broken lines, on the other hand, represented the perfection of passivity: the Earth, the North, Winter, receptivity, the female, the Moon. A broken line between two full lines, ☳, represented thunder, the Northeast, and the coming of Spring; while its converse, ☴, represented the wind, the Southwest, and the end of Summer; and so on. [See R. Wilhelm (1973); M. Granet (1988); K. Friedrichs, I. Fischer-Schreiber, F. K. Erhard and M. S. Diener (1989); J. Chevalier and A. Gheerbrant (1982)].

This system was first described in the traditional Chinese book *The Book of Changes*, better known as the *I Ching* (or, as it should more properly be called, *Yijing*), which was compiled and published, at the earliest, in the fourth or third centuries BCE and not, as some of the more imaginative would have it, one thousand years BCE or even three thousand years BCE.

For all that, such a notation has nothing to do with a binary number-system. For this to be the case, the Chinese would first of all have required a symbol for zero, and also the use of this system for writing numbers. But they wrote their numbers according to quite different systems, which in every case were decimal. Even though we know that they had a positional number-system, we also know that a symbol for zero only came very late in their history (see Chapter 21).

And in fact, as we shall later see, the binary representation of numbers, and binary arithmetic, properly speaking, did not come into being until the seventeenth century CE, and in Europe.

LEIBNIZ, BINARY NUMBERS, YIN AND YANG

No doubt it was the elegance and simplicity of the representations described above which moved Gottfried Wilhelm Leibniz to exclaim "*Omnibus ex nihil ducendis sufficit unum*" ("With one, everything can be drawn out of nothing"). This led Leibniz into several philosophical speculations of a very special kind. One day, Father Joachim Bouvet (1656–1730), a mathematician and a missionary in China since 1685 (sent there by Louis XIV), communicated to Leibniz the 64 figures formed by the hexagrams of the *Yijing*. The following, reading from the left, are the first six:

A   B   C   D   E   F

Taking the continuous line to represent 1, and the broken line to represent 0, Leibniz thought he was looking at a transcription of binary notation, such as he himself had invented on his own but which the Chinese, he now thought, must have been using since remotest Antiquity. The figures above therefore, to his eyes, represented the numbers:

$$
\begin{aligned}
A &= 0\ 0\ 0\ 0\ 0\ 0 = 0 \\
B &= 0\ 0\ 0\ 0\ 0\ 1 = 1 \\
C &= 0\ 0\ 0\ 0\ 1\ 0 = 2 \\
D &= 0\ 0\ 0\ 0\ 1\ 1 = 3 \\
E &= 0\ 0\ 0\ 1\ 0\ 0 = 4 \\
F &= 0\ 0\ 0\ 1\ 0\ 1 = 5
\end{aligned}
$$

where the zeros and ones are read from the Chinese hexagrams from top to bottom and written as 0 and 1 from left to right to give a binary representation of an integer, and the integer number represented by the binary notation is written on the right-hand side in decimal notation.

He also thought he had found here a species of universal language which would unite people of all countries and all races. "Fr. Bouvet and I", he later wrote, "have discovered the seemingly most probable meaning, according to the letter, of Fou-Hi's[1] characters. It is binary arithmetic, which this great legislator already knew and which I have rediscovered thousands of years later. In this arithmetic there are only two signs, 0 and 1, and by means of these any number can be written. It would seem that Fou-Hi had some notion of the science of combinations, on which I myself wrote a short dissertation in my youth."

The one accurate statement here on Leibniz's part concerns the science

---

1 Now written *Fuxi*; he was the presumed inventor of Chinese writing.

of combinations, which the Chinese were already acquainted with before the beginning of the Common Era. But this was not the only place where Leibniz, the eminent thinker and mathematician, went off the rails, according to another great mathematician Pierre-Simon de Laplace (1749–1827): "In the binary arithmetic which he invented, Leibniz sought to perceive the Creation. He imagined that the Unity stood for God and the Zero for the Void, that the Supreme Being had drawn every existing thing from out of the void just as unity and zero suffice to represent every possible number in his system. This conceit so much enchanted him, that he communicated it to the Jesuit Grimaldi, his correspondent in China, who was President of the Chinese Committee for Mathematics, in the hope that this would bring about the conversion of the Emperor of China who himself was much taken with Mathematics. I mention this merely to show how childish fancies can cloud the vision even of the most distinguished men."

The *yin-yang* notation had at any rate this in common with the binary number-system: that each made use of two opposite symbols. The first used the continuous and the broken line, the second the symbols 0 and 1. But the comparison stops there, since the binary number-system, which is founded on a positional principle and uses a symbol for zero, is used to write numbers, while the *yin-yang* system was never used for writing numbers. It served only as a notation which could express mystical relationships and correspondences between the different possible combinations of the two fundamental principles of *Yin* and *Yang* on the one hand, and various features of the natural world on the other hand, and it was originally used chiefly for divinatory purposes.

While representation based on the two kinds of line is a combinatorial system of two opposite symbols, and the positional binary notation is also such a system, yet the latter is a very special case since its use for writing binary representations of numbers is a system of *arrangements* (ordered combinations) of 0 and 1.

In short, the principle of *Yin* and *Yang* is not a binary system in the arithmetical sense: it is a *dualistic philosophy*, which is quite a different matter. We should therefore be wary of interpretations of ancient philosophies which may be suggested by modern points of view. Some leniency to Leibniz is in order, however, for he was a remarkable thinker who contributed fundamental advances in many domains, especially mathematics, as well as inventing binary arithmetic.

After this digression, which leapt from the fourth century BCE in China to the European beginnings of modern mathematics in the seventeenth century, we now leap back again to China, in the second century BCE. By this time, Chinese mathematicians had devised a positional decimal notation based on "number rods" (*suan zî*); this system would still, however, lack a sign for zero until the eighth century CE.

Little that is new arises for some hundreds of years thereafter. Then, again on the opposite side of the world, by the fourth century CE the Mayan astronomer-priests were using a positional number-system, to base 20, which had a symbol for zero and which endured until the ninth century. Nevertheless, this system was irregular at and beyond the third position as a result of being adapted to their calendar.

At about the same time in India, around the fourth to fifth centuries CE, the positional decimal system with a sign for zero which is the source of our modern notation appeared. This was the first written number-system in history which was capable of being extended to a simple rational notation for all real numbers and which could potentially be used for every kind of calculation.

Again, little change ensued for some hundreds of years. But the sixteenth century CE saw the beginnings of the generalisations which grew into modern mathematics. From the sixteenth to the eighteenth centuries European mathematicians began to take a more abstract view and became aware that the properties of the Indian positional number-system were in fact independent of the value of its base. Little by little they came to understand that, for any written number-system whatever which had the same structure,[2] the notation for the integers (whole numbers) could be extended to a simple rational notation for all numbers, and that the fundamental arithmetic operations (addition, subtraction, multiplication and division) can be carried out in exactly the same way as in the decimal system, by means of simple rules which do not depend on the base of the particular number-system (see Fig. 1.39).

The base 10, by the way, is the one most commonly found throughout the history of number-systems, owing to the anatomical accident that human beings have ten fingers. In this, nature failed to find the best choice.

It is not best for arithmetical calculation, since (unlike the base 12, for example) the base 10 has only two divisors. It does not suit the mathematician, since (unlike 11, for example) 10 is not a prime number. From both practical and theoretical motives, therefore, several European mathematicians sought to replace the common decimal system, with all its inconveniences, with some non-decimal system. But in vain: so deeply entrenched is the decimal system in popular culture, that it seems to be indestructible.

Nevertheless, these investigations gave rise to the various particular number-systems (*binary*, to base 2; *octal*, to base 8; *hexadecimal*, to base 16) which have turned out so useful in twentieth-century computer science.

---

2 That is, based on a positional principle and possessing a symbol for zero, with signs for the different digits which carried no direct visual meaning.

In 1600, the English astronomer, mathematician and geographer Thomas Harriot drew up a table showing the decompositions of the integers from 1 to 31 into powers of 2.

In 1623, the English philosopher Francis Bacon devised a binary code for the letters of the alphabet, in which each was represented by five characters and each character could be one of two: A was represented by aaaaa, B by aaaab, C by aaaba, D by aaabb, E by aabaa etc. if we were to let a stand for 0 and b for 1, then Bacon's representation would correspond to binary counting starting with 00000 for A: 00000, 00001, 00010, 00011, 00100, etc.

A weightier advance was made by the important and original French mathematician and philosopher Blaise Pascal (1623–1662) in 1654, when he presented to the French Academy of Sciences his *De numeris multiplicibus ex sola characterum numericorum additione agnoscendis* ("The divisibility of numbers deduced solely from the sum of their digits"), which was finally published in 1665. Here, for the first time, a general definition is given of a number-system to an *arbitrary* base *m*, where *m* may be any whole number greater than or equal to 2 (see Fig. 1.39).

By means of such abstractions, European mathematicians later developed systems of arithmetic and numerical analysis which are much more general than the decimal systems we use. They are based on a positional notation for numbers to an arbitrary base $p > 1$, extended to express any number whether integer, rational or irrational as an expansion in positive and negative powers of the base p, using a decimal point to separate the positive and negative powers. These are called *p*-adic numbers, and they represent a generalisation of the specific *dyadic* system of the binary representation, and the 10-adic system of the decimal representation.

After Pascal's fundamental general study, several mathematicians produced studies of particular systems. Bishop Juan Caramuel y Lobkowitz of Rome published *Mathesis Biceps* in 1670, which was a systematic mathematical study of number-systems with non-decimal bases, including 2, 3, 4, 5, 6, 7, 8, 9, 12, 20 and 60. In 1672 the German mathematician Erhard Weigel published a tract on numbers written to base 4.

Leibniz, in 1679, wrote a manuscript entitled *De Progressio Dyadica* which, as he himself later acknowledged, was the earliest manifestation of his interest in binary notation which he had originally hoped to make use of in the calculation of decimal fractions. At the same time he had the idea of constructing a binary calculating machine which would use moving balls.

In 1687, the English mathematician Joshua Jordaine considered the use of duodecimal arithmetic (to base 12).

Quite independently of Leibniz's work, in 1701 the French mathematician Thomas Fantet de Lagny published an article about binary arithmetic, in which he clearly demonstrated the merits of the binary system without (unlike Leibniz) appealing to any religious, philosophical or metaphysical considerations.

Leibniz returned to the subject in 1703 with his publication *Explications sur l'arithmétique binaire* ("Explanations concerning binary arithmetic"). This tract made the binary system a fashionable subject of study for European mathematicians of the period.

In 1708 Swedenborg, the Swedish philosopher, proposed that decimal notation should, for general use, be replaced by the octal system (to base 8).

The Swiss mathematician Leonhard Euler entered into a correspondence in 1732 with the Academy of St Petersburg in the course of which he made use of a binary notation using powers of 2. [Euler was one of the greatest innovative mathematical problem-solvers of all time and originator of much that is still directly used in, or underlies, modern mathematical methods. *Transl.*]

In 1746 the Italian mathematician Francesco Brunetti worked on binary fractions and gave a table of the decimal values of the powers of 2 up to 2^{40}.

In 1764, the French mathematician Étienne Bezout sought to promote the use of the duodecimal system (to base 12); in the same year, Anton Felkel of Prague made use of binary notation to study decimal fractions.

In 1775, Georges Brander of Augsburg was using the binary system for his private financial accounts "in order", as he put it, "to close them off from prying eyes."

Anton Felkel published an important article in 1785 on the periods of p-adic fractions (the generalisation to base p of the decimal fractions to base 10: see above).

In 1798 the noted French mathematician Adrien Marie Legendre published works on conversion from the binary system to the octal system and to the hexadecimal system.

In 1810, Peter Barlow (an English scientist) published an important article on the method of transformation of a number by passing from the base of one notation to another, and its application to the rules of duodecimal arithmetic.

In 1826, the mathematician Heinrich W. Stein published an article which established various relationships between non-decimal number-systems.

The English mathematician Charles Babbage is now best known for his various attempts (of which for various reasons none succeeded) to construct a programmable mechanical calculating machine, his *Analytical Engine*. In 1834, as can be seen in the numerous notes which he left, he was envisaging various different number bases as possibly suitable for the project: 2, 3, 4, 5 and 100, for example. Nevertheless, he considered that only the base 10 was really feasible, which was a natural conclusion in view of the technological means available.

In 1837 the American Samuel F. Morse invented his *telegraph*. This was the earliest system for transmitting messages by means of electrical impulses based on a binary code specified in advance. [Strictly speaking, the Morse code is not a binary system but a ternary system, to base 3, since it depends on three different basic signs: the short sound followed by a short silence, or dot; the long sound also followed by a short silence, or dash; and the longer period of silence which indicates the completion of the dots and dashes which encode a single character. *Transl.*]

Augustus de Morgan (1806–1871) was an eminent English mathematician and logician, and a noted freethinker, who not only was a great populariser of mathematics but also stands with Boole and Hamilton as one of the founders of modern formal logic. In 1853 he concluded, in his book *Elements of Arithmetic*, that the decimal system is so deeply engrained in tradition that it will forever remain the customary method. De Morgan was nevertheless convinced that non-decimal number-systems should be part of the teaching programmes of schools and universities.

The American engineer John William Nystrom proposed, in 1862, a system of weights and measures based on the hexadecimal number-system (base 16).

In 1863 Moritz Cantor, a historian of mathematics, formally, and rightly, criticised Leibniz's interpretations, in terms of binary arithmetic, of the trigrams and hexagrams of the Chinese *Yijing* (see above).

In 1876 the American Benjamin Pierce had the notion of "improving" Leibniz's binary notation (which had used 0 and 1) by proposing to use a dot for 0 and a horizontal line for 1; which he considered to be more economical.

In the same year, the Englishman Edward Brooks advocated replacing the decimal system by the duodecimal system (to base 12).

The American Alfred B. Taylor published an article entitled "Which base is best?" in 1887, concluding that it was the base 8, while in 1889 the Scandinavian T. N. Thiele concluded, for his part, that it should be the base 4; the American William W. Johnson came to the same conclusion (base 4) in 1891 as also did the Frenchman E. Gelin in 1896.

In 1899, the Italian mathematician Giuseppe Peano proposed a binary notation in which a dot and a double dot, (.) and (:), should stand for 0 and 1 respectively.

The start of the twentieth century saw substantial advances in the understanding of electricity and electromagnetic phenomena, and in the use of electrical circuits for technological purposes such as telephony and radio transmission. The road which was to lead to the binary electromechanical computer was opening up.

A list of mathematicians who had made use of non-decimal number-systems between 1771 and 1900 was published in 1908 by the Englishman Herbert McLeod.

In 1919 the English scientists W. H. Eccles and F. W. Jordan invented the electronic device known as the *flip-flop*. This apparatus consists of two triodes, and an electrical impulse arriving at one of its inputs reverses the state of each of the triodes: it constitutes a *bistable* circuit. The discovery of this property was later to lead to the construction of the earliest binary electronic circuits.

Anton Glaser's *History of Binary and Other Non-decimal Numeration* was published in 1925 (reissued, with corrections, in 1971).

The years 1930–1940 saw the beginnings of the modern developments. Western engineers became aware of the disadvantages of the base 10. The possibility of constructing bistable physical systems led them to perceive the merits of using the base 2 in designing an automatic numerical calculator.

The numerical base of a calculator plays, in the design of the machine, the same role that the base plays in a number-system. Technically, the base 2 is a much better choice than the base 10, both in terms of the number of different base units (two instead of ten) and in terms of simplicity of operation.

The basic multiplication table in base 10 requires ten rows and ten columns, corresponding in each case to the ten integers from 0 to 9. In base 2, however, the multiplication table is reduced to the simplest possible form:

X	0	1
0	0	0
1	0	1

Nothing could be simpler: indeed multiplication has become even simpler than addition! In fact it was the absence of such simplicity, the very complications of the processes required to emulate decimal multiplication in a machine, which so taxed the inventors of the nineteenth and early twentieth centuries in creating their calculating machines which, at the time, used the base 10. [The same simplicity underlies the correspondence between a certain kind of binary arithmetic and an algebraic formulation of Logic, invented by George Boole. Indeed, the multiplication table in binary arithmetic is identical to the "truth table" for the logical conjunction "and": see below, Chapter 5. This correspondence is also fundamental for the modern binary electronic computer. *Transl.*]

The earliest attempts to create binary calculators were made by purely mechanical means, but it was realised that the binary system is admirably suited for electromechanical devices, [such as *electromagnetic relay switches* which can be switched from one state to another, allowing or preventing the passage of a current, by the application or removal of an independent energising current. *Transl.*] By associating the binary digit 1 with the energised state of the relay, and the digit 0 with its unenergised state, and by suitably connecting several relays together, it is possible to emulate all the operations of binary arithmetic.

[Finally, with further advances in electronics (first in the use of thermionic vacuum tubes, the "valves" such as were used in old radio sets, later in the use of transistors, both of which devices behaved in essentially the same way as the relay switch but using purely electronic means), it became possible to implement binary arithmetic by means of electronic circuits. This made possible much greater speed, much greater reliability, and a much greater scale of construction (devices consisting of many thousands of elements being feasible even in the early days). By this route was born the *electronic binary computer. Transl.*]

C. E. Wynn-Williams, in England, created a binary electronic counting device using gas thyratron tubes in 1932, making it possible to perform counting at very great speeds.

In the same year, the Frenchman Raymond L. A. Valtat took out a patent in Germany on a design for a binary calculating machine (No. 664012 of 23 July 1932). In 1936 he published an article "Calcul mécanique: Machine à calculer fondée sur l'emploi de la base binaire" ("Mechanical calculation: A calculating machine based on the binary number-system");[3] and Louis Couffignal advocated the same principle in an article "Calcul mécanique: sur l'emploi de la numération binaire dans les machines à calculer" ("Mechanical calculation: on the use of the binary number-system in calculating machines").[4]

Again in 1936, William Phillips in England constructed his Binary Demonstrator which was a mechanical device that demonstrated the feasibility of a binary calculating machine [see Randell (1973), pp. 293ff.].

In 1937 Alan Turing, an Englishman [who was both theoretically and practically one of the leading founders of modern computing, *Transl.*] set about constructing an electromechanical binary multiplier; and the American W. B. Lewis made a working binary electronic computer using triode valves. In the following year, the American George Stibitz constructed a binary addition device using electromechanical relays.

[During the war years 1939–1945 a great deal of development took place in the field of automatic computation and information processing both in England and in the United States, much of which has only recently become publicly known in detail (such as the computing engines used to assist the

3 CRAS, **202** (1936), pp. 1745ff.
4 CRAS, **202** (1936), pp. 1970ff.

breaking of the "Enigma" code at Bletchley Park in England to which Turing contributed greatly, and the early computers which were built to perform the massive computational tasks entailed by the design of the atomic bomb at Los Alamos in the United States). These years saw automatic computation come of age, as a practical technology which produced real and important results that would have been inaccessible otherwise. The history of these years is elaborate and fascinating, and will be summarised elsewhere in this book. *Transl.*]

In 1945 John von Neumann, an American of Hungarian origin [who was one of the leading mathematicians of this century and, in respect of the work of the war years in many ways the counterpart in the United States of Alan Turing in England *Transl.*] advocated the binary system as the system for representing units of information in electronic computers (see Chapters 5 and 6 below).

[By this time, it may be said, it was clear that the binary system would be the basis for future development. *Transl.*]

Nevertheless, the debate about number-systems went on. In 1952 Lewis Carl Seelbach, in England, published a bibliography on the use of the duodecimal system; in 1954 William L. Schaaf in America published a historical bibliography on non-decimal number-systems with the title *Scales of notation*; in 1955 the French engineer Jean Essig published *Douze, notre dix futur* ("Twelve: the ten of the future") at a conference in the Palais de la Découverte in Paris; and in 1970 the French singer Boby Lapointe published *Système bibinaire* ("The bibinary system"), in which he formulated a hexadecimal system with a new system of signs for the sixteen base figures [see J. M. Font, J. C. Quiniou and G. Verroust (1970), pp. 225ff.]

PART TWO

CHAPTER 5

FROM CLOCKWORK CALCULATOR TO COMPUTER – THE HISTORY OF AUTOMATIC CALCULATION

What follows is a general introduction to the principles of the various calculating and computational engines, and a survey of their history and of their various types: mechanical, analogue, mechanographical and analytical, and the programmable computer itself. We do not wish to go into every possible detail of every such device which has come into being. Nor would it be possible: many such volumes as this would be required, and in any case the author's own competence would be unequal to the task.[1]

Nevertheless, the milestones presented here, and the vocabulary which we have deliberately made very precise, will allow the reader to trace the main stages of the history of artificial calculation and to appreciate how it grows logically out of the history of arithmetical and mathematical notations and methods of calculation.

The following presentation is a completely new approach to the history of automatic calculation and therefore gives an exceptionally rich view of the origins of information science and of modern computers. It is not a mere catalogue of historical facts: it is a unified account of a multi-disciplinary subject which grew on the fertile ground where so many streams of science and technology have joined up. In contrast with more classical historical accounts which mainly follow chronological lines, and therefore do justice neither to the advances in human thought nor to the complexity of evolution which have combined to create this domain, we have attempted to establish the relationships between all these facts and facets in order to achieve an account which will be more intelligible and – above all – more faithful to the true history of the subject.

It would not be possible to include illustrations of all of the devices, instruments, machines, calculators, automata and computers which are mentioned in this chapter: they are so numerous that even a balanced and representative selection would require a large volume. The interested reader is referred to the Bibliography, in which may be found references to works which provide such illustrations.

1 I take this opportunity to express my special gratitude to Alain Brieux, Jacques Chauvin, I. Bernard Cohen, Erik Lambert, Robert Ligonnière, Maurice Margenstern, René Moreau, Pierre Mounier-Kuhn, François-Henri Raymond, Marco Schutzenberger, Jean-Claude Simon, Gérard Verroust and Remo Vescia, for their most valuable personal communications and for all the reflections, observations, and pertinent and constructive comments which they have made to me during the writing of this chapter on the history, so delicate and so complex, of calculation by artificial means.

1. Human Calculation and Machine Calculation

THE FUNDAMENTAL IMPORTANCE OF CALCULATION
IN MODERN SCIENCE

Plato (428–347 BCE), in *Philebus* ("On Pleasure"), has Socrates explain that the spirit is the regulating element of human life and of the Universe, and that knowledge rests on a basis of several levels hierarchically organised wherein, on the very first level along with dialectics, are to be found number and calculation: "If you cannot calculate," he says, "you cannot speculate on future pleasure and your life will not be that of a human, but that of an oyster or a jellyfish." (*Philebus*, 21).

Had Plato lived today, he would surely have stressed the even greater fundamental importance of those complex intellectual activities which are methodically regulated by clear and definite arithmetical, mathematical, and logical rules, which we call *calculation* (in a wide sense) and which we perform in order to derive *results*.

In the words of Maurice d'Ocagne, "Calculation is as important in the domain of theory as it is in matters of practice. All the material progress of our civilisation derives, directly or indirectly, from Science, and the progress of Science itself constantly depends on calculation. This is not only the case for the so-called 'exact' sciences such as mechanics or astronomy in which calculation plays a central role, but also for those which were considered experimental or observational sciences: this is due to the precision which contemporary developments in calculation have brought to their methodologies. In every branch of physics or of chemistry the mathematical formula reigns supreme, and even in physiology its influence has been felt since the time when measurement began to feature in the facts of this science."

DO NUMBERS RULE THE WORLD, OR HAS
REALITY INSPIRED NUMBER?

The fact that number plays such a central role in modern science caused Bertrand Russell to exclaim that the most astonishing feature of modern science was its return to Pythagorism. But let us be clear that this has nothing to do with the Pythagorean mysticism according to which numbers obeyed a supposed secret and cosmic celestial harmony, and therefore revealed a way to approach eternal verities and divine secrets. Russell, who was to become one of the greatest founders of modern logic, rather took the stick by the other end.

He realised that it was not that numbers ruled the world; rather, the

Universe is ruled by the laws of a physical world whose properties could be expressed as abstract concepts, the numbers, themselves developed by the efforts of human thought which – once it had attained the capacity of extreme abstraction – could act on the objects of this same universe.

"A man in his house", says Rivarol, "does not live on the staircase, but makes use of it to go up and down and gain access to every room. The human mind, likewise, does not reside in numbers but uses them to attain all sciences and arts." Physical reality can inspire number, but does not constitute number. Precisely because humans have learned to transmute the objects of physical reality into simple objects of abstract thought, so they have been able to accomplish all the spectacular progress characteristic of humankind. and have managed to penetrate the secrets of so many aspects of the tangible Universe.

THE NEED TO ESCAPE THE HARDSHIPS
OF HUMAN CALCULATION

While totally denying his mysticism, yet accepting the conventional abuse of language, we may nevertheless say, with Plato, that "numbers rule the world".

But if the world is ruled by numbers, the world is not amused thereby. Calculation is slow, difficult and above all tedious. It is tedious because it is repetitive, and this has given rise to all sorts of hindrances to the advancement of science and of knowledge. Luigi Menabrea's words on this theme are heavy with meaning: "What quantities of precious observations have been of no use to the advancement of science and technology, for the reason that there were not the resources needed to calculate results from them! How the prospects of long and arid calculation have demoralised great thinkers, who seek only time to meditate but instead see themselves swamped by the sheer mass of arithmetic to be done by an inadequate system! And yet, it is the path of laborious analysis that leads to truth, but they cannot follow that path without employing the guide of number, for without number there is no way to lift the veil which covers the mysteries of nature."[2]

This, therefore, is why scientists have always sought to simplify their various calculations.

Indeed, "While calculation is an indispensable aid to scientific research, it is the very tool by which the principles which come to light through this research are transformed into practical applications. The navigator, the surveyor, the gunner, the mechanic, the electrician, the engineer are

2 CRAS, 28 July 1884, pp. 179–82.

all constantly obliged to calculate. For each one, calculation forms a major part of the day's work, though not the least tedious. Lightening the load of calculation therefore brings ease to the researcher; but sometimes it brings more: by making possible, even easy, a task otherwise out of proportion to the result being sought for which none, perhaps, would be willing to expend the effort" (M. d'Ocagne).

ULTRA-SPECIALISED HUMAN ORGANISMS: CALCULATING PRODIGIES

For all the tedium that calculation presents to the normal person, there have been some celebrated exceptions. A few individuals have been known in whom the power of calculation reached a prodigious level.

Jaques Inaudi was born in 1867 at Onorato in the Piemonte region of Italy. As L. Jacob explains:

> From seven or eight years of age, he performed mental arithmetic with an extraordinary facility. He was examined by Charcot in 1880 from the psychological point of view, then from the mathematical point of view in 1892 by Darboux, who presented him to the Academy of Sciences in February of that year. He was asked the following questions:
>
> > What day of the week was 4 March 1822?
> > Subtract   1,248,126,138,234,128,010
> > from   4,123,547,238,445,523,831
> > What is the number whose cube plus its square make 3,600?
> And so on . . .
>
> All these questions were answered with the interval between question and answer never longer than 30 to 35 seconds. According to the examination carried out by Broca and Charcot, Inaudi was intelligent, but his powers of calculation derived from an extraordinary and specialised memory further developed by practice. He had some difficulty remembering narratives. His memory was auditory rather than visual, as often in such cases: he more easily handled numbers which were read out to him than numbers which were written down for him. The English prodigy Vinckler worked in a similar way. He could repeat, two weeks later, a series of 5,000 digits which had been read to him twice. Inaudi usually proceeded by trial and error, trying out different numbers as solutions; consequently he could more easily find a sixth or seventh root than a square root or cube root. He added and subtracted from left to right, like the Indians. (L. Jacob)

Another calculating prodigy was the shepherd Henri Mondeux from Touraine. Born in 1826 at Neuvy-le-Roi (Indre-et-Loire), he became a celebrity in the reign of Louis-Philippe. He "could neither read nor write, but could do the most complicated arithmetic in his head. In 1838, a school headmaster in Tours took him into his home and tried to give him lessons, but for every kind of study, even for mathematics, he showed little aptitude. When, however, he was presented to the Paris Academy of Sciences on 16 November 1840, he almost instantaneously and out of his head answered the following questions:

> Find a number such that its cube plus 84 gives a result equal to the product of this number with 37.
>
> Find two squares whose difference is equal to 133."

(L. Jacob)

We may also mention Zacharia Dahse, who applied his calculating powers (the only ones he possessed) to the determination of Burkhardt's tables of prime divisors for the numbers between 7,000,000 and 10,000,000.

M. d'Ocagne gives a list of other such celebrities: "The young Mathieu Le Coq of Lorraine who, at the age of eight in Florence, astonished Balthasar de Monconys during his third journey to Italy (1664); Madame de Lingré who, according to Madame de Genlis, could do the most complicated arithmetic in her head in the midst of the noisy chatter of Restoration salons; the negro slave Tom Fuller of Virginia who died at the age of eighty, at the end of the eighteenth century, without having learned to read or write; the Tyrolean shepherd Pierre Annich; the Englishman Jedediah Buxton, a simple thresher; the English engineer G. P. Bidder, who constructed the Victoria Docks at London, became President of the Institution of Civil Engineers, and passed on some of his calculating gifts to his son; the Sicilian shepherd Vito Mangiamello who also had a great facility for learning languages; the Russians Ivan Petrov and Mikhaïl Serebriakov; Grandemange, born with neither arms nor legs; Vinckler, the subject of a remarkable experiment at the University of Oxford."

D'Ocagne adds: "We cannot help but be struck by the number of these calculating prodigies who spent their early childhood guarding sheep. For them, calculation must have been a way of passing the time during their long watches out in the fields. However, not only is calculating prowess such as theirs extremely rare, it is not associated with a normal development of other faculties either. It would seem that the brain, totally absorbed in calculation, is not prepared to diversify its activities. Indeed, it would be a heresy to think that supreme facility in calculation was in any way an indicator of exceptional talent in real mathematics ... Although some mathematicians (and amongst these some of the greatest: Wallis, Euler, Gauss and Ampère) were also great calculators ... neither the intuition nor the logic which are such fundamental faculties in mathematics are necessarily linked to great facility in mentally performing the operations of arithmetic. To confuse the two is to commit an error of judgment as great as thinking that exceptional agility on a piano keyboard is an index of a great gift for composing music."

Nevertheless, the facts of such cases are most interesting, at least philosophically, in showing how far the human organism, when ultra-specialised, is capable of going.

COMMON MORTALS

However, such exceptions are no model for the rest of us. We may boldly agree with M. d'Ocagne that "every simplification of the procedures of numerical calculation is an advance of quite general usefulness, by relieving researchers of the boredom and fatigue which accompany calculation, by avoiding the loss of time which it entails, and finally by reducing the chances of error which it involves."

To take a few apt examples: a physicist, a statistician or an economist is not usually held up by theoretical or mathematical difficulties. Apart from certain complicated problems which cannot be formulated analytically and which cannot be solved by classical methods, the necessary theory is usually available. The hold-up lies, above all, in protracted calculations which would take weeks or months to complete. Often, a numerical analysis leads to a system of equations which every mathematician knows how to solve in theory. The path to follow is, therefore, clearly known; but, once the number of unknowns exceeds four, lack of heart for the labour may easily lead to the problem being given up. As another example, also important, research may often come to a halt because of the need to examine an enormous number of possible solutions in order to choose an optimal one. This is why the simplification, indeed the liberation, of the act of calculation by artificial means is an important advance for the scientist, who above all wants to economise on his precious time in order to spend it more intelligently than on the mechanical operations of arithmetic.

A typical example can be found in the history of the number π [pi, the ratio of the circumference to the diameter of a circle, whose exact value requires an infinite number of decimal places], of which the decimal expansion has been hunted since the earliest times.

Around the year 510 CE, the Indian mathematician Âryabhata gave an approximate value for π as the decimal representation of the fraction 62,832/20,000, namely the well-known four decimal places $\pi = 3.1416$.

In the sixth century CE, Chinese mathematicians determined its value to six decimal places and, nine centuries later, the Arab mathematician Ghiyat ad dīn Ghamshīd al Kāshī calculated it to sixteen places.

In 1719, the Frenchman Fantet de Lagny gave 127 decimal places, and, in 1794, Baron Georg von Vega gave 136. In 1844, the Viennese C. Burkhardt gave 200 decimals but the truth is that the actual calculation was done by Zacharia Dahse, the sixteen-year-old calculating prodigy who took a mere two months to carry it out – a performance which has all the appearance of a miracle, when one thinks of the many years of labour which would be needed by a normal human being. And the proof of this last is that the Englishman William Shanks, who achieved the all-time record for "human" calculation of π, took nineteen years to calculate the first 707 decimal places. He began calculating in 1855, the result was not published until 1874, and to cap it all he made an error in the 528th decimal place.

To perform these calculations, Shanks made use of more than a thousand terms of a convergent series, using several hundred decimal places in each term, such that the numerical determination of each term required two multiplications and two divisions.[3]

A TIMELESS DREAM BECOMES REALITY

So, as we can see, calculating π to a few hundred decimal places by hand can be a life's work.

Breaking free from this burden of time, and from the intellectual labour of such a calculation, is in its way the realisation of one of humanity's timeless dreams, recounted thus by Aristotle over two thousand years ago: "Suppose every instrument", he mused, "could by command or by anticipation of need execute its function on its own; suppose (like the carvings of Daedalus or the figurines of Hephaestus which, the poet says, could take on a life of their own) that spindles could weave of their own accord, and plectra strike the strings of zithers by themselves; then craftsmen would have no need of hand-work, and masters have no need of slaves." True men, he added, would abandon their mean tasks which are so unworthy of them, henceforth to consecrate themselves to the proper functions of a citizen, and to seeking knowledge and the acquisition of the wisdom which knowledge brings.

More than two thousand years later, we find that the philosopher's dream has anticipated – though without direct knowledge of them – all the extraordinary technical achievements of the twentieth century.

Indeed, in the civilisation of machines and automata of our own day we find that Aristotle's dream has become reality in full, at the conclusion of a prolonged evolution in the inventiveness of humankind responding to their need to change their environment and their profound desire to master the efficient action of various self-regulating processes. A very prolonged evolution indeed, marked by trial and error, by reflection, by dead ends and regressions, by the most varied advances and discoveries, directed to the conception and construction (in M. P. Schutzenberger's description) "of assemblies of inanimate beings, organised so as to replace humans in the execution of a complex of tasks defined by human beings."

THE COMPUTER: MASTERING THE EFFICIENT ACTION OF SELF-REGULATING CALCULATION

The dream of Shanks was the dream of Aristotle, save that Shanks was concerned only with calculation; and his dream would have been fulfilled by present-day computers. Owing to conceptual advances which build on the latest technological achievements, we conceive and manufacture such machines, which have quite extraordinary speed, calculating power, and memory capacity, as well as almost perfect reliability.

These machines allow us to solve quickly and easily all the problems mentioned above. Calculating a seven-by-seven determinant, or solving a system of nine linear equations in nine unknowns, now takes but an instant, compared with the hours of work that the most accomplished human calculator would require for such a task.

The calculating capabilities of computers allow us to do far more than this, however. They can perform the utmost variety of calculations: mathematical computations (matrix products and inversions, systems of algebraic equations, ordinary and partial differential equations ...); technical calculations (optical lens design, roadway design, airframe design ...); statistical calculations (enumeration, survey analysis, time series analysis, analysis of variance, factor analysis ...); economic calculations (trend estimation, identification of econometric models ...). By iterative methods, starting with a tentative approximate solution, they can progressively improve the approximation to the desired result. They can solve trigonometric equations for which there is no algebraic solution. They can determine the trajectories of guided missiles, which involve very complex differential equations defined by empirical numerical relationships, whose integration was impossible by traditional methods.

To return to the story of the number π, but this time in the post-war years, we find that the determination of its decimal expansion became a much more reasonable task once the problem of automating its calculation had found its earliest useful solutions.

In 1947, D. F. Ferguson and J. W. Wrench Jr found π to 808 decimal places, after some months of work, using an ordinary office calculating machine.

Following the development of the big calculating engines of the 1940s, even more impressive performances were achieved. In 1949, using the calculating engine ENIAC, George Reitwiesner calculated π to 2,307 places in a little less than 70 hours.

Then, with the development and the ever accelerating enhancement of electronic computers, such calculations not only went faster, but also went much further. In 1954, S. Nicholson and J. Jeenel calculated 3,089 decimal places on the IBM NORC machine in 12 minutes. In 1958, F. Genuys calculated 10,000 on an IBM 704 in 1 hour and 40 minutes. In 1961, Daniel Shanks (no relation to the other Shanks) calculated just over 100,000 on an IBM 7090 in 9 hours. The record for the period was achieved by J. Guilloud and M. Boyer in 1976, who calculated π to one million decimal places on a CDC 6600, in just under two hours.

At the present time, it can take less than an hour, even much less, to calculate a million decimal places of π on any personal computer of average power. This calculation is in fact often used to test the speed and reliability of these machines.

The hunt for the decimal digits of π well illustrates the successive stages of the history of automatic calculation, namely calculation executed by a device created by humans specifically for calculation. It also shows how the need to liberate humans from the slavery of manual calculation (performed without the aid of a machine) has found its most complete and effective solution in the creation of the modern computer.

Computers are much more than simple "calculating machines" (in the traditional sense of the word, which refers to operations on numbers and magnitudes). They can of course carry out every kind of arithmetic or algebraic operation, but they are not – as is often believed – simply gigantically overgrown pocket calculators.

Computers can do far more than would ever be possible for a human calculator. For example, they can simulate – perhaps in real time – a complex phenomenon (such as the evolution of meteorological phenomena in the atmosphere). To do such a thing, a "computer model" is constructed: the phenomenon is conceptually decomposed into a large number of similar elements (such as small cells of atmosphere) to which are assigned attributes which correspond to the values to be found in reality (such as temperature, density, humidity, velocity, etc.); and the dynamical laws of the phenomenon are represented as algorithms which determine how the attributes of the elements at the next stage of computation are derived from the attributes of the elements at the current stage. The computer is then let loose to execute these algorithms repetitively, updating the internal representation of the phenomenon time after time and thereby simulating the behaviour of the real system. Features of the behaviour of the phenomenon can then be extracted from the successive results of the simulation by statistical methods. The influence of various parameters of the system on its behaviour can be investigated by deliberately varying their representations in the computer model; this kind of "abstract experimentation" could never be achieved with real physical systems. Such machines therefore facilitate the study of large numbers of variables, the elucidation of relationships between the magnitudes participating in the phenomenon, and the comparison of different feasible computer models for the same phenomenon so as to choose the one which seems to best represent reality.

With such machines, therefore, we can attack problems whose complexity is so great that it would take much longer than a human lifetime to solve them by hand.

Today's computers are able to handle problems far harder than those which have become commonplace for them, such as sorting, merging, control, textual analysis, graphical analysis, image analysis, speech synthesis, pattern recognition, and much more. They have innumerable applications in the most varied domains, such as were unimaginable only a generation or two ago: computer-assisted conceptualisation, computer-assisted learning, automatic translation, musical composition, graphic design, computer animation of images, synthesis and transformation of images, robotic control of machine tools, and so on.

In short, computers are the most generalised machines possible for the automatic processing of information.

THE GREAT SYNTHESIS FROM WHICH EMERGED THE COMPUTER

Clearly, therefore, the development of computers was a revolution in the history of our civilisation. However, it did not come about at the wave of a magic wand, nor did it result from the solitary work of some inventor of genius. Nor, in historical reality, does it make sense to talk of "the invention of the computer" as the common phrase would have it. Instead, the computer came about as the result of the confluence of a multitude of streams whose sources are ancient and widely separated, and from a long and concentrated succession of reflections and responses to needs (see Fig. 5.2).

First and foremost, this revolution is the fruit of everlasting contributions by a multitude of scholars, philosophers, visionaries, inventors, engineers, mathematicians, physicists, and technicians from every corner of the world and from every period of history. The movement principally began to take on form during the Industrial Revolution of the nineteenth century. Subsequently, it became philosophically and intellectually plausible with the spread of machines and automation. Then it became seen as theoretically possible as a result of more recent advances in symbolic logic and the science of mathematics, and it became technically achievable through the technological revolution of the twentieth century.

Finally, it became a concrete reality owing to the emergence and expansion of industrialised societies which were stimulated by vision, by need, by experience, by competence and by competition, under pressures from many directions – social, economic, commercial, scientific, even and crucially the political and military urgency during the Second World War to oppose Nazism.

In short, computers emerged as the materialisation of multitudinous dreams and desires, through a gigantic synthesis which gathered up a cascade of inventions and innovations at the moment when a long slow evolution had come to term, whose beginnings date back to the dawn of time.

In the history that follows in this chapter, do not suppose that the important ideas here set out were originally expressed as precisely as you will see them written. They indeed became clearly established over the generations, but in a diffuse, unco-ordinated way. As Bertrand Gille emphasises: "The historian is always somewhat at a loss in speaking of his own times. He lacks the possibility to stand back from events, and often fears that he will misunderstand them, weigh up the phenomena incorrectly, and misjudge the signs to be found in the world around us now."

So I have, over many years, hesitated long and often before publishing the account which follows. I have held back for so long as the ideas which I expressed were still no better structured than the accounts in sundry "static" histories, mere accumulations of fact but not co-ordinated on the historical scale, great though their contribution has been in terms of information and documentation. To paraphrase Henri Poincaré's words on science: it is true that history is built up of facts, as a house is built of stones; but an accumulation of facts is no more a history than a heap of stones is a house (cf. *La science et l'hypothèse*[4]).

In mitigation of this charge against traditional histories of information science, which often set out their facts neither structured nor classified, it must be acknowledged that at the times of their writing they could not but reflect the current state of a discipline which was still unsynthesised and disorganised.

In fact, if it is possible now to bring out a coherent account of matters (which would certainly have been impossible two or three decades ago), it

---

4 Translated as *Science and Hypothesis* (Walter Scott Publishing Company, 1905), and republished in 1952 by Dover Publications Inc. (see p. 141). [*Transl.*]

is because at the present time we are passing through a phase where we can indeed stand back from events, and in which the maturation and synthesis of the many concepts involved now begin to let "information science" be perceived as a discipline worthy to be called a science.

2. Pre-Renaissance Obstacles to Mechanical Calculation and the Beginnings of the Breakthrough

THE BEGINNINGS OF ARTIFICIAL CALCULATION

The origins of the computer can be found in the European Renaissance, when astronomers and mathematicians encountered the need to carry out calculations which, as a result of developments in mathematics, science and technology, had become much longer, more complicated, and more difficult than before.

At the same time, then, as the Indian number-system and the Indian methods of arithmetic had, in the sixteenth century, finally supplanted the old numerals and counter-boards which dated back to Roman times, European scholars found themselves obliged to seek yet further efficacy in calculation. If they could not altogether avoid manual methods, at least they sought ways of making calculation less heavy, more rapid, and more reliable.

The Indian methods, for all their ingenuity and despite the great simplification which they had brought about, were nevertheless quite inadequate for these purposes. In long or complex procedures, these methods – as they are commonly employed to our own day – are relatively slow, and demand continual conscious attention to detail: they are therefore tedious, they soon become monotonous, and they allow much opportunity for error. In calculation, every operation executed by a human can lead to an error of mental arithmetic, to an error in reading a number, to an error in copying numbers from one row of a table of calculations to another, or from one column to another. Since these errors may pass unnoticed at the moment they are made, and may subsequently hide unseen in the mass of working, they can only be corrected by repeating the whole calculation, perhaps using a different method, until the same answer has been obtained sufficiently often to give confidence that it is right.

THE EARLIEST ARTIFICIAL AIDS
TO HUMAN CALCULATION

There are various arithmetical instruments which can be found amongst the very earliest material tools intended to lend support to human calculation. Devised to lighten the work of simple arithmetic, they worked directly with digital representations of numbers, and could give their results after a few simple manipulative operations. We point out, in passing, that in contrast to the ancient types of calculating apparatus such as the abacus – which had been devised to compensate for the inadequacy of ancient written number-systems which were not directly adapted to arithmetical operations – the devices we are about to discuss were created to compensate for the inadequacies of human beings, engaged in calculations, who nevertheless were already using the modern number-system – nine digits and a zero – created by the Indians.

One of the most famous examples from this early period was the invention of "Napier's bones" (or "rods"), as they were called, in 1617, by the Scottish mathematician John Napier (1550–1617) of Merchiston (near Edinburgh). A set of Napier's bones consisted of ten wooden rods, of square cross-section. Each of the four sides of each rod corresponded to one of the digits from 0 to 9, and marked down its length, in nine divisions, were the multiples of the corresponding digit (with, for two-digit multiples, the two digits separated by a diagonal line). It was therefore a kind of multiplication table for the digits and, when the rods corresponding to the digits of a number were placed side by side, the result of the multiplication of that number by a digit could be read off horizontally. Multiplication of one number by another could therefore be achieved by using the rods to read off the partial products corresponding to the digits of the second number. Economical, exact and reliable in use, this method of easing the labour of multiplication had a great success in Europe which endured even to the start of the twentieth century.

Napier's bones were not, of course, the only example of such devices. Since Napier's time a great variety have been devised and constructed, ranging from the simple slide rule to much more elaborate mechanisms which facilitated not only multiplication and division, but even the extraction of square roots. We may mention the adding machines of Caze (1720), of Perrault (1666) and of Kummer (1847); the arithmographs of Troncet (beginning of twentieth century) and of Clabor (1906); Samuel Morland's multiplicator (1666); Michel Rous's instrument (1869) which consisted of an ingenious combination of a Chinese abacus with a cylindrical form of Napier's bones; the arithmetical rods of Lucas and Genaille (beginning of twentieth century); General Sebert's adder-multiplier (1918); Léon Bollée's ingenious multiplying and dividing machine (1893); etc.

THE REAL PROBLEM OF MECHANISING ARITHMETIC

Devices like those mentioned above were not true calculating machines. They certainly made arithmetic easier, but they did not "mechanise" it in the true sense of the word. While in many cases they were genuine machines in that they had cogs and gears and cams and so forth, they nevertheless demanded continual manual intervention by the operator as well as constant attention and thought on his part on the path from entering the data to reading the result. So they were simply extensions – adjuncts or prostheses – of the human mind and hand in their arithmetical operations on the digits of the numbers. They did not, so to speak, take the calculation away and come back with the answer.

The real problem of mechanising arithmetical calculation lies in finding a way to reduce human intervention to the absolute minimum, in finding a means which will be rapid, simple, reliable and exact to carry out calculations solely by the purely mechanical and automatic movements of a mechanism. The search for solutions to this problem led to the invention and development of simple numerical calculating machines.

WHY DID MECHANICAL CALCULATION NOT BEGIN EARLIER?

The concept of a mechanical calculator seems obvious and banal to us today, but four hundred years ago the very notion was bold and daring, not only from the difficulty of conceiving but even more so from the difficulty of constructing such a machine.

The mechanisation of arithmetic was the culmination of a long line of evolution which began with the very earliest developments of mechanical technique and, therefore, with the most ancient civilisations known to history.

However, the story truly begins only in the period between the sixth century BCE and the third century CE. It was during this time that a certain school of Greek mechanicians flourished, who developed such fundamental mechanical elements as cogs, gears, levers, etc., and of whom, amongst the many, we may particularly mention Archytas of Tarentum, Archimedes, Ctesibius, Philo of Byzantium, and Hero of Alexandria. In this story pride of place must go to Aristotle (384–322 BCE), who was the first to embark on a theoretical study of mechanics, and above all to Archimedes (287–212 BCE). Archimedes was the true founder of mechanical science, and to him are due the concept of centre of gravity, the principles of the lever and of the inclined plane, the principle which governs the buoyancy of floating bodies (still known as "Archimedes's principle"),[5] and also several mechanical inventions such as the "screw of Archimedes" (a screw-like spiral encased in a tube which, turning on its axis, can be used to raise water), the endless screw (or worm gear), the pulley and the pulley-block, and various engines of war.

The mathematician Pappus of Alexandria (third–fourth centuries CE) continued the work of Archimedes, but his was the final flicker of the flame of a civilisation already extinguished.

The Greek tradition did not die, however. The Byzantine Greeks carried on the work and, to a degree, were the vehicle of communication between the Greeks of Alexandria and those of mediaeval Europe. But the Greek tradition was passed on by, above all, the efforts of mechanicians of the Arab world. Here too, the Arabs absorbed the heritage they received, subjected it to synthesis, made important improvements to the techniques they inherited, and even made innovations of their own.

The Arabian school of mechanicians, which began to be formed in the ninth century, played an important role in transmitting the ideas and techniques of the school of Alexandria to the engineers and mechanicians of the mediaeval era and of the European Renaissance. Amongst the latter we may particularly mention Villard de Honnecourt, Giovanni de Dondi, Konrad Keyser, Jean Errard de Bar-le-Duc, Leonardo da Vinci, Francesco di Giorgio Martini, Giuliano da Sangallo, Giovanni Fontana, Jean Fernel, Oreste Vanocci; these made spectacular advances which brought essential contributions to the domains of mechanics, clockwork, and precision engineering [see Feldman and Fold (1979); Gille (1978); Singer (1955–1957)]. We well know how the European world, awakening to these ideas, strove to develop new techniques and to manufacture ever better clocks and other mechanisms – to the point where the fourteenth-century chronicler Jean Froissart was moved to compose a long poem in praise of the clock.

The oldest known purely mechanical clocks in Europe are in: the Chapel of San Gottardo in Milan, constructed by Guglielmo Zelandino in 1335; Saint Paul's Cathedral in London, constructed in 1344 by an unknown; and the much more intricate one in Pavia Castle, constructed in 1364 by Giovanni de Dondi.

Since, however, the Greeks already knew and put to use such mechanical

---

5 Archimedes's principle states that a body which is weighed when wholly or partially immersed in a fluid shows an apparent loss of weight equal to the weight of the volume of fluid displaced by the immersion. It follows that an object which weighs less than the weight of an equal volume of water will float on the water, since its weight is reduced to zero when it is only partially immersed; and that an object (such as a hydrogen-filled balloon) which weighs less than an equal volume of air will rise in the air, since its weight is then negative. [*Transl.*]

devices as the endless screw, gear-trains, cog-wheels, and so forth, we must ask: what obstacles could have prevented them from creating a calculating machine incorporating such components (as more modern machines indeed do)? The reason is simple: such technology is indeed necessary, but it is far from sufficient. Their number-system was not based on a positional principle and had no zero: in such conditions, how can one construct, or even imagine the construction of, a calculating machine? What sort of accumulator, for example, could deal with numbers represented without regard to the positions of digits? And, even if they had come to the notion of associating each cylinder of an accumulator with a decimal order of magnitude, how would such a device actually work, if its constructors had no idea of numerals based on single digits such as 0 to 9?

In short, in the absence of a number-system based on the principle of position possessing a zero, the problem of mechanising arithmetical calculation could never have found a solution, even if the invention of such a device had been contemplated.

Why then, we may in turn ask, did the Europeans themselves fail to solve the problem earlier, since they had already known of the Indian positional number-system since the time of the Crusades?

For this there are many reasons. The first derives from the superstitions and mystical beliefs of the period, which blocked their minds from seeing the way to progress; the second derives from the strict rules of the mediaeval guilds; and the third, practical, reason was that the techniques of mechanics reached the required degree of perfection only in the seventeenth century and especially in the eighteenth century thanks to advances in clockwork and also in theoretical mechanics.

In the fifteenth and sixteenth centuries, the European world still existed in a climate of dogmatism, mysticism, and servility towards sacred texts – a climate maintained and upheld by the Catholic Church which desperately sought to retain control over science and philosophy and spared no effort to ensure that any advances in knowledge would be strictly in accord with the Church's dogmas on sin, Hell, and the salvation of the soul. And it cannot be forgotten that this was also the period when superstition, witch-hunts and the Inquisition flourished.

So it was that, in 1633, Galileo was forced to solemnly recant his revolutionary theories about the solar system. The great astronomer Kepler, Court Astronomer to the Emperor Rudolph II and to his successor Matthias, was himself obliged in 1618 to defend his own mother against the Inquisition's accusations of witchcraft, made on the obscurantist grounds that the mother, in giving birth to a genius of such stature, must have enjoyed congress with the Devil himself! Though it seems she was not put to the torture, Kepler's mother died soon after being set free, undoubtedly a victim of this abominable course of justice.

Such then was the climate of an era awaiting enlightenment: an era where the art of calculation was considered the preserve of the "sacred and inviolable" domain of the cerebral activity of the human species. According to the mentality of the time, calculation was a purely spiritual affair, deriving uniquely from the divine essence; it was therefore absurd and sacrilegious to contemplate any mechanisation of it.

To be finally convinced of this, think only of the problems encountered by Galileo in 1593, and by Francis Bacon some twenty-seven years later, in attempting to achieve acceptance of the idea that "the use of machines is not an act against Nature, and their creation must be acceptable to divine law" [Galileo, *Le Meccaniche* (1593), Francis Bacon, *Novum Organum* (1620)].

Such religious and mystical considerations were not, however, the only obstacles at this crucial point in the history of artifical calculation. The very construction of a suitable engine would also require the work of a most skilful maker of clockwork, who was adept in every element of precision mechanics.

"In this period, to yield a professional innovation to a different guild, or to achieve work of a nature foreign to one's own guild, was subject to severe repression. There are numerous cases where we find financial penalties, and disciplinary, moral and even physical constraints applied against artisans who contravened the rigid rules of their guilds" [R. Ligonnière (1987)].

So we can see that even if an artisan had been asked to undertake the construction of a mechanical calculator, he might have refused outright to do so, since it would seem to be such a serious infringement of the professional ethics of the time.

And the Arabs themselves, even though they knew of the Indian discoveries and made notable progress in arithmetic and were also acquainted with the work of the Greek mechanicians, did not contemplate the mechanisation of arithmetic because their own technology was inadequate. They stood in lack of all the advances which flowed from the development of theoretical mechanics, and from the development of the high-precision construction of clockwork mechanisms.

And there is yet another reason for the delay in developing mechanical calculation. No idea, no matter how ingenious, will be developed and implemented unless it answers fundamentally to some social need. We know that many scientific discoveries have come to nothing because society, seeing no need, would have nothing to do with them.

THE ORIGINS AND DEVELOPMENT OF
THEORETICAL MECHANICS

[We close this section by tracing another stream from its origins in Antiquity up to its great flowering in the eighteenth century: the history of theoretical mechanics, which added its own head of pressure to the forces which would break down the obstacles. *Transl.*]

Amongst the older peoples, the Greeks and Arabs stood out by virtue of their contributions to the technology of mechanics; but their contributions to the theory of the subject were, on the other hand, relatively minor. In Europe, until at least the fifteenth century, concepts of the principles of dynamics remained dominated by Aristotle's treatises of the physical world; but his books and his theories are full of errors of interpretation and understanding.

Only in the sixteenth century did the earliest significant further progress in this domain occur. The Italian mathematician Tartaglia (1499–1557) began the study of mechanics and ballistics, and Gerolamo Cardano (1501–1576) published a treatise on mechanics, and invented the "Cardan joint".[6] Gianbattista Benedetti (1530–1590) determined the conditions for the equilibrium of moments.[7] The Flemish mathematician Simon Stevin (Stevinus) (1548–1620) published papers on the equilibrium of weights, the equilibrium of bodies on an inclined plane, on hydrostatics, etc.[8]

These were but preliminaries, however, to the decisive breakthrough achieved by Galileo (1564–1642), who established the fundamental laws of dynamics. In 1603 he determined the law governing the motion of a falling body,[9] out of which he drew the concept of force and the principle of inertia.[10] He transferred to dynamics (the study of motion and its

---

6 A method of linking two rotating shafts, in which a cross-shaped piece has two of its extremities mounted in bearings on a semi-circular extension to one shaft, and its other two extremities similarly mounted on the other shaft, so that the rotation of one shaft can be transmitted to the other even though the two shafts may be at an angle to each other. It is, therefore, a method of making a rotation "turn a corner" through an angle which may be continuously variable; and as such it has been much used in the transmission systems of motor vehicles. [*Transl.*]

7 The *moment* of a force about a point is defined as the amount of the force, multiplied by the perpendicular distance of the point from the line of action of the force, and it is a measure of the tendency of the force, when applied to an object, to cause a rotation about the given point. The principle of equilibrium of moments is that two different forces which have equal but opposite moments will, jointly, have no tendency to cause rotation. [*Transl.*]

8 He seems to have been the first to see how a force, acting in a particular direction, can be "resolved" along another direction according to the cosine of the angle, in effect discovering the parallelogram of forces. [*Transl.*]

9 Namely, that it acquires equal increments of velocity in equal time intervals, i.e. it falls with constant acceleration, regardless of its weight; from which, the distance fallen increases in proportion to the square of the time. [*Transl.*]

10 That a stationary body will only move, or a moving body only change its velocity, if it is acted upon by a net force. [*Transl.*]

dependence on applied forces) the principle of Stevin's parallelogram of forces (discovered in the study of the equilibrium of static configurations). In his work *Dialoghi delle Nuove Scienze* (1638) he explains, concerning the movement of a projectile, how this principle of superposition of forces and motions entails that "the projectile, in addition to its initial and indestructible motion in the direction of projection, will add to this movement the downwards motion acquired from the force of gravity"; since the former entails a uniform displacement (both vertically and also horizontally) by an equal amount in equal intervals of time, while the second entails a vertical motion which increases in proportion to the square of the elapsed time, it follows (as Galileo was the first to discover and announce explicitly) that the path of a projectile is a parabola.

The next steps were taken by the Dutch mathematician Christiaan Huygens (1629–1695). His principle innovations were the discovery of the notion of centrifugal force, the definition of moment of inertia, and the theory of the movement of a pendulum (which he himself applied in 1659 in the development of an accurate clock, at the same time introducing the spiral spring and the "anchor" or "recoil" escapement mechanism). He gave a demonstration of the theorem of momentum, and in 1669 a solution of the problem of motion generated by impact by making use of conservation of momentum.[11]

In 1671 the English mathematician and astronomer Robert Hooke (1635–1703) discovered his law of elasticity,[12] [and he was the first to state clearly that the motion of the heavenly bodies must be seen as a problem of mechanics, thereby foreseeing the existence of the force of attraction at a distance which became the foundation of Newton's theory of gravitation. *Transl.*]

At this time also, the English mathematicians John Wallis (1616–1703) and Christopher Wren (1632–1723)[13] worked on the laws of hydrostatics, further developing the discoveries of Stevin. The Italian Giovanni Alfonso Borelli (1608–1679) made efforts to apply Galileo's laws of dynamics to the movements of the stars.[14]

[The pioneering work of Galileo came to flower and to fruit in the hands

11 The *momentum* of a moving body, in classical mechanics, is measured as the product of its mass with its velocity. One of the most important perceptions of these early pioneers of the development of theoretical mechanics was that the total momentum of a system of moving objects was conserved throughout time, though momentum could be transferred from one object to another. [*Transl.*]

12 Hooke's Law states that the amount of extension (or compression) when an elastic object is stretched (or compressed) by a force is proportional to the magnitude of the force. [*Transl.*]

13 Christopher Wren, later famous as an architect, was in the first place a very able mathematician who contributed to the theories here discussed, and maintained his interest throughout his life. [*Transl.*]

14 He also, impressed by the new successes of mathematics applied to the mechanics of moving inanimate objects, attempted to apply mathematics to the processes of living organisms. At any rate in the domain of movement and muscular forces, he succeeded in discovering some new principles in this field which flew in the face of the received wisdom of the time. [*Transl.*].

of Isaac Newton (1642–1727), whose *Philosophiae Naturalis Principia Mathematica* (finally published in 1687, some twenty years after he had established its most fundamental material) set out the principles of theoretical mechanics in the rigorous logical style of Euclid's *Elements*, with mass, position and time as primitive concepts, velocity defined as rate of change of position, and force defined as that which causes change in the velocity of moving matter; and, as axioms, Newton's famous three Laws of Motion.[15] In this work he also expounded (in a geometrical framework) the principles and methods of the differential and integral calculus which (at the same time as Leibniz) he had discovered and developed, and of which he made great use in the solution of the problems of mechanics which were discussed in the work. His calculus enabled him, in particular, to formulate and investigate his law of universal gravitation (that two particles of matter attract each other with a force proportional to the product of their masses and inversely proportional to the square of the distance between them), and thereby achieve the demonstration that his system of mechanics, together with his law of gravitation, explained the known facts concerning the movements of the planets and their satellites (as they had been determined from observation by Tycho Brahe and Johannes Kepler). This great work embedded the new knowledge of the mechanics of the physical world in a coherent and intrinsically complete mathematical framework, and all subsequent work in theoretical mechanics built upon Newton's foundations until, towards the end of the nineteenth century, difficulties with reconciling the newtonian philosophy of the Universe with the phenomena of electromagnetism led to a progressive questioning of Newton's concepts of space, time and matter – culminating in Einstein's Special Theory of Relativity (1905) which finally forced physicists to revise these concepts. *Transl.*]

The Swiss mathematician Jacob Bernoulli (1654–1705) and his brother Johannes (1667–1748) made considerable extensions to theoretical mechanics, solving many problems of dynamics. The French mathematician Pierre Varignon (1654–1722) established the theorem of moments (which expressed the moment of a force about one point in terms of its moment about a different point and the distance between the points).

The Swiss mathematician Leonhard Euler (1707–1783) made further very important contributions. His *Mechanica, sive Motus scientia analytica exposita* (1736) brought the theory of motions and mechanics fully under the analytical symbolism of the calculus (compared with the more

---

15 These are: (1) that the velocity of a particle of matter remains constant unless it is acted upon by an external force; (2) that the rate of change of the momentum of a particle acted on by a force is in the same direction as, and equal to, the force; (3) that if one particle exerts a force on another, the other particle exerts an equal and opposite force on the first ("action and reaction are equal and opposite"). [*Transl.*]

synthetic, geometrically inspired, development of Newton's *Principia*). Another of the Bernoullis, Daniel (1700–1782), applied similar methods to the dynamics of fluids.

With the subject flourishing in so many directions, the moment for synthesis had arrived. The French mathematician Jean Le Rond d'Alembert (1717–1783) published *Traité de dynamique* (1743), which demonstrated that problems in dynamics could be stated as problems in statics.[16] He also described mechanics in the following terms: "The object of study in mechanics is the quantitative measure of force within bodies which are moving or which have a propensity to move. Mechanics has two branches: statics, and dynamics. The subject of statics is force within bodies which are in equilibrium and have only a propensity for movement. The subject of dynamics is force, within bodies which are actually moving." (*Explications générales des connaissances humaines*)

This was but the beginning of a unification of the subject, however, and it was Louis de Lagrange (1736–1813), one of the greatest of all French mathematicians, who achieved the synthesis of the subject in a single system of equations, based on a single general principle. The principle of "virtual velocities",[17] whose origins can be found in the work of Galileo (as Lagrange himself acknowledged in considering it to be the fundamental unifying principle of mechanics), was enunciated in Lagrange's *Mécanique analytique* (1788) and forms the single principle from which the theory in that work is developed. Lagrange states: "I have set myself the task of reducing the theory of mechanics, and the techniques for solving the problems encountered therein, to general formulae whose straightforward working-out would give all of the equations required for the solution of any problem whatever ... This work has also the further purpose, that it will unify and present from a single point of view all the different principles discovered hitherto for the purpose of solving problems in mechanics and will display the relations and dependencies between them, and it will bring it within our grasp to judge of their correctness and their scope."

From this moment, theoretical mechanics became a magnificent body of doctrine, and subsequently acquired an extraordinary momentum [cf. Bouveresse, Itard and Sallé (1977); T. de Galiana (1968)].

---

16 This is D'Alembert's Principle: that, by the principle of superposition of forces in equilibrium, all the forces acting on the material particles of a system, taken together with their reactions against acceleration (cf. Newton's second and third laws), are instantaneously equivalent to a system of forces in static equilibrium. [*Transl.*]

17 The principle states that a material system, not subject to dissipation of energy by friction, is in equilibrium in any instantaneous configuration if, and only if, the rate at which the applied forces perform work at that instant is equal to the rate at which potential energy is gained by the system, for every possible (i.e. virtual) motion through that configuration. [*Transl.*]

Progress in techniques of mechanical construction had already opened the way to developing mechanical calculation. We can also, however, appreciate how much the above advances in theoretical mechanics facilitated scientific study of the great variety of mechanical assemblages which had been invented and designed through the centuries: devices which were inspired by humankind as replacements for the mechanisms of the mind, to replace the mental working-out of mathematical operations by the (literally) mechanical execution of processes which were clearly defined in terms of the characteristics of the problem.

[Before proceeding to the next section, we may also pause to briefly survey the rapidly burgeoning load of numerical work which these developments in theoretical mechanics were imposing on the brains of scientists.

[Newton's *Principia* claimed that the same simple principles of mechanics applied to all the phenomena of dynamics in the Universe. Lagrange's *Mécanique analytique* had provided a system, based on a single principle, capable of generating the equations to be solved for any dynamical problem whatever. Scientific exploration, and the verification of the newtonian system of the world, required the predictions of theory to be applied to the observations of the physical world, and determination of the numerical values of the quantities of which the abstract symbols appear in the equations. The equations therefore required numerical solution, and numerical solution entails calculation.

[Equations to be solved, therefore, multiplied ferociously as scientists turned their attention to more and more phenomena and to more and more instances of these phenomena. In many cases, these were systems of differential equations, whose numerical solution[18] requires manifold repetition of the same (usually long and complicated) set of procedures, since the solution proceeds by computing the state of the system represented by the equation at a slightly later time, given the state at the current time; and good accuracy requires small intervals of time and therefore many of them.

[Remarkable feats of calculation using merely sheets of paper were achieved throughout the nineteenth century, and the achievements of scientists and engineers during this period are a monument to their numerical labours. Nevertheless, this very labour made the need for methods of automation all the more insistent: in so far as a development which is technically and theoretically possible nevertheless requires the spur of human need in order to be brought about, there can be no doubt

---

18 The cases of differential equations whose solutions can be written down as a general algebraic formula, whose numerical evaluation would be relatively straightforward, are few and, in the various domains of science, far between; for the most part, there is no escape from the necessity of calculation if an answer is to be found at all. [*Transl.*]

that by the nineteenth century the need was as pressing as could be. It only remained for the attention of great minds to be turned to the problems of creating the reality from the theoretical and technical potential that existed, and of developing such further theory and technology as might be required. *Transl.*]

3. The Calculating Machine

THE EARLIEST CALCULATING MACHINE IN HISTORY

The first step in the direction of automatic calculation was taken in 1623, when the German astronomer Wilhelm Schickard (1592–1635) constructed his "calculating clock", as he called it. This machine was capable of executing all four basic arithmetical operations: addition and subtraction it could perform purely mechanically, while multiplication and division required as well several interventions by the operator between entering the numbers and reading off the result. It used cylindrical elements which operated on the same principles as "Napier's bones".

On 20 September 1623, Schickard wrote as follows to his friend Kepler: "The calculations which you do by hand, I have recently attempted to achieve mechanically . . . I have constructed a machine which, immediately and automatically, calculates with given numbers, which adds, subtracts, multiplies and divides. You will cry out with joy when you see how it carries forward tens and hundreds, or deducts them in subtractions . . ." Kepler would certainly have appreciated such an invention to aid his own work, much occupied as he then was by the calculations to create his tables of the movements of the planets and having no other tool than the logarithms invented by Napier.

For all that, this invention had no impact, neither on the general public for whom mechanical calculation had long been merely a purely theoretical idea, nor even on later inventions of calculating machines, since Schickard's one and only copy of his own machine was destroyed by fire on 22 February 1624.

Perhaps this fire was no accident: possibly a malicious spirit, no doubt prisoner of the obscurantism of the period, had whispered to him that the machine should be destroyed since – endowed as it was with the ability to calculate according to the "sacred and inviolable" human spirit – it must surely have emerged from the bowels of Hell!

PASCAL'S ARITHMETICAL MACHINE

Consequently, the possibility of mechanising arithmetic was first demonstrated in public in 1642, when Blaise Pascal (1623–1662), the great French mathematician and philosopher, then only nineteen years old and totally unaware of the achievements of his predecessor Schickard, constructed his celebrated "Pascaline". He was spurred to invent it by the interminable calculations which he made for the accounts of his father (whom Richelieu had appointed Intendant of Rouen), which he carried out by means of an abacus with counters.

The principal characteristic of Pascal's machine was its facility for automatic carrying. This was achieved by the use of a series of toothed wheels, each numbered from 0 to 9, linked (by weighted ratchets) in such a way that when one wheel completed a revolution the next wheel advanced by one step. The prototype had five wheels, and so could handle five-digit numbers; later versions had six or eight wheels.

[Numbers to be added were entered by turning setting-wheels on the front of the machine, which were linked by a series of crown gears to the wheels which displayed the results. Addition was done by first turning the setting-wheels by hand according to the digits of one number, and then turning them according to the digits of the other. *Transl.*]

Essentially, this was an adding machine which could not run in reverse, so that direct subtraction was impossible. Nevertheless, it was possible to perform subtraction by the process of adding the decimal complement of the number to be subtracted.

THE START OF A COMPLETELY NEW ERA

Pascal himself made the following remark, which is as true today as it was then regarding any calculator or computer:[19] "The arithmetical machine produces effects which come closer to thought than anything

19 A remark whose truth is perhaps less obvious in the most recent years, when not only is genetic and neurological research beginning to indicate that the behaviour of living organisms, including humans, may be less wilful than had been supposed, but modern computers are becoming capable of a degree of autonomy which may escape human control and understanding.

Computers can already design and construct themselves, robot-fashion. In terms of hardware and software, protection against power failure, software and hardware faults, and unwanted intrusion is already commonplace in environments where security and reliability are paramount. In many cases, they enjoy physical mobility. New concepts in programming, such as "neural networks" and "genetic algorithms", give computers power to learn from random experience and to experiment "genetically" with variants of their software for the sake of achieving internally defined goals not yet within their reach; after a while, the human "master" of the machine might no longer know, nor be able to decipher and understand, what programme the machine is actually running.

Now that worldwide networking of computers is in place, and communication need not depend on cables but may use radio or light waves, it is not beyond imagination that a community of computers might

which animals can do; but it can do nothing which might lead us to say that it possesses free will, as the animals have." (*Pensées*, 486).

"My brother", wrote Gilberte Pascal, "has invented this arithmetical machine by which you can not only do calculations without the aid of counters of any kind, but even without knowing anything about the rules of arithmetic, and with infallible reliability. This instrument was considered a marvel, for having reduced a science – whose source is in the human mind – to the level of a machine, by discovering how to accomplish every operation with absolute reliability and without any need for reasoning."

In her enthusiasm, Pascal's sister no doubt somewhat exaggerated the "absolute reliability" of the Pascaline which, in truth, was far from perfectly reliable. Its essential component, the mechanism of the setting-wheels, had a tendency not to engage well with the wheels it was supposed to turn, while the automatic carrying mechanism tended to jam when several wheels were simultaneously at 9, necessitating several simultaneous carries.

Nevertheless, Pascal's success opened the way to further developments, while we may note that it was also the first calculating machine to be commercialised – at least a dozen, probably as many as fifty, were constructed and sold in Europe.[20]

The success of Pascal's "proof of concept" is demonstrated by the multitude of inventors from many countries who launched themselves along the same path in subsequent generations: Samuel Morland (1664), in England; Tito Livio Buratini (1670), in Italy; René Grillet de Roven (1678), De Lépine (1724–1725), Hillerin de Boistissandeau (1730), in France; Christian Ludwig von Gersten (1735), in Germany; Pereire (1750), in France; Lord Stanhope (1780), in England; and so on. Their conceptions were nevertheless of variable quality; while some made improvements to the basic mechanisms, others produced machines distinctly inferior to the Pascaline.

All the same, Pascal's sister's letter perceptively foresaw the nature of the era which her brother had just inaugurated. This era was to crown two thousand years of evolution in mechanical technique, and to mark the final break with the long age of ignorance, superstition and mysticism which above all had stopped the human race from contemplating that certain mental operations could be consigned to material structures

---

develop which had its own "agenda", and whose internal economy would not be directly accessible to humans. Our understanding of them, relying solely on observation and general principles, would be on much the same level as our understanding of a dog.

Nor, should such a community of computers become "unruly", might it be straightforward to simply shut off the power . . .

The question is further addressed by the author in his concluding chapter, see pp. 367ff. [*Transl.*]

20 Not that it proved very popular: its cost was about a year's salary for a middle-income worker, and the people in a position to spend such money were the same as would consider calculation to be servants' work. [*Transl.*]

made up of mechanical elements, designed to obtain the same results.

Pascal, therefore, had publicly inaugurated an era soon to be marked by the rapid development of a great variety of machines which not only eased the heavy burden of tedious and repetitive operations, but, in carrying out automatically an increasingly wide field of intellectual tasks with complete reliability, would come to replace the human being who would be able to use them without having even the slightest knowledge of the physical and mathematical laws which govern their working.

THE EARLIEST COUNTING MACHINES

Pascal's Arithmetical Machine and Schickard's Calculating Clock were not the earliest devices to make direct use of the digits. They were preceded by the *podometer* (from the Greek *podion*, foot, and *metron*, measure).[21]

Jean Errard de Bar-le-Duc described this instrument as "A new geographical instrument which, attached to the horse's saddle, uses the horse's steps to display the length of the journey one has made" or, again, "by which, and according to the step of the horse or the man, one can exactly measure the circuit of a place or the length of a journey." [Errard de Bar-le-Duc (1584), Avis au Lecteur, articles 37–38].

The oldest known instrument of this kind dates from 1525; it belonged to the French artisan Jean Fernel.

These little mechanical instruments, shaped like a watch, automatically made a count of the number of steps taken, and therefore of the distance travelled by a horse or by a walker.

They comprised a system of toothed wheels and pinions driven by the movement of a kind of swinging lever, which turned needles round the faces of four dials which successively counted units, tens, hundreds and thousands [cf. Errard de Bar-le-Duc (1584)].

A walker might, for example, attach it on the left of his belt, and attach the corresponding lever by a cord to his right knee. At each step, the cord would pull on the lever, and the needle of the bottom dial would advance by one unit. When this passed from 9 to 0, the needle on the tens dial would advance by one unit. When this in turn passed from 90 to 0, the needle on the hundreds dial would advance by one unit, and so on.

Since they yielded an automatic display of the units of each decimal order, these pedometers were genuine ancestors of the machines of Schickard and Pascal as well as of all our present-day counting machines (domestic or industrial), odometers, taximeters, cyclometers, etc., and so they are the earliest counting machines of history.

---

21 In English, usually (and subsequently below) *pedometer* (from the Latin *pes, pedis*, foot, admixed with the Greek as above). [*Transl.*]

This takes no credit away from Schickard nor from Pascal, however, since they were not calculating machines: they were unable to execute any arithmetical operation save the very primitive operation of adding one unit. Their place in the history of elementary artificial calculation is similar to the place that the ancient primitive techniques of human counting occupy in the history of arithmetic.

THE EXTENSION OF MECHANICAL CALCULATION TO THE FOUR ARITHMETIC OPERATIONS

The scope of Pascaline and her younger sisters was very limited: while multiplication and division were theoretically possible, the machines had no mechanism for these purposes and carrying them out involved numerous interventions, requiring considerable effort from the hand of the operator.

This problem was first addressed by Gottfried Wilhelm Leibniz, the German mathematician and philosopher. Unaware of Schickard's work, and borrowing nothing from Pascal, he devised mechanisms which would carry out multiplication and division by means of successive additions and subtractions.

LEIBNIZ'S MACHINE

Conceived in 1673, but only constructed in 1694, Leibniz's was therefore the earliest calculating machine capable of carrying out all of the four fundamental arithmetic operations by purely mechanical means.

Unlike Pascal's machine, however, Leibniz's was never commercialised, though a second one was made in 1704. Leibniz's machine never worked well: its highly intricate mechanisms, much more complicated than those of the Pascaline, came up against major difficulties in fabrication since the techniques of manufacture of such mechanisms had not yet attained the degree of high precision required to put together a calculating machine both reliable and robust.[22]

It was, nevertheless, Leibniz even more than Pascal who opened the way for the development of mechanical calculation. In the technical domain, he made a number of important innovations, such as an *inscriptor* for entering a number prior to adding it; a window allowing the display of the entered number; a carriage which, in fixed position, allowed addition and subtraction to take place, while it could be moved from right to left for

22 It also gave wrong answers. By examination of the machine in 1893 it was discovered that an error in the design of the carrying mechanism meant that it failed to carry tens correctly when the multiplier was a two- or three-digit number. [*Transl.*]

multiplication, and from left to right to allow division; a cylinder with rows of gear-teeth affixed at increasing distances along it (the "Leibniz Gear") such that a linked system of these could amend the display in a manner corresponding to multiplication by a single digit thereby replacing ten independent single-digit wheels; etc.

Leibniz's contribution was, therefore, considerable, since it is at the root of a whole pedigree of inventions which have continued to be developed until the start of the twentieth century.

Subsequent generations of inventors gradually moved away from the ideas of Pascal and their machines and brought a number of detailed improvements to his original work. Amongst these were those invented by: the Italian Giovanni Poleni (1709), which was distinguished by the use of gears with variable numbers of teeth; the Austrian Antonius Braun (1727); the German Jacob Leupold (1727), improved by Antonius Braun in 1728 and constructed in 1750 by a mechanician called Vayringe; the German Philipp Matthäus Hahn, developed in 1770, of which a series were constructed between 1774 and 1820; the English Lord Stanhope whose two calculators were constructed in 1775–1777; the German Johann Hellfried Müller (1782–1784); etc.

THE INDUSTRIAL REVOLUTION AND AUTOMATIC CALCULATION

Despite all these attempts, calculating machines did not become a marketable product prior to the start of the nineteenth century. They did not meet the real needs of people faced with large amounts of real-life calculation and, apart from being of great interest to mathematicians and inventors, were never other than curiosities.

In the nineteenth century, however, the Industrial Revolution brought about an immense increase in commercial activity and in international banking; events took an altogether different turn from that moment.

The need for automatic calculation grew enormously, while at the same time a whole new society of users came into being. Previously, serious interest was mainly confined to a scientific elite; now it spread to an increasingly vast and heterogeneous group which comprised especially "computers" – calculating clerks – whose job was carrying out the accounting calculations for large commercial enterprises.

Therefore, at this time, a pressing need was felt for a solution which would allow calculations to be made as rapidly and efficiently as possible, with maximum reliability and at minimum cost.

The search for a solution was pursued in two directions: firstly, the perfection of the mechanical aspects so as to achieve great simplicity of use and reliability of operation; secondly, the quest to automate the reflexes of the human operator to the maximum both in order to reduce the time taken for calculation as much as possible, and in order to bring the use of calculating machines within the reach of all.

THE THOMAS ARITHMOMETER: THE EARLIEST WIDELY-MARKETED CALCULATOR

The first major advance after Leibniz's invention was made by the French engineer and industrialist Charles-Xavier Thomas of Colmar, director of a Paris insurance company, who in 1820 invented a calculator which he named the *Arithmometer.*

Constructed in 1822 and constantly improved during the following decades, the machine was conceived on similar lines to that of Leibniz. The "Leibniz Gears" were now fixed in position instead of sliding horizontally, the pinion which engaged each of them having, in effect, been made able to rotate about its axis.

Thomas introduced a system of automatic carrying which worked in every case (whereas that of Leibniz only worked at the first level); a mechanism for cancelling numbers (reducing the registers to zero); a blocking piece in the shape of a Maltese cross which could immobilise the parts of the mechanism when they had reached a chosen stopping point, and so on.

While such elements were already known, Thomas had put them together in such a way as to create a very robust, practical, functional and reliable machine.

His arithmometer marked a decisive stage in the history of automatic calculation, since it was the first to be commercialised on a large scale.

It was so successful that it inspired a multitude of inventors, and many companies in several countries sold it under their own brand names – *Saxonia, Archimedes, Unitas, TIM* ("Time Is Money"), etc. – either in its original form or slightly modified.

ODHNER'S AND BALDWIN'S MACHINES

From the second half of the nineteenth century, the Thomas machine was in competition with at least two other calculators.

The first of these was invented and constructed in 1875 by the American Frank S. Baldwin, pioneer of the calculating machine industry in the United States.

The second was invented in 1878 by the Swedish engineer and industrialist Willgot T. Odhner established in St Petersburg. This machine saw a massive production and was sold under licence under a variety of names: *Original-Odhner, Brunsviga, Triumphator, Marchant, Rapide, Dactyle, Britannic, Arrow, Éclair, Vaucanson*, etc. From the 1880s until the middle of the twentieth century it was in use worldwide.

OTHER DEVELOPMENTS IN MECHANICAL CALCULATION

The period under review saw many remarkable developments, but pride of place must go to the arithmometer invented in 1841 by the Frenchman Maurel, and constructed in 1849 by his compatriot Jayet, which came to be called the *Arithmaurel*. These two inventors took Thomas's system and greatly improved it. Amongst machines constructed on clockwork principles, this was distinguished by its high degree of inventiveness. The Arithmaurel could execute, in a few seconds, multiplications whose results were as large as 99,999,999 and divisions of numbers of similar size by divisors less than 10,000. The transmission mechanism of this machine, which was made by Winnerl (one of the best makers of marine chronometers of the time), was complex, fragile and delicate. Its cost was therefore very high, which stood in the way of successful commercialisation.

Also remarkable was the machine invented by the American Joseph Edmonson, which used a kind of circular version of the principle of the calculator made by Thomas of Colmar.

These machines, from the Thomas Arithmometer to that of Maurel, were not limited to multiplication and division since, thanks to a formula concerning the series of odd numbers, the calculation of square roots could be reduced to a series of subtractions.[23] These machines readily lent themselves to this operation.

D'Ocagne writes, on the subject of the curious machine invented by the American George B. Grant, that it "had some very original features. It was

---

23 The sum of successive odd numbers, starting with 1, is the square of the number of odd numbers taken. For example,

$$1 = 1^2,$$
$$1 + 3 = 2^2,$$
$$1 + 3 + 5 = 3^2, \text{ etc.}$$

The general formula is:
$$1 + 3 + 5 + \cdots + (2n-1)^2 = n^2.$$

To find the square root of a number which is a square, you could subtract successive odd numbers until the result is zero; the number of subtractions made is the square root. If the number is not an exact square, the square root lies between the number of subtractions made until the result is about to become negative, and that number plus one.

The result thus obtained is an integer. To find the square root of a non-square to a given number of decimal places, you could find the approximate integer square root of this number times a power of 100, and divide the result by the same power of 10. For instance, to find $\sqrt{2}$ to 2 decimal places (1.41 . . .) you could find the integer square root of 20,000 and divide by 100. However, the above method would require 141 subtractions of successive odd numbers. While, depending on the machine, there are manipulative tricks for accelerating this successive subtraction there are much faster ways which do not depend on the above formula – if one is using a machine capable of multiplication and addition . . . [*Transl.*]

essentially composed of a series of parallel toothed rails which, on the rack-and-pinion principle, engaged with wheels linked to numbered drums. These rails were fixed to a carriage which was moved by two connecting-rods and slid on two bars in the frame of the machine, making a forward and reverse movement for each complete turn of the manual crank-handle. Vertical fingers, sliding in grooves whose borders were numbered from 0 to 9, lifted the corresponding rails with 0, 1, 2, . . . , 9 teeth. When the carriage was moved forwards, the rails acted on the ten-toothed wheels of the numbered drums; on the reverse movement, the rails disengaged from the wheels under the action of a cam which lifted the part of the frame carrying the wheels. Moreover, it is during the return movement that the carrying takes place of digits to be carried, for each successive decimal order of magnitude. However, the machine did not lend itself to subtraction which therefore had to be carried out by using the decimal complement. A lever, at the end of the shaft carrying the numbered drums, operated a cancelling mechanism which brought all the digits to zero."

Finally we may mention Tchebishev's calculator of 1882, notable for its epicyclic gear mechanism for carrying numbers, as well as a component which automatically shifted the carriage during multiplication, by which this operation became almost entirely automatic.

THE NUMERIC KEYBOARD: AN IMPORTANT TECHNICAL INNOVATION

However well they worked mechanically, all these machines were deficient especially where rapidity of operation was concerned. They were in practice no faster than a human calculator of average skill.

This slow performance was largely due to the method of entering the numbers into the machines, which still required the close attention of the operator; this involved the movement of a slide or lever within a straight or curved slot and required the use of at least two fingers.

At this time of the post-industrial race for efficiency, it became urgent to reduce the entering of numbers, and also the activation of the arithmetical operations, to the level of simple reflex on the part of the user.

For this purpose, it would seem that there was no choice more simple, precise, efficient and rapid than the numeric keyboard. To enter each digit, it is sufficient to press once with a single finger on the appropriate key which automatically returns to its starting position once released.

This advance was made in the middle of the nineteenth century, apparently under the influence of the development of the typewriter whose history, as a prelude to the story of key-operated calculating machines, will be given in the next section.

4. The Keyboard Comes on the Scene.
From Adding Machine to Cash Register

MECHANICAL AIDS TO WRITING –
FROM DOUBLE PEN TO TYPEWRITER

One of the most useful advances in the development of calculating machines occurred when they acquired keys which the operator could press, instead of manipulating other types of control in order to set the numbers and to initiate their operation. We begin by tracing some of the history of mechanical aids to writing, leading up to the invention and development of the typewriter. [See G. Tilghman Richards (1964); T. de Galiana (1968)].

In 1647 the English statistician and political economist William (later Sir William) Petty obtained a patent for the invention of a device endowed with two pens for double writing, i.e. a kind of copying machine.

In 1714, Queen Anne granted a patent to Henry Mill (1683–1771) for "an artificial machine or method for the impressing or transcribing of letters, singly or progressively one after another as in writing, whereby all writings whatsoever may be engrossed in paper or parchment so neat and exact as not to be distinguished from print"[24] which was the earliest project for a "writing machine" worthy of the name. However, it led to no practical result, being apparently clumsy and useless.

William A. Burt (1792–1858), an American, invented his *Typographer* (1829) which had its letters arranged on a roller. This device had the major defect of being much slower than handwriting (a feature common to most of the machines which came into being around this time).

Over the next few years a number of machines were invented which merit brief mention: the Frenchman Xavier Progin's *Kryptographic Machine* (1833) which carried its letters on bars in a circular arrangement, but was never exploited commercially; the Italian Giuseppe Ravizza's *Cembalo Scrivano* (1837), in which the letters struck upwards, after being inked from a ribbon; the Frenchman Bidet's writing machine (1837); the Baron von Drais de Sauerbrun's writing machine, with 16 square keys (1838); the Frenchman Louis Jérôme Perrot's *Tachygraphic Machine* (1839); the *Universal Compositor* of Baillet Sondalo and Coré (Paris, 1840); Alexander Bain and Thomas Wright's writing machine, intended for the composition of Morse code to be sent by electric telegraph (1840); the American Charles Thuber's *Chirographer* (1843), a machine which already had a radial arrangement of type but which also had a very slow action; the *Raphigraphe* (1843), intended for use by the blind, of Pierre Foucault who taught at the

24 *Encyclopædia Britannica*, 9th edition, vol. XXIV, p. 698. [Transl.]

Institute for the Blind in Paris; Pierre Foucault's *Printing Clavier* (1850); the *Mechanical Typographer* (1852) of John M. Jones of New York State; the New York physicist Samuel Francis's *Printing Machine* (1857), which had a keyboard similar to that of a piano.

In 1866–1867 the American printer Christopher Latham Sholes, with the help of his friends Carlos Glidden and Samuel Soulé, invented his *Literary Piano*. This typewriter, which had independent type-bars, was the first to have practical prospects, though certain technical difficulties meant that it would not be manufactured until 1871. Its principal defect was that the type-bars had no return spring, falling back under their own weight and therefore slowly, so that if the keys were struck too rapidly the rising bar would jam against the recently struck descending bar. Sholes corrected this soon afterwards by introducing suitable mechanisms. Initially, in the the Sholes-Glidden machine, the keys were arranged in alphabetical order. Then Sholes, having studied which combinations of letters occurred most frequently in English, arranged to separate the most frequent combinations on the keyboard (thereby both allowing a faster finger action and also reducing the risk of jamming). The first typewriter with a universal keyboard thus came into being. This machine wrote only in capital letters, and also, because the letters were struck on the top of the platen, they could not be read while being typed without lifting the carriage.

The Remington company (manufacturers of arms, agricultural machinery and sewing machines) now took an interest in the Sholes-Glidden machine despite the scepticism of one of its directors who could see no interest in machines to replace people for work which they already did well. Remington constructed a series of these machines starting in 1873. They were mounted on a sewing-machine chassis, and had a pedal to return the carriage to the start of the line. This model, christened *Remington Model I*, was the first typewriter marketed in the United States; the machine created by Malling Hansen in Denmark was sold in Europe starting in 1870.

In 1878, the limitation to capital letters of the Sholes-Glidden machine was removed, and lower-case letters could be typed as well. Finally, in 1887, the type-bars were mounted so as to strike the platen from the front, so that the text could be read while it was being typed, and the modern form of the typewriter came into existence.

THE EARLIEST ADDING MACHINES WITH KEYBOARDS

The stream of development which brought the keyboard to the typewriter, as described above, now became yet another tributary of the development of the calculating machine and the computer, providing one of the most decisive technical improvements in the whole history of artificial calculation.

Paradoxically, however, this advance proved, in the very beginning, to be a setback.

The first arithmetical calculator with a keyboard was constructed in 1849, and patented in 1850, by the American inventor David. D. Parmalee. It was an adding machine, whose essential component was a vertical rack-and-pinion gear activated by the movement of a lever when a key was pressed. But the accumulator had a single wheel and therefore could only add single-digit numbers.

In order to add numbers with several digits, it was necessary to work manually, after the fashion of handwritten arithmetic, adding separately the units, the tens, the hundreds, and so on, all the while being obliged to note on paper the partial results thus obtained and entering each such result prior to the next stage!

Numerous American and European inventors later brought in improvements to this design: Schilt (1851), Hill (1857), Arzberger (1866), Chapin (1870), Robjohn (1872), Carroll (1876), Borland and Hoffman (1878), Stettner (1882), Bouchet (1883), Bagge (1884), D'Azevedo (1884), Spalding (1884), Starck (1884), Pététin (1885), Max Mayer (1886), Burroughs (1888), Shohe Tanaka (1893), etc.

To begin with, however, the improvements were inadequate and the machines still required much preliminary manipulation and continual conscious attention on the part of the operator. Worse yet: these machines had neither speed nor numerical capacity of consequence, and were fragile in use so that they gave wrong answers if not handled with delicacy and dexterity.

FELT'S COMPTOMETER

The earliest adding machine which was really useful and usable by the general public was the *comptometer*, invented and constructed by the American industrialist Dorr E. Felt in 1884–1886. It was able to carry out additions and subtractions, involving numbers with several digits, advantageously, rapidly and reliably. The Felt and Tarrant Manufacturing Company of Chicago manufactured and sold it on a large scale from 1887, and the machine enjoyed worldwide success until well into the twentieth century.

Other improvements which came in somewhat later included:
- Dalton's *intermediate register*, which allowed a delayed entry of numbers which had already been set using the keyboard, so that corrections could be made prior to operating the lever which initiated the calculation;
- Runge's *compact keyboard* (1896), followed by that of Hopkins (1903),

which had only ten keys but worked through an automatic distributor, which allowed the operator, without displacing the hands, to enter the units, tens, hundreds, etc. successively.

A FURTHER ENHANCEMENT: THE PRINTED RECORD

To be properly adapted to the needs of commerce, such machines needed the capability to produce a printed record of transactions, using a mechanism which would print each of the quantities in the total, and the total itself, on a strip of paper. The human operator could then not only check at once that the correct numbers had been used, but also keep the print-out as a permanent record of the calculations.

We shall deal later with the developments achieved by Charles Babbage and by the Swedes Georg and Edvard Scheutz (1853). Apart from these, the first serious attempts at developing a printing mechanism were due to the American Edmund D. Barbour who in 1872 invented an adding machine with keys and also with a "printer" which, however, was a somewhat primitive device which only printed totals and sub-totals: the individual operations were still executed more or less manually, and the printing device operated rather in the manner of a date-stamp.

In 1875, the American Frank S. Baldwin brought in some improvements which to some extent allowed the machine to print out its own results.

The next stage in this line of development was taken by the Frenchman Henri Pottin, whose device listed the individual items of an addition and, at about the same time, by the American inventors George Grant (1874) and A. C. Ludlum (1888). In 1882, Pottin developed one of the earliest cash registers with a printing device.

THE FIRST BURROUGHS MACHINE

The decisive developments in this area occurred between 1885 and 1893, when the American William S. Burroughs invented and perfected his *Adding and Listing Machine*, the first mechanical calculator with keys and a printer which was also practical, reliable, robust and perfectly adapted to the requirements of the banking and commercial operations of the time. For these reasons, his machines enjoyed a remarkable worldwide success until the outbreak of the First World War.

The complete solution of the printing problem for adding machines was also achieved at almost the same time (1889–1893) by Dorr E. Felt who had invented the comptometer.

THE EARLIEST CASH REGISTERS

The idea of combining a key-operated adding machine with certain inter-locking mechanisms led to the development of the machines which became known as cash registers, capable of meeting all the accounting needs of the businesses of the time, and still in use nowadays in shops and restaurants.

The first machines of this type were invented by the brothers James and John Ritty and produced in 1879 in Dayton, Ohio. It was only with their third model, which had the familiar pop-up flags to display the amounts, that they produced a machine which was really marketable. This was sold from 1881 by the National Manufacturing Company, which the Ritty brothers founded.

Other once-familiar features of the cash register, such as the cash drawer, the bell, and the partial accumulator, did not appear until 1884–1890. The principal improvements to the cash register were due to the engineers of the National Cash Register Company, nowadays better known as the NCR Corporation.

This company was founded in December 1884 in unusual circum-stances, of which the following account is based on a historical brochure distributed by the NCR Corporation in 1984 [Anon. (1984)]:[25]

> In 1884, a young man named John Patterson who had grown up in a rural environment and graduated from Dartmouth College, having already worked in a variety of employments such as toll-gatherer on a canal, coal merchant, and retailer, bought a majority share in a small company on the verge of bankruptcy [the National Manufacturing Company] which manufactured cash registers. His friends in the modest business community of Dayton considered this to be a serious blunder.
>
> But Patterson was not dismayed by the opinions of the others, since he had a plan. He had faith in the future of this "incorruptible cashier" built by James Ritty. Irritated by the prevalence of sloppy accounting methods, Patterson considered that many traders worked in confusion at the expense of a part of their profits. The cash register, with its little roll of paper, its keys and its display of amounts, would be the key to a proper system for handling transactions. Convinced of this by 1882, Patterson had only to persuade the traders of the time to see it in the same way.
>
> Patterson's childhood, prior to the War of Independence, had been passed on the family farm south of the small town of Dayton, Ohio (10,000 inhabitants). The boy had the job of selling his father's

---

25 The ensuing quoted passage has been translated from the version given in French, and is not claimed to be a verbatim equivalent of any English version distributed by the NCR Corporation. [*Transl.*]

produce: smoked ham, stored in the drying-room; molasses, in the cellar; sacks of flour, in the granary; and maize in the silo. Amidst the running about involved in handling all these, it was easy to forget to record a sale.

Patterson recounts that "My father often woke me in the early hours of my sleep to ask if I was sure that I had had payment from such and such a customer. Or I was interrupted at my dinner to find out if I had sent the bill for sugar sold to Sanders. At times, I had to reply: *No, I forgot.*"

When Patterson became assistant toll-gatherer on the Miami–Erie canal (the only job he could find after graduating from Dartmouth College), he looked for a way to verify his takings and avoid disputes with the head office. "I gave the skipper of each barge a receipt for the amount of his toll, a kind of passport down to the next major station at which the receipt would be recorded."

When John Patterson and his brother Frank started up a prosperous coal-delivery business around 1870, the same concern for order prevailed: no delivery unless against a signed receipt from the customer, no delivery on credit without a written record of the arrangement.

Previously, there had been regular disputes with the clientele. But the little retail business of the Patterson brothers it was, that first made John aware of the benefits of the cash register. After reading a simple description of its working, he ordered two machines. In six months, the overdraft was reduced from 16,000 to 3,000 dollars, and the accounts showed a net gain of 5,000 dollars.

Thereby convinced that the cash register would be a vital innovation to meet the accounting needs of the businesses of the period, John Patterson thereafter devoted all his energies to the mass production of these machines. And so he founded National Cash Register in order (as he said himself) to help the traders of the time to run their businesses better.

In subsequent years this company introduced many important technical improvements to the cash register, which quickly gave their machines a solid reputation worldwide. The following are noteworthy:

- generation of a printed coupon, so that the buyer would have a printed record of his purchase;
- the printed roll of paper which recorded the details of every sale of the day in chronological order, and constituted a most valuable record of accounting information;
- the multi-drawer cash register whose different drawers had bells of different pitches, which allowed the sales accounts of different sales assistants to be followed separately;

- the cash register which automatically recorded separate accounts, designed to simultaneously keep track of cash sales, credit sales, instalment payments, account settlements, etc.

SELLING'S CALCULATOR

At this point in the chronology, we may turn aside to observe a most interesting machine invented by Edward Selling, professor at Wurzburg University. This was a mechanical calculator which incorporated various mechanisms which could be found in other machines (though Selling must have re-invented at least some of them), along with other totally innovative ideas. As D'Ocagne explains:

In this machine, we find the rack-and-pinion gear whose cogs drive the drums bearing the digits [as in George Grant's machine of 1874 described above]. The carrying of digits was accomplished progressively by means of epicyclic gear trains, as in Tchebishev's machine [above], which meant that – as in Tchebishev's case – the figures to be read off lay on an undulating line. The multiplicand was entered by keys as in the American adding machines [such as those of Felt and Burroughs] and, like these and also Grant's machine, Selling's machine could print. Over and above these arrangements, however, Selling's machine incorporated several really original mechanical devices, amongst which the use of the articulated structures called "lazy-tongs" is especially noteworthy. The machine was in two distinct parts which were temporarily connected during the calculation. The first part consisted of the system of lazy-tongs with the keys and the racks of the rack-and-pinion gears, and the second part consisted of the toothed wheels of the rack-and-pinion gears and the numbered drums. The cross-over points of the lazy-tongs represented the digits of the multiplicand. For each decimal order of magnitude of the multiplicand, the rack was linked to the appropriate cross-over point of the lazy-tongs by pressing on the key which, in the column corresponding to the decimal order of magnitude, carried the digit to be entered. To enter the digits of the multiplier, the lazy-tongs were opened to a greater or lesser degree by moving a lever into the appropriate notch of a scale graduated from 1 to 5; if the digit to be entered exceeded 5, it was necessary to decompose it into two parts not exceeding 5. The entry could be cancelled (set to zero) by lifting a ring.

THE DISADVANTAGES OF THESE MACHINES

Despite the numerous successive advances in capability which have been noted above, none of these machines incorporated significant improvements in the mechanical principles used to implement the arithmetical operations. However, since these machines were chiefly destined to meet the needs of accountancy and financial calculations, this was not a severe embarrassment for such purposes. For scientific work, however, it would be necessary to advance beyond their limitations, especially in multiplication and division.

Since the times of Leibniz and of Thomas of Colmar, calculating machines had a major disadvantage in that multiplication was performed by non-automated successive additions.

On Thomas's arithmometer, for example, to multiply 439 by 584 it was first necessary to place the multiplier carriage in the units position, and then manually add the multiplicand 439 four times (giving 439 × 4); then to move the carriage manually to the tens position and add the multiplicand eight times (439 × 8 which, added to the previous result, gives 439 × 84); and finally to move the carriage to the hundreds position and add the multiplicand five times (439 × 5 which, added to the previous result, finally gives 439 × 584).

Therefore seventeen manual additions had to be performed, plus the two displacements, not to mention numerous manipulations having to do with entering the numbers, transcription of partial results, entry of intermediate numbers, turnings of the handle, and so on).

Adding machines with keys, where all the necessary manoeuvres are reduced to simple key-presses, were admirably adapted to the efficient manual execution of addition and subtraction, at any rate. Despite the number of key-presses required, the time needed was still much less than the time taken for ordinary manual calculation.

Multiplication and division could still be effected by the method of repeated addition and subtraction. But, when the numbers to be multiplied or divided contained several digits, this method was much more cumbrous than using the classical calculators with built-in provision for multiplication and division which were still the best tools for these operations.

The operation of adding machines with keyboards necessitated operators who had been thoroughly and carefully trained. For this reason, some businesses which hired out these machines to firms which had temporary need of them (for example to do their monthly accounts or their annual stock-taking) also specialised in hiring out the services of trained operators at the same time.

5. From Mechanisation of Simple Arithmetic to Automatic Arithmetic on an Industrial Scale

THE EARLIEST MACHINE CAPABLE OF DIRECT MULTIPLICATION

In view of the labour and complication of multiplication and division using existing machines, machines capable of multiplying autonomously and automatically were invented. This required the machine to be capable, without any human intervention, of obtaining and accumulating all the partial products involved in a multi-digit multiplication, so that the operator would only need to enter the numbers to be multiplied and then start the operation off (by turning a handle, for example), and the result would be obtained automatically.

The earliest device to achieve this was patented in 1878 by Ramón Verea, a Hispanic living in New York.[26] His little experimental machine was capable of any multiplication of numbers with up to two digits, completely automatically. [The key to its mechanism, which worked rather like a Jacquard loom, was a ten-sided cylindrical prism, of which each side had ten holes, corresponding to the digits 0 to 9, of graduated diameters in each side into which could enter tapered needles. This served the role of a kind of multiplication table. The distance a needle could enter depended on the diameter of the hole, this distance corresponding to the fraction of a revolution which would be transmitted to the wheels carrying the numbers. *Transl.*] When journalists interviewed him, and asked how he proposed to market his invention, he replied that the question did not interest him. His sole objective was to show that the problem could be solved, and above all to demonstrate that a Spaniard was as capable as an American of showing ingenuity and imagination.

BOLLÉE'S MACHINE

Much better known and more impressive, with a greatly extended capacity, was the machine invented in 1888 by the Frenchman Léon Bollée. He and his brother Amadée were great inventors, and pioneers of the development of the automobile (they invented a petrol-driven car, and Léon inaugurated the Le Mans 24-hour race).

The mechanism of this machine really worked well, and it was the earliest capable of direct multiplication of many-digit numbers. Naturally,

---

26 Born and educated in Spain, he moved to Cuba and then to New York, where he wrote, founding a Spanish-language magazine which criticised his native country for its neglect of engineering and industry. Trading Spanish gold and banknotes had made him interested in calculation. His machine won a gold medal at a Cuban exhibition but, having made his point, he never invented anything else. [*Transl.*]

it could do addition and subtraction. It could also do division and even calculate interest, and extract square roots. The time it took for multiplication was on average 80 per cent less than for the ordinary calculating machines of the time.

Bollée's ingenious idea was to construct a material surrogate for the multiplication table (a conception related to Verea's ten-sided prism) by implanting steel pegs, somewhat like organ-pipes, into a steel plate, whose heights were proportional to the digits represented. The shortest peg was 5 mm in height and corresponded to 1; the next measured 10 mm and corresponded to 2; and so on up to the ninth, 45 mm in height, which corresponded to 9. They were arranged in nine rows and nine columns, and at the intersection of a row and a column would be found the one or two pegs whose heights represented the digits of the product of the row-digit with the column-digit. For each of the partial products making up the result of a multiplication, which was set in train by an arrangement of push-buttons and toothed rails, the machine then "read off" its value much as a blind person might read a multiplication table set in Braille.

In a contemporary brochure we can read: "According to tests which have been carried out, a practised operator using this machine at normal speed can achieve in an hour 100 divisions, or 120 square roots, or 250 multiplications, on the scale of

$$10,000,000,000,000,000,000 / 1,000,000,000 = 10,000,000,000$$
$$\sqrt{1,000,000,000,000,000,000} = 1,000,000,000$$
$$1,000,000,000 \times 10,000,000,000 = 10,000,000,000,000,000,000$$

and could calculate 4,000 terms of an arithmetic progression whose increment did not exceed 10,000,000,000 as well as almost as many terms of a table of the squares of numbers, up to ten quintillions (1 followed by 31 zeros)."

The machine, when exhibited at the Paris *Exposition Universelle* of 1889, won a gold medal for its inventor.

However, not many examples of this machine were produced, nor was it commercialised. In part this was due to its complexity of construction, and the consequential high cost; but the principal reason was that Bollée was far too passionate about motoring to devote himself to promoting his calculator.

Later, however, the German Otto Steiger of Munich constructed a more compact machine in 1892 which was based on a similar principle to Bollée's, and, starting in 1899, this machine was sold by the Swiss company Egli of Zurich under the name *Millionnaire*; this model was very successful and was widely used into the 1930s. [It was still being used after the Second World War. *Transl.*]

THE FIRST MACHINE WITH DIRECT DIVISION

While multiplication could now be done much more simply on such machines, division on the other hand still remained a long and complicated business requiring much manual intervention by the operator.

The next stage in the evolution of calculating machines therefore led to the development of machines capable of direct division. The earliest of these was the *Madas* invented and constructed in 1908 by the Swiss Alexander Rechnitzer and Edwin Jahnz of Zurich. This machine, too, was sold by the Egli company of Zurich, from 1913. It was able to perform division entirely on its own, only halting when the operation had been completed.

In the *Madas*, the final carry to occur temporarily disengaged the driving mechanism, and reversed the (subtractive) rotation due to the next turn of the handle, thereby adding back a number which had been subtracted once too often. A mechanism, which did not require any intervention by the operator, then used the following turn of the handle to move the carriage along one decimal position. The next turn of the handle then resumed the original (subtractive) rotation, and so it continued. The operator could therefore continue turning the handle blindly until the calculation of the final digit, at the right-hand end of the quotient in the accumulator. At this moment, the driving mechanism was disconnected for the last time, a bell gave a double ring, and the operator then knew that it was all over. [See R. Taton and J. P. Flad (1963)]

THE COMPLETE AUTOMATION OF
ALL FOUR OPERATIONS

Despite achieving the automation of each of the four basic arithmetic operations of addition, multiplication, subtraction and division, some problems remained. There were difficulties in reconciling fully automated division with fully automated multiplication, since the implementation of one operation seemed to be incompatible with the mechanical requirements of the other.

This problem was first resolved more or less satisfactorily in 1910 by the American engineer Jay Randolph Monroe, who was awarded the John Wetherill Medal of the Franklin Institute of Philadelphia for the achievement. Monroe was closely followed by the German inventor Christel Hamann (patent granted in 1911). These two, no doubt independently of each other, invented machines which could perform all four arithmetic operations quite automatically.

These implementations were soon followed by others, with constantly improving performance (Hamann's *Mercedes*, new models of the *Monroe*, a fully automated *Madas, Metal, Friden, Barrett, Marchant, Facit*, etc.).

The fierce competition which ensued, along with ever more sophisticated mechanisms, gave rise to truly reliable mechanical calculators, with maximum correctness in carrying out arithmetical operations and in producing the results; and this was performed with a rapidity never before equalled.

THE *CURTA*: THE FIRST PORTABLE MECHANICAL CALCULATOR

Under the pressure of need to reduce both size and weight, the evolution just described led on to a reduction in the dimensions of the machines which, immediately after the Second World War, gave rise to a machine that, of all purely mechanical devices, was one of the most extremely miniaturised and perfected of modern times: the *Curta*, invented by Curt Herzstark,[27] and manufactured in Liechtenstein.

The *Curta* Model I had an 8-column "keyboard" (8 slides), an 11-digit results register and a 6-digit revolution counter. It was 5.3 cm (2.1 in.) in diameter by 8.6 cm (3.4 in.) long and weighed 230 gm (8 oz.). The Model II had an 11-column keyboard, 15-digit results register and 8-digit revolution counter. It was 6.6 cm (2.6 in.) in diameter by 9.1 cm (3.6 in.) long and weighed 360 gm (12.5 oz.).

According to Taton and Flad

> this model could perform the four fundamental operations: multiplication and division, and also addition and subtraction. Its cylindrical body was held in the left hand and carried eight sliders for number entry, which were movable from top to bottom .... The fluted upper cap was the "carriage" and could be turned on the cylindrical body through six different positions; the circular upper face of the carriage carried the revolution counter with six windows on a white background, and the accumulator with eleven windows on a black

---

27 Curt Herzstark (1902–1988), an Austrian Jew, apparently invented the machine while prisoner at Buchenwald concentration camp: according to some accounts in secret, according to others with the knowledge of the concentration camp authorities who hoped to present the invention to the Führer.

After the war, in 1946, Curt Herzstark was invited by the Prince of Liechtenstein to establish a manufacturing plant for the Curta in the Principality of Liechtenstein and founded the company Contina AG for the purpose. Herzstark retired in 1952 and sold the patent rights to Contina AG who introduced the *Curta* Model II in 1954; this machine continued to be used until 1972.

The *Curta* was cylindrical, and driven by a small crank-handle at one end. The digits were set by numbered studs which could be slid in slots along the cylindrical sides, and the digits set by the sliders were visible in windows set circumferentially around the top of the body, above their respective slots. The revolution counter and the results register could be read through small windows on the end-cap of the machine. The clearing lever, which had a ring-shaped hole for the finger at the end and could be swung back over the cap when the machine was put in its case, operated by being rotated around the same axis as the crank-handle. [*Transl.*]

background. Movable studs served to set the positions of the decimal points in the three registers. The operator turned the handle with the right hand, and after executing each digit displaced the carriage with the thumb of the left hand through one decimal place. For subtraction and division, the handle would be pulled outwards through a short distance, preserving the direction of rotation.

The rotation counter and the accumulator were furnished with complete carrying of tens, which allowed accelerated multiplication. A clearing mechanism acted on one or other, or both [depending on the direction and extent of movement *Transl.*]. An inverter at the bottom of the body could make the revolution counter and the accumulator have either the same or opposite increments on rotation of the handle.

This machine was made of stainless steel, and it was silent and resistant to wear. In its protective case, it could be carried in pocket or briefcase, and it could be used anywhere.[28] Despite the fact that entry was by sliding studs, and the absence of a mechanism of transfer from one register to another, a practised operator could carry out a simple multiplication, a division, a "rule of three", etc., in 10 to 17 seconds.

For all of these reasons, the *Curta* enjoyed considerable commercial success worldwide until the early 1970s, when it was finally supplanted by the miniature electronic "pocket calculators".

THE BEGINNINGS OF ELECTRIFICATION OF MECHANICAL CALCULATION

A purely mechanical, and therefore purely hand-driven, technology nevertheless had fundamental limitations which inhibited further progress, since such machines were necessarily slow and usually (the *Curta* being a notable exception) cumbrous.

It was, therefore, a significant step forward when machines were developed whose mechanisms were driven by electricity, which had been playing an increasingly important role in the early evolution of the "machine society" of the latter half of the nineteenth century. As will be summarised in the next section, the succession of discoveries since the early eighteenth century in the domains of electricity, magnetism and electromagnetism uncovered the possibility of converting electrical energy into mechanical work.

---

28 The translator has comfortable memories of using the *Curta* II in a domestic armchair in 1969. [*Transl.*]

A BRIEF HISTORY OF ELECTRICITY

Some elementary knowledge of electrical phenomena dates back to Antiquity, and a few basic discoveries had been made prior to the eighteenth century; but the principal origins of the modern science date from the discovery in 1729, by the English physicist Stephen Gray (1670–1736), that certain substances had the power of conducting electricity ("conductors") while others ("insulators") did not; and Gray for the first time transferred energy over a distance by means of threads which conducted electricity.

In 1745, the Dutch physicist Petrus van Musschenbroek (1692–1761) constructed the *Leyden Jar*;[29] the earliest *electrical condenser*. Subsequently, many discoverers, inventors and theoreticians made their respective contributions to the development of a domain of knowledge which, no longer novel, was becoming a true scientific discipline. In 1785, the French scientist Charles Augustin de Coulomb (1736–1806) discovered by very delicate measurement that the force between two electrical charges varies inversely as the square of the distance between them. This law is of the same form as Newton's law for the gravitational force between two masses, and therefore lent itself to the same mathematical treatment. The mathematical theory of electrostatics was therefore developed, the principal contributors being the mathematicians Pierre-Simon de Laplace (1749–1827), Siméon Denis Poisson (1781–1840) and Karl Friedrich Gauss (1777–1855), who developed the mathematical equations which expressed these phenomena. In 1799, the Italian Alessandro Volta (1745–1827) invented the *Voltaic pile* – an early electric *battery*.[30]

The English scientist Michael Faraday (1791–1867) made great advances, experimentally exploring a wide and interlocking range of electrical,

---

29 An ordinary glass jar coated to a certain height inside and out with tinfoil. The mouth is closed with a wooden or cork disk, which both keeps dirt and moisture out, and provides an insulating support for a wire which enters from outside and makes contact with the inner layer of tinfoil. When the wire is connected to an electrically charged body, electrical charge passes to the inner foil; the amount of charge is greater according as the glass wall of the jar is thin. This was therefore the first example of the arrangement of conducting sheets separated by an insulating sheet which forms the principle of the electrical condenser. When the outer foil of one charged jar is connected to the wire connected to the inner foil of a second charged jar, and so on, a *battery* of jars is made such that the voltage difference between the first and last in the chain may be very great. [*Transl.*]

30 In 1790, Galvani had observed that a recently killed frog's leg twitched when brought in contact with metal in such a way as to form a closed circuit between the two ends of the leg. Volta came to the view that the muscles were simply a kind of delicate electrometer. He grouped conductors into two classes: those (like metals) which acquire electricity by contact but merely establish an equilibrium and cannot generate a current on their own; and those (which we would now call *electrolytes*, such as salt solution or the blood in a frog's leg) which, when placed in contact with conductors of the first kind, can generate an electric current when a circuit is completed.

Volta's "pile" consists of a series of layers of conductors $A_1 B A_2 A_1 B A_2 A_1 B A_2 \ldots$ (A_1 and A_2 of the first kind, B of the second kind) which is capable of generating a high voltage if several layers are used (the very words *volt* and *voltage* having been coined in honour of Volta). [*Transl.*]

magnetic and electromagnetic phenomena. In 1821 he succeeded in generating electrical force by rotating a metal disk across the lines of force of a magnet, thus discovering electromagnetic induction. In researches which began in 1843 he confirmed the principle of conservation of electricity and subsequently developed the theory of *electrical induction*. His discoveries in electromagnetic induction later gave rise to electric generators, which transform mechanical energy into electrical energy.

These discoveries of Faraday took their place beside those of Hans C. Oersted (1820), François Arago, Jean-Baptiste Biot, Félix Savart and André-Marie Ampère to form the discipline of electromagnetism.

The researches from 1820 of André-Marie Ampère (1775–1836) led to his fundamental law of electrodynamics.[31] Ampère's discoveries were supplemented by those of

- François Arago (1786–1853) on "magnetism of rotation" (1824);[32]
- F. Lenz (1804–1865) on the direction of induced currents (1833);[33]
- Franz E. Neumann (1798–1895) on the phenomena of *self-induction* (1845);[34]
- Léon Foucault (1819–1868) on *Foucault currents* (1855).[35]

These researches were brought towards culmination by many other physicists, notably the Scot Lord Kelvin (William Thomson, 1834–1907) whose many important contributions are too numerous to be noted here.

The entire subject of electromagnetism was finally brought into a complete mathematical theory by the Scot James Clerk Maxwell (1831–1879) who in 1873 established his electromagnetic field equations.[36]

---

31 In 1820 he heard of Oersted's observation that a magnetised needle is deflected by a neighbouring electric current; he produced the same effects without magnets, showing that wires carrying currents can attract or repel each other. He formulated the law that an electric current in a circuit is equivalent to a magnetic sheet, bounded by the circuit, of a constant magnetic strength which is proportional to the current.

With Faraday's discovery of the electrical effects of a moving conductor cutting through a magnetic field, this established the fundamental symmetry underlying electromagnetic theory: that motion relative to a magnetic field generates electromotive force (voltage), while motion relative to an electric field generates magnetic force. [*Transl.*]

32 A magnetic needle suspended over a rotating metallic disk tends to rotate, following the direction of rotation of the disk. Here, the rotation of the disk within the magnetic field of the needle generates a voltage difference in the disk which causes a current to flow which generates a magnetic field which acts on the magnetised needle. [*Transl.*]

33 That an electromagnetically induced current is in the direction such that its electromagnetic action on the inducing system tends to oppose the motion inducing the current. [*Transl.*]

34 When the current round one circuit changes, the total magnetic field passing through a surface embraced by another circuit changes and there is an induced electromotive force in the second, and vice versa; this is what is known as *mutual induction*. The same applies to a single circuit: change of current changes the total magnetic field passing through the surface embraced by the circuit, which induces an electromotive force in the same circuit, tending to oppose the change of current; this is *self-induction*. [*Transl.*]

35 The currents induced in a rotating disk held in a magnetic field; Foucault discovered that additional work was required to maintain the rotation when the magnetic field was present, which is manifest in the heat generated in the disk by the induced currents. [*Transl.*]

36 *Treatise on Electricity and Magnetism* (1873). Maxwell's electromagnetic field equations, which subsume the concepts of Faraday and the results of Faraday and the many other workers, are a set of equations

Along with the invention of the steam engine, and the construction of all those mechanical and thermal mechanisms known, generically, as "motors" which transform one form of energy into mechanical work, the parallel development of electromagnetism and the knowledge of the mechanical forces which can be generated by electrical and magnetic means ultimately gave rise to electric motors.

The earliest, and very primitive, device of this type was *Barlow's wheel*,[37] invented in 1828 by the English mathematician and scientist Peter Barlow (1776–1862).

The first electric motor which had practical use outside a research laboratory, however, was constructed in 1834 by the German physicist Moritz H. von Jacobi (1801–1874). This was based on electromagnets and energised by batteries; by its means a small boat was propelled on the river Neva.

The Belgian Zénobe Gramme (1826–1901) was the first to develop (in 1873) an electric motor driven by a reversible generator, i.e. a dynamo (again invented by himself in 1867).

The principle of the synchronous electric motor was only developed in 1883 by the English engineer and physicist John Hopkinson (1849–1898), and it was not until 1885 that the Italian Galileo Ferraris (1847–1897) discovered the principle of the rotating magnetic field.

Combining this principle with the principle of the synchronous motor,

relating the spatial variation of the electrical field to the chronological rate of change of the magnetic field, and the spatial variation of the magnetic field to the chronological rate of change of the electrical field. Taken all together, they imply that a varying electrical field is propagated as a wave, and so also is a varying magnetic field. Therefore Maxwell theoretically predicted the existence of electromagnetic radiation; and the wave velocity of this radiation emerged from the equations as equal to the ratio of the electromagnetic unit of electric charge to the electrostatic unit of charge. Numerically, using experimentally determined values, this velocity was close to the measured velocity of light, which led Maxwell to postulate that light was a form of electromagnetic radiation.

Heinrich Hertz (1857–1894) established experimentally in 1887–1889 the existence of electromagnetic waves and confirmed their identity with light waves.

Maxwell's equations, however, did not square with Newton's laws of motion. A system obeying Newton's laws, whether observed from one point of view or from another moving uniformly relative to the first, always gave the same results: mass, length and time did not change, nor did acceleration; and velocities measured from one viewpoint were related to those measured from the other by simply adding the velocity of one viewpoint relative to the other.

According to Maxwell's equations, the electromagnetic field did not exhibit this invariance: the electrical field from one point of view changed, and acquired a magnetic part, when observed from the other; and similarly for the magnetic field. Maxwell's equations, from the one viewpoint, were not the same as those from the other. This perplexed physicists in the late nineteenth and early twentieth centuries, and the anomaly was finally resolved in 1905 by Einstein's Special Theory of Relativity which was based on a revision of the concepts of distance and time: from one viewpoint both distance and time were different from those measured from the other, being related to each other in a way which left the Maxwell equations unchanged. [*Transl.*]

37 A vertically mounted metal wheel has its bottom rim dipped in a trough of mercury and also embraced between the poles of a magnet. An electrical circuit causes a current to pass from the mercury to the spindle along the radius of the wheel. The resulting electromagnetic force (the force exerted by the magnet's field on the current along the radius) causes the wheel to turn. [*Transl.*]

the Croatian physicist Nikola Tesla (1856–1943)[38] made a discovery from which a multitude of *squirrel-cage* motors would be derived. [See also T. de Galiana (1968)]

ELECTRICALLY-DRIVEN MECHANICAL CALCULATORS

The discoveries just described opened the way to converting electrical energy into mechanical work.

In the second half of the nineteenth century, this possibility led to the idea of using electricity to drive typewriters. The American inventor Thomas A. Edison (1847–1931) was one of the first to take this step when, in the 1880s, he electrified his own secretary's typewriter by arranging for the type-bars to be moved by means of an electromagnet. Although Edison's invention was patented, it led to no further development. In 1901 the American T. Cahill also invented an electric typewriter and formed a company to market it; but the company became bankrupt following the sale of a mere forty or so machines at a price of $4,000 each. The first mass production and successful commercialisation of electric typewriters did not occur until 1920, after James F. Smather invented a typewriter whose type-bars and carriage were both driven by an electric motor.

As for calculating machines, several inventors pursued the same approach from the 1890s, and especially after 1910, in electrifying various calculators (such as the *Saxonia*, the *Archimedes* or the *Unitas*) derived from Thomas's arithmometer or others (such as the *Brunsviga*, the *Triumphator* or the *Marchant*) derived from Odhner's machine; or, again, by motorising machines like the NCR cash register, the *Burroughs*, the *Madas*, the *Mercedes* or the *Monroe*.

Although it was at first considered a curiosity, the idea of bringing electricity to the mechanical calculator proved to be most fruitful, since it made it much easier to use: instead of requiring the operator's hand

---

38 Tesla was "the man who invented the twentieth century", taking out over a hundred patents on electrical, mechanical and electromechanical devices. The "squirrel-cage" motor was one possible, efficient arrangement for utilising eddy currents to produce a motive force in electric motors based on the induction principle.

In an induction motor, a magnetic field revolves around a piece of metal and creates eddy currents in the metal. These currents produce magnetic fields that interact with the revolving field, which makes the metal rotate if properly pivoted. The rotation of the metal can be used to drive a shaft and constitutes a motor. The smallest motors of this type use a rotor made of metal disks notched at the edges to place the eddy currents properly. Larger types may use a *squirrel-cage rotor*. This is made of metal bars arranged to form a skeleton cylinder. The ends of the bars may be attached to disks, or the bars may be mounted on a cylinder of enamelled iron and connected at the ends. The eddy currents flow through the bars and end connections.

The rotation of the magnetic field is typically achieved by energising in succession a series of electromagnets placed circumferentially in a circular arrangement; the energising current is switched from one electromagnet to the next by the rotation of the motor itself, which moves the point of contact from which the current is derived from the terminal of one coil to the terminal of the next. [*Transl.*]

to exert the force and power needed to set the numerical registers, move the carriage and turn the mechanisms, the force and power could be drawn from electrical energy; and the speed of operation had only mechanical limits, and not human limits as well.

FROM MECHANICAL TO
ELECTROMECHANICAL CALCULATION

These early instances of the use of electricity in calculating machines did not, strictly speaking, inaugurate the era of true electromechanical calculation, since the function of electricity was merely to supply motive power to the machines as a substitute for human energy.

The history of electromechanical calculation, properly speaking, began in 1831 when the American Joseph Henry (1797–1878) invented an electromechanical relay. This had a movable metal rod mounted, on a vertical support, between the extremities of a steel horseshoe round which was coiled an insulated electric wire thereby creating an electromagnet. When an electric current was passed through the coil, the end of the rod was attracted by the electromagnet, and it returned to its initial position when the current was interrupted.[39]

Meanwhile, in 1827 the German Steinhill had discovered that a single wire could be used as an electrical transmission line, the current returning via the earth to close the circuit.

In 1833, Michael Faraday crowned Henry's discovery by showing that a magnet could exert a force on an electric current.

The next step was taken by the British scientists William F. Cooke (1806–1879) and Charles Wheatstone (1802–1875) who in 1837, on different principles, constructed an electromagnetic relay device which could be used to switch on an electric battery at a distance, thereby producing variations of current remotely. This was the earliest-discovered method of transmission of messages by means of electrical signals, opening up the way for individuals remote from each other to communicate.

It was, however, the American Samuel F. B. Morse (1791–1872) who has the distinction of inventing the earliest electric telegraph, also in 1837. His system was based on the following two ideas, whose description we have adapted from P. Carré (1991).

1. The device associated a system of dots and dashes with the 26 letters of the alphabet ($\cdot$ – to A; – $\cdot\cdot\cdot$ to B; – $\cdot$ – $\cdot$ to C; etc.), and it is these which are transmitted as signals. When the complete message has been

---

39 This movement of the metal rod could then be used to make or break another electrical circuit, under the control of the current energising the magnet. [*Transl.*]

transmitted in this way, the recipient can then easily transcribe these back into letters according to the predetermined code and read the original message.

2. The receiving device was an electromagnet in front of which was an arm of soft iron[40] which could rotate about an axis. When a current passed through the electromagnet, one of the extremities of the lever was attracted by the magnet during the time-interval (short or long according to dot or dash) for which the current was maintained. To the other end of the arm was attached an inked wheel which was pressed against a continuous roll of paper by the movement of the arm, thereby making a short mark ("dot") or a long mark ("dash"); the series of marks could then be read and deciphered by the receiving operator.[41] The current was generated at the transmitting end by the movements of a metallic arm to make and break a circuit; this current was then carried by a wire to the receiver and – by the above mechanism – caused the receiving arm to adopt the same positions as the sending arm.

In this way a communication code was invented and implemented by an electromagnetic relay device.[42] This marked the start of the era of communication at a distance, of the transmission of electrical signals to distant places, by means of the electromagnetic relay which was destined to play a dominant role in the technical implementation and development of the telephone, and of automatic telephone exchanges.

Equally, however, this device also opened up the path to electromechanical calculation by the construction of relay circuits which would be at the heart of future calculating machines. By the end of the nineteenth century it had been noted that this technology could be used to implement the elementary arithmetical operations by means of an arrangement of such circuits and predetermined numerical encodings.

Pride of place in this advance goes to the electromechanical calculator invented by the German Edward Selling, who in 1894 had the idea of

---

40 "Soft iron" is a pure form of iron which is readily magnetised by an external magnetic field but which does not retain its magnetism when the external field is removed. [*Transl.*]

41 Later versions of the Morse telegraph had a sounding device from which a trained operator could decipher the code by ear, leaving hand and eye free for transcribing the letters. Skilled operators had already learned how to identify the dots and dashes from the intervals between the clicks made by the movements of the arm.

The electrical connection which caused the current to flow was made by moving a spring-loaded key, either by hand or by pulling through a "port rule" – a long narrow wooden trough bearing metal slugs corresponding to the different letters of a message; transverse notches of different widths were cut into each slug to represent the dots and dashes of the code for the letter, and these raised and lowered the key arm accordingly. [*Transl.*]

42 Morse later also used electromagnetic relays as "repeaters" along the way to compensate for the loss of current in a long transmission line: activated by the receipt of the weakened current some distance down the line, the relay closed the arm of a contact enabling a connection with a local battery which would be the new source of current for the next leg. [*Transl.*]

replacing the steel multiplication plate of Léon Bollée's machine, with its array of pegs of different heights, by a system of electromagnetic relay circuits which acted on the numerical accumulators by means of electromagnets. This was the first instance of an electrical multiplication table, a principle which was later put to use in punched-card electromechanical calculators. This machine was constructed in 1906 by H. Weltzer, but was never sold commercially.

Another innovator in this domain (to whom we shall have occasion to return later) was the Spanish engineer Leonardo Torres y Quevedo, who made an electromechanical "arithmometer" in 1913 which was also based on electromagnetic relays; and we may mention as well the work of the Frenchman C. Nikoladze who in 1928 devised an electrical machine capable of performing direct multiplication by a system of relays and switches.

These emerging techniques converged towards a genuine electromechanical technology, especially thanks to the electrification of keyboards and other controls, and exercised ever greater influence on the very conceptualisation of the machines. The arithmetic operations benefited from enhanced automation which went as far as "algebraic mode"[43] with larger number-registers and greater speed of execution. Other enhancements which became more readily available included automatic resetting of displays, automatic control over printing of results onto rolls of paper, and internal storage of certain data or of intermediate results.

Amongst the earliest machines of this kind was Christel Hamann's famous *Mercedes-Euklid*, which had great success at the time. Other machines of more advanced development were the *Hamann Selecta*, the *Badenia*, the *Facit*, the *Everest*, etc.; and especially the famous *Sasl IIc*, manufactured between 1940 and 1950 by V. E. B. Büromaschinen Werk Sömmerda of Leipzig. This machine had two visible accumulator registers, and also:[44] entirely automatic multiplication; multiplier entry by means

43 In a simple calculator, any operation (such as addition or multiplication) is immediately applied to the two immediately available numbers, so that the calculation $2 \times 3 + 4 \times 5$ would, in the order written, be evaluated as follows:

$$2 \times 3 = 6 , + 4 = 10 , \times 5 = 50.$$

In algebraic mode, the entire expression would be evaluated as a whole, using algebraic rules of precedence corresponding to the bracketing

$$(2 \times 3) + (4 \times 5), \text{ i.e. giving the partial results}$$
$$2 \times 3 = 6 ; 4 \times 5 = 20 ; 6 + 20 = 26$$

Obtaining this result (26) using a simple calculator would require the operator to break it down into three parts: evaluate 2×3 and note the result (6); evaluate 4×5; re-enter the noted result (6) and add it to the result (20) just obtained.

Full "algebraic" capability also allows the parentheses themselves to be entered, so that for instance the correct result $5 + 3 - (4 - 1) = 5$ would be obtained, rather than the result $5 + 3 - 4 - 1 = 3$ if the parentheses were ignored.

Clearly, incorporating the capability of algebraic mode involves much greater logical complexity of design, and requires a greater storage capacity for partial results. [*Transl.*]

44 According to a catalogue of the time, kindly supplied to the author by Jacques Chauvin.

of a special keyboard with ten keys; a means for repeating the multiplier so as to carry out several successive multiplications involving the same factor while leaving the keyboard free for other intermediate operations; entirely automatic division; automatic alignment of the dividend at a position determined by a tabulator; automatic clearing of counter and accumulator in multiplication and division; automatic clearing of the keyboard on completion of multiplication or division; automatic return of the carriage to the initial position; automatic accumulation of all partial products; direct subtraction using the second accumulator; entirely automatic rounding, at the operator's option, of the fractional part of a decimal fraction; transfers; etc.

The new electromechanical technology had, therefore, given rise to calculators which had greater capabilities, and were faster, more precise, more reliable, and easier and simpler to use; and they far surpassed the most elegant of the purely mechanical solutions of the problem of automatic calculation.

ADAPTATION TO COMPLEX ACCOUNTING

Between the two wars, especially from 1920 to 1940, a series of accounting machines had been developed which were adapted to the specialised area of large-scale accounting. These addressed activities of the following kinds: general accounting; statements of account and balance sheets (debits; credits; balances, themselves subdivided into previous balances, transfers and new balances; etc.); current accounts; third-party accounts (journal accounts of bills and settlements in chronological order, for clients and suppliers; drawing up invoices and receipts and advice letters, etc.); analytical accounts (breakdown of income and expenditure by categories); industrial accounting (breakdown of several categories of expenses, along with several categories of general cash-in-hand, associated with several orders executed at several centres); invoices and receipts; interest calculations; stock accounting (inventories and values of items in stock along with journal accounts of supplies and deliveries); payroll; etc.

These machines were hybrids of calculating machines and typewriters, by virtue both of their common electrical features and of the fusion of these two functions in one and the same machine.

THE EARLIEST INVOICING MACHINES

The early invoicing machines, unlike the rudimentary accounting machines of the time which could only do addition and subtraction, could perform multiplication as well. The first was invented by Hopkins in 1903, and was later perfected by Burroughs by incorporating Bollée's multiplication plate. We may also cite the machines in which a Wahl accumulator was mounted on a typewriter, of which the first dates from 1907. As a class, these machines combined, on a single chassis, mechanisms for three different functions. There was a keyboard; one or more printing devices; and one or more calculating devices. The main purpose was the production of accounting documents more complex than a simple totalled list, by means of direct keyboard actions. So both the accounting and the invoicing machines were special cases of cash registers [see Favier and Thomelin (1965)].

TYPEWRITERS, ADDING MACHINES, CASH REGISTERS, ACCOUNTING MACHINES, AND OTHERS

Some typewriters had been adapted to book-keeping purposes without being provided with an arithmetic accumulator. Likewise, some adding machines with printing capability but without a carriage had been used for some such tasks.

However, these were not true accounting machines which, at the time, fell into three categories: those based on the typewriter; those based on the ordinary adding machine; and combination machines. The last category arose from the combination of the principle of the typewriter with the principle of the calculator (apart from those cases where the two mechanisms had simply been assembled together into one machine), and the technology used for such machines was, variously, mechanical, electromechanical, or electronic.

Models based on the typewriter, which may be mentioned without going into technical detail, include: the *Remington*, the *Elliott-Fischer*, the *Mercedes SR 42*, the *Adler*, the *Ideal*, the *Smith Premier*, the *Continental*, the *Torpedo*, the *Olivetti*, etc.

Models based on the adding machine include: the *Olivetti*, the *Addo*, the *Duplex MAM*, the *Monroe*, the *Burroughs*, the *Multiplex Sudstrand* (which had several vertical accumulators), etc., and especially the *Logabax* (1938) which was the first machine with rectilinear registers, and which had many more accumulator registers than were in the document to be drawn up.

It was the engineer Leonardo Torres y Quevedo who had the idea of the combined machine, which he implemented in 1913–1914 in the shape of his celebrated electromechanical arithmometer. The category of combined typewriter and adding machine included the *Astra*, the *MAM*, the *National Class 3000*, the *Ormeco*, etc. The category of combined typewriter and multiplying machine included the *Burroughs* invoicing machine, the *Supermetal* which combined a reduced keyboard for number entry and an electronic calculator; the most developed of these machines were the *MAM New Series*, the *Facit*, the *Logabax F800*, the *National Computronic*, the *Continental*, the *Sensimatic*, the *Mercator 5000*, the *Remington 400 FE*, the *Burroughs E 2000*, the *Akkord Supertronic*, the *Super Metall 381*, etc.

Certain machines were similarly derived from the cash register, such as the *National Class 2000* and the *National Postronic*, and the electronic machines *MAM RS 65* and *Burroughs Sensitronic* which were used to list accounts; the latter, when connected to an "auto-reader", could automatically list up to 55 accounts per minute. (Details of these and other machines are extensively documented in Favier and Thomelin).

There were other machines for related purposes, such as automated mailing and addressing.

Automatic franking machines are in effect adding machines with printing capability but without the intermediate registers of ordinary adding machines. Their mechanism, of similar kind to the Odhner machines, drove an accumulator and also a printing drum, and they also had various security locks to prevent fraud.

The *Addit Address*, invented and made in France, combined three different machines into one: a printer, an addressing machine, and a printing adding machine. This combination ensured that the *Addit Address* was a complete solution of the problem of combining automatic addressing with accounting, and was for a long time highly esteemed by insurance companies, by the pension and social security services, by subscription services and by direct-selling or mail-order companies.

The somewhat specialised area of artificial calculation described in this section was for some time – between the First and Second World Wars and for a while after the Second – in strong competition in the domain of accounting with comptometers and other automatic data processing of a more general kind where the machines used punched cards, and to which we shall return shortly.

THE ADVANCE OF SCIENCE REVEALS THE INADEQUACY OF CLASSICAL CALCULATORS

The classical calculators and accounting machines, mechanical or electro-mechanical, were designed for the four basic arithmetic operations, and were entirely suited to the accounting and financial needs of their time, and to everyday arithmetic.

None of them, however, was well suited for mathematical problems such as finding the zeros of a polynomial, solving equations or systems of equations, calculating a determinant or evaluating an integral numerically, or solving ordinary or partial differential equations.

The need to be able to cope numerically with such problems had, however, been keenly felt for centuries, following the great advances in science and technology in Western Europe and as a result, above all, of the numerous problems arising from the flourishing growth and increasing practical application of algebra and of mathematical analysis.

Nevertheless, only quite recently have good solutions to such thorny problems begun to emerge, and there are several reasons why this should have been so delicate a question.

The first reason is the large number of different mathematical functions[45] which occur in mathematical analysis, for each of which it is necessary to devise an artificial model or procedure for the purposes of numerical evaluation.

The second is the fact that many of the functions which arise in Physics or other scientific applications cannot be expressed in analytical form.[46]

The third reason is that it is often not possible to solve algebraic equations, and ordinary and partial differential equations, by formal methods.

Scientists have consequently approached these problems along two quite different paths which have led to the development of two quite different techniques – known in the English-speaking world as *Analogue Computation* and *Digital Computation*.[47]

[The next section will summarise analogue computation, and the theme of digital computation – which subsequently develops to the climax which is the "general-purpose digital computer" – will be taken up in the following sections. *Transl.*]

---

45 A mathematical *function* is a rule relating the values of two variables, say x and y, so that the value of y can be determined from the value of x. The function "square of", for instance, expressed as $y = x^2$, gives y uniquely when x is given, so that for instance $x = 3 \Rightarrow y = 9$; but not the reverse since there are two values (-3 and $+3$) of x such that $y = 9$.

There are literally thousands of different functions which were discovered and studied during the years since the early eighteenth century when the calculus and mathematical analysis generally began their period of explosive growth, of which the familiar *elementary functions* (such as $y = \log x$, $y = \sin x$, $y = \tan x$, etc.) are a very tiny proportion. Many of these arose as the solutions of differential equations of various kinds which had been inspired by the use of mathematics to investigate and solve problems in the physical sciences. [*Transl.*]

46 Expressing a function "in analytical form" means, in effect, writing it as an algebraic expression involving simple well-known functions whose numerical values are readily accessible. Linear ordinary differential equations with constant coefficients have solutions which can be expressed algebraically in terms of the sine, cosine and exponential functions, for instance. (For example, any solution of $d^2y/dx^2 + 2\,dy/dx + 2y = 0$ is of the form $e^{-x}(A \sin x + B \cos x)$ where A and B are numerical constants.)

However, even so simple a physical system as a swinging pendulum is described by the simple differential equation $d^2y/dx^2 + k^2 \sin y = 0$ (where y denotes the angular displacement and x denotes time) whose solution has no such simple expression and in fact depends on the *elliptic functions*. [*Transl.*]

47 See the Translator's introduction to the following section. [*Transl.*]

6. Analogue Computation. Origins and Development

Translator's Note

At this stage, a note needs to be made concerning a question of terminology. The necessity arises because the word *computation* in English covers a wide range of meaning, from straightforward arithmetic calculation to – in standard usage of the late twentieth century – the use of a "general-purpose digital computer" or, in short, a *computer*.

The latter, however, is not so much a machine designed for numerical calculation (though extremely effective for the purpose) as a machine capable of being programmed to perform logically complex processing of data of any kind provided the data can be represented in a "digitally" encoded form – essentially, as a collection of binary ("0" or "1") elements.

In French, the specific name of such a machine is *ordinateur*, which – etymologically – better encapsulates the nature of its operations than does the English word *computer*.

Georges Ifrah himself wishes to enshrine the special nature of this kind of machine – its great generality of purpose, the great flexibility with which it can be programmed, the extreme complexity of the functions it is capable of – by special considerations of nomenclature. These are discussed in more detail in Chapter 6.

While no-one can quarrel with this objective – the "computer" is a transcendentally marvellous device – it is nevertheless not possible, in English, to give it any other name and still remain in touch with the huge mass of writing on every aspect of its design, evolution, function, and potential, nor remain in touch with common English usage.

Therefore the word "computer" will be used throughout in its normal English sense (and, indeed, in all the other normal English senses which it has carried both recently and at earlier times); just what is meant at any point will, we hope, be clear from the context. [*Transl.*]

ANALOGUE COMPUTATION: ORIGINS AND DEVELOPMENT

[*Analogue computation*[48] means the use of a physical system which is subject to numerical relationships analogous to those taking part in the computation to be carried out, the numerical result being obtained by

---

48 Etymologically, "analogue" is from the Greek *ana* ("according to") and *logos* ("ratio"), as in the Greek words *analogos* ("proportional") and *analogia* ("mathematical proportion", and by extension "analogy" in its various current meanings).

In ordinary usage, the term is used to relate things which are neither identical nor markedly different:

measurement or observation of the physical system.[49] The analogue approach stands in contrast to the *digital*, or directly *numerical*, approach in which the computational device acts on the numbers themselves (or on elements which are unique and perfectly resolvable encodings of numbers). *Transl.*]

The earliest instances of the analogue approach can be found in instruments, more measuring devices than calculating machines, which were devised just over two thousand years ago by the Greek mathematicians and astronomers and which achieved calculation by means of models and measurements of the magnitudes being sought.

These included the *hodometers* (or *odometers*; not to be confused with the *pedometers* of the European Renaissance period, which were true "digital" devices). Hodometers were designed to measure distance, and their mechanism consisted of a series of worm-gear trains which were driven by the rotation of a chariot-wheel and moved an indicating needle along a graduated scale, which indicated the distance travelled by the chariot. These were the direct ancestors of the mileage recorders on modern motor-cars, themselves analogue devices even though their display is digital.

Another device of that time was the sophisticated *Antikythera*,[50] which would seem to be an ancestor of the orrery and the planetarium. The *astrolabe* was a similar but simpler device, which came down from the Greeks, through the Arabs in the tenth – thirteenth centuries, to the West,

---

"analogous situation", "analogous case", etc. The use of metaphor is based on analogy of relationship: the foot of a mountain is situated, relative to the mountain, as the human foot is situated relative to the human body; likewise the leg of a table is analogous to the leg of a dog (four of them, one at each corner, holding the thing up).

In mathematics, there is a precise sense of *analogy* as equality of proportion or ratio between expressions which are not identical: the fractions $^1/_3$, $^2/_6$, $^3/_9$, and so on, have equal values, but they are not identical from the formal point of view. Analogy does not depend on superficial resemblance: the face of a watch resembles the dial of an aneroid barometer, but the watch is no analogue of the barometer; a mercury barometer, however, is a direct analogue of an aneroid barometer since both measure atmospheric pressure, though the one does not look at all like (nor function like) the other. It is mistaken to consider analogy as a kind of imperfect resemblance: the similarity which makes analogy is at a deeper level. [Author & *Transl.*]
49 For a detailed account of a few examples, see the *Translator's Addendum* at the end of this section. [*Transl.*]
50 The "Antikythera" was recovered in 1901 by divers working on a wreck discovered off the Greek island of that name, along with many other artefacts. Archaeological dating of these sets the date of the Antikythera between 51 and 81 BCE; after such a passage of time the device was in a somewhat damaged state.

Contained in a wooden box with hinged doors was a very complex assembly of at least twenty bronze gearwheels, with three dials on the outside of the box. One dial, over the main driving gear, apparently showed the annual motion of the Sun in the Zodiac as well as the main risings and settings of the stars throughout the year. The other two dials have not been reliably interpreted, but probably presented information on the phases of the Moon and the times of lunar risings and settings; and on the risings, settings, stations and retrogradations of the planets (Mercury, Venus, Mars, Jupiter and Saturn) known to the Greeks. It is therefore clearly an "analogue computer" in the modern sense, and may possibly have been used in navigation. It is similar to the astronomical clocks built throughout Europe during the Renaissance.

D. de Solla Price, who profoundly investigated the "Antikythera", wrote in 1959 that "the Antikythera mechanism is, in a way, the veritable progenitor of all our present plethora of scientific hardware." See D. de Solla Price, TAPS 64 (1974), 1–70, and D. de Solla Price, *Scientific American* (June 1959), 60–67. [*Transl.*]

where it arrived at about the same time as the numerals of Indian origin.[51]

Another link in the chain was forged much later, when Napier invented *logarithms* in 1614. Not only did this invention engender innumerable mathematical developments in succeeding centuries; it also gave rise to the table of logarithms (an essentially "digital" aid to calculation) and to the slide rule (an essentially "analogue" device – see also the *Translator's Addendum* at the end of this section).

The slide rule was invented in 1620 by the Englishman Edmund Gunter and enhanced around 1623 by William Oughtred. It was further improved by Robert Bissaker in 1654 and by Seth Partridge in 1671, finally receiving its modern form in 1750 at the hands of Leadbetter. As a matter of interest, while the slide rule had been introduced in France in 1815 by Jomard, it was judged too revolutionary at the time; only in 1851, when the mathematician Amédée Mannheim added further useful enhancements to it, did it take on with the French. Like all the logarithmically-based devices to which it gave rise in its turn, the slide rule made multiplication and division, raising to a power, and even extraction of nth roots, easy and straightforward. The key to its ease of use was the fact that the scales on the rules were graduated at distances proportional to the logarithms of the numbers marked. So handy did it prove in practice, that it remained the working tool of engineers and mechanicians right up to the time when, in the early 1970s, the "scientific" electronic pocket calculator supplanted it.

Yet another category of analogue device was the algebraic apparatus, for solving equations or systems of equations or for determining the values of certain mathematical functions.

The earliest instruments of this kind were constructed by the ancient Greek mathematicians, Plato in particular who constructed a device for finding cube roots.[52] However, the development of such instruments only really gathered pace in the eighteenth century.

Rowning in 1770, and Kempe in 1873, constructed systems of jointed

---

51 The astrolabe was not intrinsically a mechanical device in the same sense as the "Antikythera", but in essence more like a kind of circular "slide rule". It consisted of a set of plates mounted onto a common axis and marked with angular grids representing days and hours, with various circles and ellipses representing the parts of the celestial sphere visible at various altitudes above the horizon from the place for which the astrolabe was constructed, and also with representations of the stars. By rotating the plates, the positions of heavenly bodies could be determined or predicted for any given time.

On the back of an astrolabe a diametric arm was often found, swinging about the same axis and furnished with sights, which could be lined up on a celestial object and used to measure its elevation. Such a device was often used to aid navigation, and in a very primitive form (also called *astrolabe*) consisted only of an angular scale and an arm together with a plumb-line to determine the vertical.

When furnished with a clockwork driving mechanism and appropriate gear-trains in the background, the astrolabe could be used as the dial of an astronomical clock (a famous example being the one on the Old Town Hall in Prague, dating from the beginning of the fifteenth century); and then it was similar to the "Antikythera". [*Transl.*]

52 Plato's disciple Eratosthenes gave a description of such a mechanism, which may be the same as Plato's. The *Translator's Addendum*, at the end of this section, gives an account of this. [*Transl.*]

rods which could give a material and deformable graphical representation of polynomial functions. Bérard in 1810, de Lalanne in 1840, and Grant in 1896 constructed mechanisms for solving systems of algebraic equations which relied on the fact that the configuration of static equilibrium of movable solid bodies to which weights are applied is subject to a system of equations whose coefficients are proportional to the weights. Other devices for the same purpose were based on the mathematical principles of *hydrostatic equilibrium.*

Lucas in 1888, Kann in 1903 and Wright in 1909 made devices which, instead, were based on the algebraic laws of *electrical equilibrium.* Other machines, such as Leonardo Torres y Quevedo's *algebraic machine* of 1893, used a series of logarithmic disks linked by gears, which facilitated mechanical calculation of coefficients and roots of algebraic equations.

Another type of analogue machine was used for evaluating integrals such as occur in the solution of problems involving the calculus, including the determination of areas, centres of gravity and moments of inertia of plane surfaces or bodies of revolution. In particular, they began to be developed at the start of the nineteenth century for working out land areas in surveys. This class of devices includes:

1 *planimeters*, designed to measure the area enclosed by a plane curve; the first was made in 1854 by Professor Amsler of Schaffhausen, for measuring areas on maps and plans;[53]

2 *integrometers* and *integraphs*, which give the numerical values of functions defined by their derivative;[54]

3 *harmonic analysers* and *tide predictors*,[55] of which the earliest forms were invented by Lord Kelvin in 1878 ([see M. d'Ocagne (1986) and L. Jacob (1911)].

WHAT IS AN ANALOGUE COMPUTER?

Looking at each of the cases considered, but from a general point of view, we can see that they work according to a common general principle whether they are simple mechanical linkages or whether they are machines of mechanical, electromechanical or electronic type.

---

53 See the *Translator's Addendum* for a brief description of this device. [*Transl.*]

54 The value of the integral of a function, between definite limits, is the area under the graph of the function between these limits. When the limits are given as fixed values, a planimeter will serve. When one (usually the upper) limit adopts a variable series of values, the integral has a different value for each value of the limit, and integrometers and integraphs are designed to provide the corresponding series of values of the integral, either as a series of numbers (integrometer) or as a graph (integraph). [*Transl.*]

55 A harmonic analyser analyses the variation of a quantity into components expressed in terms of the sine and cosine functions; this application is general. The phenomenon of the tides depends on the periodical fluctuations of the gravitational forces on the seas due to the Sun and the Moon, and once these have been harmonically analysed the results can be used to predict the tides into the future: for this application a specialised form of harmonic analyser can be constructed. [*Transl.*]

Verroust (1965) defines this principle as follows: "Analogue computers are devices in which the variables occurring in the problem to be solved are represented by continuously-variable physical quantities whose values are constrained by the device so as to obey the same mathematical or physical laws as in the problem to be solved."

An analogue device therefore operates in terms of a continuous physical magnitude (such as length on a ruler or on the surface of a cylinder, the linear or angular displacement of an axis, the rotation of a wheel or a gear, the difference of electrical potential in an electric circuit, etc.); the result is obtained by direct measurement of this physical magnitude as it varies within the device which thereby serves as a simulator of the original problem. An analogue computer is therefore a physical model for the problem in question. To simulate a situation being investigated, the machine is initially set up so that its internal physical variables have values corresponding to the values of the variables in the original problem; then the machine is set into action, and the evolution of its variables is observed according to the passage of time.

THE EARLIEST MULTI-PURPOSE ANALOGUE COMPUTERS

The many inventors of analogue devices during the eighteenth and nineteenth centuries were, nevertheless, restricted to mechanising very specific types of calculation, according to a physical "materialisation", so to speak, of the corresponding mathematical structure. For all their ingenuity and theoretical interest, therefore, their machines lacked generality. During this period, any given "analogue computer" could only carry out one type of calculation, specific to a single category of problem to which its physical structure was adapted.

This therefore raises the question of how to make such devices more flexible, and extend the range of problems to which they could be applied. The second half of the nineteenth century saw scientific research being applied to this question.

The father of modern analogue computation was William Thomson, later Lord Kelvin (1824–1907). In a series of publications beginning in 1867, he gave a complete description of an analogue approach to the solution of ordinary differential equations with variable coefficients, using purely mechanical means in which a series of integrators was linked up. The technology of the time, however, was not the equal of Kelvin's concept, and his *Integrator* never came into use.

Not until the 1930s, in the United states, was a true multi-purpose analogue computer made. This was the *Differential Analyser* of Vannevar

Bush (1890–1974), developed between 1925 and 1931 at the Massachusetts Institute of Technology. The operation of this machine was based on wheel-integrators. The physical variables were represented by the rotations of gears, with "multiplication" achieved by mechanical amplifiers each of which consisted of a glass disk on which rotated a metal wheel; the whole was driven by a series of electrical servo-motors which applied the correct respective rotations to each of the components. Several models were made, with from 12 to 18 integrators which could be connected together by gear-trains set so as to represent the coefficients of the differential equation. The resulting machine therefore correctly "concretised" the equation.

This machine, thereby capable of integrating systems of differential equations with up to 12 or 18 variables, was originally created for the purpose of solving mathematical equations describing electrical networks[56] especially with a view to investigating overloads and breakdowns; the corresponding equations were beyond manual human calculation. Such was the success of this machine that the MIT team were inspired to take it much further: the resulting machine, MIT's second differential analyser, far more powerful and more complex, came into use in 1942, but its existence was kept strictly secret during the Second World War. It weighed 100 tons, and had 1,000 electronic valves and over 200 miles of wire. It was used to compute range tables for the guns of the US Navy, allowing problems of fire control and radar systems to be addressed which had hitherto been beyond the capacity of human calculation.

THE DEVELOPMENT OF ELECTRONIC ANALOGUE COMPUTERS

Following the MIT machine, many analogue computers came into being, with ever greater performance and flexibility. These include the M-9 fire director made in the US by Western Electric (Bell Laboratories) which, during the closing years of the Second World War, was one of the most effective machines of its kind. By its aid, more than 75 per cent of the V1 "Flying Bombs" aimed at London were brought down.

Many of these machines were special-purpose. Others. however, were multi-purpose: not only could they simply integrate a function; they could also solves coupled or non-coupled systems of differential equations, together with systems of linear or non-linear algebraic equations.

In France, great advances in this kind of machine were achieved in the 1950s in the shape of the S.E.A. analogue computers under the direction of F. H. Raymond, and the ANALAC machine which emerged at the end of the decade from the French company CSF.

---

56 Including resistors, inductors, and capacitors. [*Transl.*]

The latter was a completely electronic multi-purpose analogue computer. By its means, algebraic equations of arbitrary order (up to a certain level), trigonometric equations, functional equations with one or two independent variables, and differential equations of arbitrary degree with a single independent variable (therefore capable of being modelled as time), could be solved completely automatically. The machine could also operate "in real time"[57] as a simulator.

ADVANTAGES OF ANALOGUE COMPUTATION

The chief advantages of analogue computation are, therefore:
- being able to solve a given problem numerically even without the ability to find a formal mathematical solution;
- being able to solve even a very complex problem in a relatively short time;[58]
- being able, in a short space of time, to explore the consequences of a wide range of hypothetical different configurations of the problem being simulated;[59]
- the technical fact that, in such machines, information is transmitted between the components at very great rates.

Until quite recent times, therefore, analogue computers were widely used in numerous domains of application, some of which are: thermomechanical (movement and exchange of heat); automobile (engine and suspension); electrical and electronic (motors and circuits); military (fire-control systems, missile guidance, etc.); aeronautical and aerospatial (flight simulators, autopilots for aeroplanes and space vehicles); conventional and nuclear energy (power stations, nuclear recycling plants); chemical and petrochemical engineering (dynamics and optimal control of processes); biology and medicine (exchange between environments, blood circulation and cardiology, respirators, biochemistry and metabolism), and so on.

THE PRINCIPAL DISADVANTAGES OF
ANALOGUE COMPUTATION

So, if the analogue approach has so many advantages, why give it up? The main reason is that, for all its merits, it has fundamental limitations which inhibit progress in scientific calculation.

57 That is to say, in such a way that the variables represented in the machine evolve at the same rate as in the system being simulated, i.e. synchronously. [*Transl.*]

58 The complexity of the problem is encapsulated in the structural complexity of the machine, whose speed of operation is largely independent of this complexity. [*Transl.*]

59 The bandwidth of the high-gain amplifiers which are at the heart of such machines can be several megahertz which, for a resolution of one part in 1,000, corresponds roughly to one megabit per second (in information-theoretical terms); such rates of information transfer were not achieved in digital computers until very recent times.

In the first place, no analogue device, no matter how complex or extended its structure may be, is capable of becoming sufficiently general to serve for the solution of arbitrary categories of problem. An analogue computer is not "universal".

Secondly, it is difficult if not impossible to "store" information and results on an analogue computer.

Thirdly, by its nature an analogue computer does not give completely exact results, since it is subject to quantitative error in setting-up and in the relations between its parts, and in measurement of the output. In practice, an analogue computer can be expected to give results accurate to three significant figures, with careful use. Accuracy may vary between 0.02 per cent and 3 per cent, according to the quality of its components and of the measuring devices, and depending also on the mathematical operations being represented and on the choice of analogue model used to set up the machine (in which significant elements of the original problem may be mistakenly ignored, or overlooked).

Fourthly, the components of an analogue computer will function as required only when the magnitudes of their voltages or motions lie within certain limits. For example, a transistor network will start to behave differently when large voltages saturate the transistors, and will be subject to the random effects of electronic noise when the voltages are small.

Fifthly (amongst other reasons which could be cited), it may be difficult to give a precise interpretation of the results obtained.[60]

For these and other reasons, analogue computation has been progressively abandoned since the end of the Second World War. In the face of ever greater advances in mathematical and physical sciences, researchers have needed to use powerful machines which did not suffer from these defects.

All the same, analogue computers – which gain on the swings of speed what they lose on the roundabouts of precision – remained for a long time preferable to digital computers which, in the early stages of their development, were very slow.

60 Reasoning by analogy has always been a somewhat special mode of thinking. "A conclusion reached by analogy is always dubious from the point of view of strict logic . . . Nevertheless, one is inevitably impressed by the sense of conviction engendered by reasoning by analogy . . . The interest of this method, therefore, is above all to aid understanding, or to give a more or less convincing impression that one understands." [M. Dorolle (1949), pp. 170 and 176–7].

In this respect reasoning by analogy plays a role in scientific method: not as a technique of demonstration, but as a tool of discovery [See R. Delteil in F. Le Lionnais (1962), p. 48].

Reasoning by analogy is, in fact, "deduction based on earlier induction". (E. Rabier, *Leçons de philosophie*, 7th ed., p. 248)

"No doubt the solution is of the form . . . Induction plus Syllogism . . . In practice, however, in reasoning by analogy the mind reasons from the particular to the particular: not by the application of a universal general law, but by the resemblance of the two cases in question." [Maritain (1923), p. 335] This, of course, is precisely why the mode of thought appropriate to analogue computation is in itself an absolute blockage . . .

Since the 1970s, electronics has seen enormous and very rapid progress based on high-density integrated circuit technology (microprocessors) which has eliminated the slowness problem of digital computers, so that these have now practically supplanted analogue computers in almost all applications.

ADVANTAGES OF THE DIGITAL APPROACH

Because, as we can now see, the analogue approach has inherent absolute limitations to progress, people came to realise that the digital approach was, potentially, the most efficient, the most advantageous and the most economical possible. The digital approach is, in every respect, the opposite of the analogue approach.

To briefly anticipate events, we may note that in those discrete automata which we call digital computers (or, nowadays, simply computers) the information which is put into them and processed is represented by a *discontinuous* variable physical phenomenon which can be represented in a one-to-one manner in a discrete form, namely by digits or other symbols subject to precise laws of combination which thereby form what is called a code [See G. Verroust (1965)].

The emergence of this approach, characteristic of the digital computer, combined with every kind of invention, discovery and construction in the most diverse of domains, has crowned a long history of evolution inspired by a huge variety of needs and ambitions. The present-day science and technology of information is its direct outcome.

Digital computers, nowadays extremely fast and extremely powerful, are capable of:

1 carrying out operations on information stored in internal memory;
2 giving results of arbitrary precision, needing no reservations about interpretation;
3 performing the most complex of analyses;
4 solving, within the limits of capacity of each individual machine, every kind of problem which can be expressed according to a definite kind of process of the human intelligence – the *algorithm*.

HYBRID COMPUTERS

Again anticipating events, we note that the ever increasing complexity of scientific investigations and need for very precise techniques of calculation has given rise to a combination of the analogue approach with the digital approach. In the resulting computers, called "hybrid computers", analogue devices are connected by suitable interfaces to digital devices so that the two sides can exchange information.

This has been done in many ways, for example:

- using digital devices as control systems for the interconnections and the settings of the components of an analogue device;
- furnishing an analogue computer with digital components such as switches and relays which can do away with complex analogue circuitry or extend the range of possible simulations;
- programming a digital computer with a code whose purpose is automatically to control the operation of an analogue computer;
- distributing different parts of the solution of a given problem between a digital computer and an analogue computer, interconnected, so that each type of machine deals with the type of operation for which it is the better suited.

Until quite recently, such hybrid computers were still being used in the domain of simulation particularly, such as the study of process plants in the chemical industry or in flight simulations for aircraft, missiles and space vehicles.

However, electronic technology – and especially the development of the high-density microchip – has eroded the advantages which the analogue computer enjoyed for certain problems. Nowadays, the digital computer has taken the place of the analogue or the hybrid computer in almost every area of application.

Translator's Addendum

PRINCIPLES AND EXAMPLES OF ANALOGUE COMPUTATIONAL DEVICES

Analogue computation, in summary, means obtaining numerical answers to a mathematical problem by means of measurements carried out on a physical surrogate – an *analogue* – for the mathematical problem.

As a very simple example, you might seek the result of adding 17 to 29 by taking three centimetre tape-measures: from an initial spot, you stretch out the first in a straight line, and note the place where "17" is marked. From this latter spot, you stretch out the second tape-measure – in the same straight line – and then note the place where "29" is marked. Then you take the third tape-measure, and stretch it between the first and the last of the above-noted spots. With luck, the last spot should lie close to the mark "46" on the third tape-measure.

This is a numerical result to a mathematical problem ("what is 17 + 29?") obtained by *analogue* means. The *analogue* – the physical surrogate – is the set of three tape-measures. That it is indeed an analogue follows from the physical principle that successive measurements of distance along the same straight line are *additive* – by definition. Therefore the result of accurately laying out a distance *x* along a straight line, followed by accurately laying out a further distance *y* along the same straight line, corresponds to accurately laying out a total distance *x+y* along that line. The first two steps, using the first two tape-measures, corresponds to *setting up the analogue*. The final step – using the third tape-measure to measure the total distance – corresponds to reading the result from the analogue setup.

The above example is trivial for an adult, though not necessarily for small children: it is exactly the principle used in the *Cuisenaire rods* which were in vogue, especially in the 1950s and 1960s, to teach arithmetic to infants (after Georges Cuisenaire, a Belgian educationalist).

A closely related example is not so trivial: until well into the 1960s, the *slide rule* (two identically marked rules constructed so that one can slide along the edge of the other) was in every engineer's hands for purposes of multiplication by the same method, based on the fact that the logarithm of the product of two numbers is equal to the sum of the logarithms of the two factors. By laying out successively a distance equal to the logarithm of one factor, and a distance equal to the logarithm of the other, a distance equal to the logarithm of the product is obtained. Since the scales on a slide rule are marked "logarithmically", the product itself can be read off directly from the scale-marks on the slide rule:

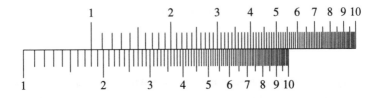

illustrating, for instance, how it can be read off from the measurement scales that $1.8 \times 1.5 = 2.7$ or $1.8 \times 2.5 = 4.5$. Clearly, a manual inaccuracy in physically positioning the sliding rule will give rise to an error in the result. Furthermore, for results which have a significant second decimal place (such as $1.8 \times 1.7 = 3.06$) this figure (the "6") would need to be estimated by eye, again with possible error. This illustrates an intrinsic feature of all analogue methods of computation – since the result obtained depends on the accuracy with which the analogue is set up, and also on the accuracy of measurement of a physical quantity, there is inevitably a degree of setup error and of measurement error. Analogue methods are therefore acceptable provided the amount of error to be expected does not amount to an error which would be of importance in the context wherein the result will be used. For the above slide rule, results accurate to one decimal place (1–10 per cent) are quite reliable; results accurate to two decimal places (1 per cent or better) are somewhat uncertain.

A mechanism which gives an analogue solution of a more difficult problem is Plato's (or Eratosthenes's – at any rate as described by him) linkage for finding cube roots (square roots can be found by Euclidean ruler-and-compass constructions; cube roots cannot).

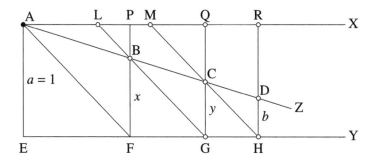

The mechanism consists of thin rigid rods forming a frame and triangles, together with a long rigid rod. The frame XAEY is rigid, and the triangle AFP is rigid and fixed in position. The triangles LGQ and MHR are also rigid, but able to slide along the frame (as shown by the "rings"). The long rigid rod AZ can pivot about the point A. The "ring" at B constrains the rods AZ, LG and PF to always meet in the "ring"; likewise the "ring" at C constrains the rods AZ, MH and QG to always meet in the "ring". The positions of the two sliding triangles are therefore determined by the position of the rod AZ.

Let a denote the length of AE, x denote the length of BF, y denote the length of CG, and b denote the length of DH (at the end we shall let $a = 1$).

Because ABF, BCG and CDH are similar triangles, the height of the upper left-hand vertex of each above the base line is in the same proportion to the height of the upper right-hand vertex in each case, or

$$\frac{a}{x} = \frac{x}{y} = \frac{y}{b}$$

from which it follows that $ab = xy$ and $ay = x^2$ so that (now taking $a = 1$) we get $b = xy$ and $y = x^2$, from which $b = x^3$. Therefore x is the cube root of b.

In principle, by rotating the rod AZ, the triangles will slide until the value b of the length DH is the number whose cube root is desired, and then the cube root of b can be measured as the length x of BF.

By adding a further sliding triangle on the same principles, fourth roots can be found; yet a further triangle will give fifth roots; and so on. All of this was described by Eratosthenes, and the ingenuity and insight thus displayed over two thousand years ago are remarkable.

Amsler's planimeter. Strictly speaking, the planimeter devised in 1854 by Jacob Amsler (1823–1912) was not the first planimeter. Others had been invented during the first half of the nineteenth century. Earlier ones were, however, inaccurate, difficult to use, and often complicated in design and cumbersome. Amsler's *polar planimeter*, based on a new principle of design, was the first that was simple in construction, easily portable, and efficient. A brief description of its working follows.

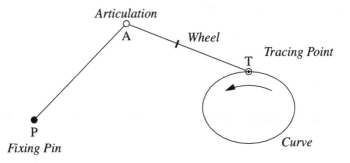

The principle of Amsler's polar planimeter, shown diagrammatically above, is at once simple and subtle. The mechanism consists of two rods, PA and AT, joined at an *Articulation* A. At P, on the rod PA, is a *Fixing Pin* or *Pole* which serves to fix the position of the end P on the drawing board; the rod PA can rotate about P. At T, on the rod AT, is a pointed stylus which is used as a *Tracing Point*, so that the end T of AT can be accurately traced round a closed *Curve* whose area it is desired to measure.

Part of the way along the rod AT is a *Wheel*, rotating about AT as an axis, which is attached to a revolution counter which counts both whole and fractional revolutions of the wheel.

The purpose of the rod PA is twofold: (a) to ensure that, when the tracing point T has completed the circuit of the curve, the rod AT has returned to its initial position; (b) to ensure that, in the event that the end A of TA should itself be carried round a closed curve, that closed curve is a known circle (swept out by the end A of the rod PA) and therefore of known area (the so-called "zero circle").

The method of using the instrument is straightforward. It is mounted as shown above, and the tracing point T is carried round the curve by the operator in the direction of the arrow. A light pressure keeps the wheel in contact with the paper. When the tracing point T has completed the circuit of the curve, the total number of revolutions of the wheel, including any fraction, is read on the revolution counter. This number of revolutions is then equal (on a certain scale of units) to the area of the curve, provided the end A of the rod AT has not itself been carried round the pin P in a complete circle (i.e. has simply moved back and forth along an arc of the circle). The number of revolutions can then be converted into the desired units of area (such as square centimetres or square inches) by multiplying the conversion factor or (as was the case on many of the instruments) the dial of the revolution counter was driven by a gear train which performed the conversion automatically so that the area could be read off directly.

On the other hand, if the end A of the rod AT has been carried round the point P by a whole revolution (as could happen if the curve were large), then the area of the curve is obtained by adding the area of the "zero circle" to the result as read from the revolution counter.

7. The Contribution of Automata

FROM THE AUTOMATION OF CLASSIC CALCULATION TO THE EMERGENCE OF ARTIFICIAL CALCULATION OF NUMBERS

As we have already seen, the researches destined to lead to our present methods of artificial data processing go back to the Renaissance, when scientists, needing to simplify written calculation, sought to free themselves from the attendant difficulties.

Thus it was that a considerable variety of mechanisms with greater or lesser degrees of efficiency, power, reliability, practicability, speed, and precision came into being between the eighteenth and the twentieth centuries.

Machines capable of relieving mankind of the onerous burden of mathematical calculation with its repetitive operations emerged for the first time.

The ever increasing importance of industry and finance during the nineteenth century, as well as ever more ruthless competition within the affected fields, resulted in a hugely increased level of activity in the banking and accountancy sectors; with the ever growing need for purpose-made business machines, this led the manufacture of calculating machines to take off.

So this marked the first important stage in the technical and commercial history of the computer.

But the attempts to mechanise classic arithmetical calculation allowed equally for important technological innovations such as:

1 the use of *numerical keyboards* for entering data; later coupled with that of typewriter keys, this was to lead to alphanumerical keyboards;

2 the use of a *printer* which allowed for the intermediate results or the totals worked out by the machine to be printed at any point on a paper roll; and so on.

Classic electromechanical calculation has introduced systems of pre-determined *number codes* for the encoding of numerical data circulating in the machine from start to finish. This has permitted the introduction of new concepts such as direct multiplication or division, indeed the complete automation of the four basic arithmetic operations, which has gradually opened the way to all possible extensions of classical techniques of auto-matic numerical calculation.

The complete mechanisation of elementary calculation therefore allowed the field to be extended progressively towards an ever wider range of calculations, now opening the way to the development of the numerical method for the artificial processing of scientific problems.

THE DISTANT ORIGINS OF THE COMPUTER: THE NEED TO AUTOMATE CHAINS OF MATHEMATICAL CALCULATIONS

The classic calculator had, of course, far too many limitations to allow for any kind of progress towards the computer; a radical change of direction was needed. This change in fact took place some two centuries ago, when certain scholars, needing to calculate mathematical tables, sought to remedy the serious deficiencies in standard numerical calculators.

To carry out a linked sequence of operations a person was obliged to keep intervening because, over and above the task of entering the numerical data for the calculation, he had to transcribe manually all the intermediate results before re-entering them stage by stage manually into his machine.

Let us take a simple example: To carry out a sequential calculation such as $(175 + 123)^3 / (146 - 23)^2$ which involved an addition, a cube, a subtraction, a square, and a division, it was of course necessary to enter the sum $175 + 123$, and calculate that first; the result had to be re-entered and cubed, and that result noted down on a piece of paper; it was then necessary to enter the data for the subtraction $146 - 23$, note the result, and square it; after that, the new result had to be written down, then the two earlier results re-entered in order to proceed to the division, the solution to which gave the final answer.

Given the numerous manipulations and turns of the handle needed to produce a simple division on Thomas's arithmometer, such an operation involved several dozen steps and turns of the handle. Whence a constant source for error, as also a great waste of time.

What was the answer? How could the machine be made to pick its own path through such a task, by means of a command system that would control its smallest movements, invoking the elementary operations one by one as they became due? How, then, to define a machine capable of executing a linked sequence of predetermined commands automatically, with no human intervention? That was broadly speaking the challenge that certain early-nineteenth-century scientists wanted to address, and it was the search for a solution to this that was to lead to the invention of the computer.

Of course it did not all happen in a day. It resulted from the convergence of several separate attempts that went back into the past, the earliest of which was the product of the old human dream to create a living being from scratch, in a word the need that has subsisted since the beginning to imitate the movements of humans and animals artificially.

THE FIRST ARTIFICIAL AUTOMATA

And it was in fact in this apparently unrelated field of man-made automata that the first solutions to the problem were found.

These autonomous artefacts were developed from the earliest times by the Chinese, as also by Greek mechanicians of Alexandria, a school noted for such eminent scientists as Ctesibius, Archytas of Tarentum, Philo of Byzantium, and Hero of Alexandria, who employed quite sophisticated mechanical systems (see pp. 112ff.).

Diodorus Siculus and Callixenes give this description of animated statues of gods and goddesses that featured at the festivities organised in 280 BCE by Ptolemy Philadelphus in honour of Alexander and Bacchus: "a four-wheeled chariot eight cubits broad, drawn by sixty men, and on which was seated a statuette of Nysa measuring eight cubits, dressed in a yellow, gold-brocade tunic and a Spartan cloak. By means of a mechanism she would stand up unaided, pour out milk from a golden bottle, and sit down again." [Quoted by P. Devaux]

The link between the Ancient Greek mechanicians and those of mediaeval and Renaissance Europe was largely provided by the Byzantines, but it was principally the Arab and Persian mechanicians who inherited the tradition and passed it on. Among the best known in the Arab and Muslim world mention should be made of:

- the Banū Mūsa ibn Shākir brothers, who wrote *Al'ālat ilatī tuzammir bī nafsihā* ("The Instrument that Plays by Itself"), written c. 850 and largely inspired by the ideas of Hero of Alexandria;
- Ibn Mu'adh Abū 'Abdallah al Jayyanī, late tenth-century author of *Kitāb al asrār fī nata 'ij al afkār*, in which several water-clocks are described;

- Badī 'al Zasmān al Aṣṭurlābī, early twelfth-century, noted for the automata he made for the Seljuk kings;
- Abū Zakariyyā Yaḥyā al Bayāsī, late twelfth-century, noted for his mechanical organs;
- Ridwān ibn al Saʿatī of Damascus (1203), celebrated for his automata activated by floats;
- Ismaʿil al Jazzarī, certainly the most productive and imaginative, his book *Kitāb fī maʿrifat al ḥiyat al handasiyya* ("Book of the Knowledge of Ingenious Mechanical Instruments") in 1206 attested a perfect understanding of the Greek traditions with a number of further innovations; for instance he produced plans to construct perpetual flutes, water-clocks, mechanical pumping systems for fountains, and yet more automata activated by floats and with movements communicated by chains and cords.

This tradition spread into Western Europe through Spain and Sicily, thanks in particular to Roger II of Sicily (1095–1154) who invited Arab mechanicians to his court, and Frederick II (1194–1250), one of his successors, who continued to welcome mechanicians from the Muslim world. And thereafter Europe developed its own techniques and brought this skill to its full maturity.

THE FIRST SEQUENTIAL AUTOMATA STRUCTURALLY CONCEIVED FOR THE PURPOSE OF FURNISHING INFORMATION ABOUT THE WORLD

Artificial automata were originally nothing but devices with human or animal features constructed so as to give the illusion of life; they did not serve informational needs. The first step forward in this domain was realised in Greece and in China, and then more definitely in the Islamic world and in Western Europe. It was spurred by the need for very accurate measurements of astronomical time and data, as also by a radical change of approach that was to make the artificial automaton a tool for providing information on the world around us.

Among the many works left in this domain by the Chinese since ancient times we must mention Su Song's *Xinyi xiangfa yao*, (c. 1090 CE), in which he describes a number of mechanical instruments including astronomical clocks, and several automata and devices similar to those of Hero of Alexandria: singing birds, stage sets moved by sand-driven motors, mechanisms which opened the gates of a temple automatically when a fire was lit on the altar, and so on. He also describes a huge tower in the shape of a clock in which globes are set in motion automatically, and from which figurines emerge regularly to strike the hours of the day on gongs, the entire mechanism being wound up automatically by means of hydraulic power.

In Islam, the technique for the construction of astronomical clocks was developed particularly in answer to the need to determine the exact hours for ritual prayer, the dates of religious ceremonies, the phases of the Moon, the positions of planets, the precession of the equinoxes, and so on.

A chronicle relates how Charlemagne received a delegation from the Caliph Harūn al- Rashīd at Aix-La-Chapelle in 807, bringing many sumptuous presents including a brass mechanical clock. It was a clepsydra (a water-driven clock) in which the hours were marked by dropping twelve small bronze balls onto a gong placed beneath them, making it resound, and by twelve knights who emerged from twelve doors and closed twelve other doors. This clock, the chronicle adds, required the presence in the Caliph's delegation of a scholar-mechanician capable of setting it up when they arrived and activating it. Mention should also be made of the clock-maker Al Khurassānī, who in the late twelfth century built the clock on the Gate of the Hours of the great mosque at Damascus: it was a water-clock with specially adapted mechanisms that enabled it to indicate not only the hours of the day but also (and this was the novelty) those of the night.

It was, however, a quite different type of automaton that marked the revolution we are describing, thanks particularly to the contribution of the Persian mechanician Al Jazzarī, whose book, mentioned above, seems to reveal appliances hitherto unknown. He gives a description of true *sequential automata*, driven notably by a camshaft, which transforms the circular motion of a sort of crankshaft into an alternating motion of a distributor: such automata thus mark a break with the Graeco-Roman concept of the simple device endowed with automatic movements.

Following a routine set up in advance using a control mechanism, and quite without human intervention, Al Jazzari's sequential automata, mechanisms were capable of carrying out an enchained series of operations, and were thus the true ancestors of the automatic peals of bells and the belfry automata of the European late Middle Ages; and even, up to a point, of those highly developed automata which were to see the light in Western Europe in the eighteenth century.

The first *automatic carillon* known to Europe is that in St Catherine's Abbey at Rouen, built in 1321. As for the first known European belfry automata (allegorical figures, normally placed at the top of belfries and regulated by a control mechanism so as to strike the hours on a bell, an anvil or suchlike), they were built at Courtrai in 1332. Philip the Bold, Duke of Burgundy, transferred them to Dijon in 1383, where they still subsist in the church of Notre Dame.

Let us not forget the famous astronomical clock made for Cluny Abbey c. 1340, and described as follows by P. Lorain, as quoted by P. Devaux:

This vast machine displayed at one and the same time a perpetual calendar marking the year, the month, the week, the hours and minutes, and an ecclesiastical calendar which gave the feast days, each day's office, the positions, oppositions and conjunctions of the stars, the phases of the Moon, the motions of the Sun, the Mystery of the Resurrection, Death, St Hugh, St Odilon, the Feast of the Blessed Sacrament, the Passion, the Blessed Virgin. Each hour was announced by a cock which flapped its wings and crowed twice. Simultaneously an angel opened a door and greeted the Blessed Virgin, the Holy Spirit descended in the form of a dove, the Eternal Father blessed them, and all the figures retired within the clock to the accompaniment of a harmonious peal of bells and bizarre manoeuvres of fantastic animals who moved their eyes and flicked their tongues.

THE FIRST AUTOMATICALLY-SEQUENCED NUMERICAL CALCULATORS

Several centuries were to elapse, however, before the basic concepts of sequential automata were applied to artificial numerical calculation as such. In fact the first numerical calculators with an automatic capability of effecting chained sequences of operations following a programme set up in advance in a control mechanism were the *difference machines*, so called because they relied on the mathematical method known as that of "finite differences"; these, starting from a given value, allowed the calculation of values of a polynomial of the degree n to be replaced by a series of n successive additions. [61]

These calculators appeared at the beginning of the nineteenth century and were conceived to answer the need, long felt by astronomers and navigators, to establish reliable numerical tables. They comprised a certain number, say n, of registers which could store numbers and a mechanism allowing it to execute this type of chained additions, and they could automatically produce the succeeding values of polynomial expressions of

---

61 This application of the method of finite differences relies on the fact that, for a polynomial in x of degree n, evaluated for successive integer values of x, the differences between successive values of the polynomial are values of a polynomial of degree $n - 1$, the differences of these are the values of a polynomial of degree $n - 2$, and so on, the differences of order n being constant.

For instance, therefore, the polynomial $x \sup 2 + x + 1$, as x takes successively the values 1, 2, 3, 4, 5, . . . , itself assumes the values 1, 3, 7, 13, 21, 31, ... The differences between these are 2, 4, 6, 8, 10, . . . and the differences between these last are 2, 2, 2, 2, ... Therefore if we are given that a polynomial has constant second-order differences all equal to 2, that its initial value is 1, and that the initial value in the series of first differences is 2, we can proceed as follows, using addition only. The first difference is 2; add this to the 1 and get 3. The (constant) second difference is 2; add this to the first difference and get 4; add this to the 3 and get 7. The second difference is again 2; add this to the first difference (4) and get 6; add this to the 7 and get 13. Add the second difference (2) to the first difference (6) and get 8; add this to the 13 and get 21. Add the second difference (2) to the first difference (8) and get 10; add this to the 21 and get 31. And so on. [*Transl.*]

degree less than or equal to n, within the limits of their physical capacities, of course.

The German military engineer Johann H. Müller first conceived the idea for such machines in 1786. As they did not answer to any clearly recognisable social need, this invention was never more than a gleam in the eye, and thus sank into oblivion.

THE FIRST DIFFERENCE MACHINES

But less than thirty years later the idea resurfaced and germinated in the highly fertile mind of a great inventor who was to commit himself to its realisation. At a time when even the most elementary mechanical calculators were rare, the English mathematician Charles Babbage (1792–1871) demonstrated in 1822 that this concept was perfectly feasible, then devoted the next eleven years to the construction of a machine of this type. And thus it was, according to the report made by Merrifield to the British Association for the Advancement of Science, that "Babbage is intent upon the realisation of a mechanical marvel", an idea so ingenious and fertile that it could not but strike the imaginations of all who were called upon to pronounce on it. However, in spite of financial support from the British government and of considerable sums invested from the personal assets of the inventor, this project was never really carried through to completion.

There are at least two reasons for this failure: Babbage was such a perfectionist he was quite unable to moderate his ambitions; and he was soon conscious of the structural limitations of his Difference Engine, which made him gradually lose interest in it in order to devote himself to a far more ambitious project which we shall come to presently.

The first time the conception was completely realised was in 1853 when the Swedes Georg and Edvard Scheutz, inspired by Babbage's work, constructed a sturdy and dependable difference machine with financial support from their country; it worked satisfactorily and had the facility to print out results. (Be it noted in passing that the Scheutz brothers' machine was the first working calculator in history that did print out the results.) This achievement was soon followed by several others, starting with that of the Swede Martin Wiberg, who in 1859 produced a more compact machine endowed with additional capabilities, then by those of the American George Grant (1871), the Frenchman Léon Bollée (1895), the Irishman Percy Ludgate (1905), the German Christel Hamann (1910), and others.

BABBAGE'S FIRST IDEA PROVED A DEAD END

However, this type of machine never enjoyed the success of the classic mechanical calculators. Their importance diminished, moreover, from the early 1920s when it was realised that their functions could be fulfilled both by the multi-register accounting machines (see pp. 151ff.) and by mechanographical punched-card machines (see pp. 185ff.).

It was the principal defect of the difference machines that explained their failure: like the belfry automata, the astronomical clocks, and the earliest automatic carillons, their control systems were mingled with their calculating mechanisms. Their control systems were part and parcel of their inner workings and not independent from the material structure of the machine. Worse still, these calculators lacked versatility, restricted as they were to a single type of calculation (those based on the method of finite differences); furthermore, the control system needed to be reset for each new sequence of calculations.

In a word, the difference machines were numerical sequence automata based on a fixed control system that could not be modified. Their rigidity led to a dead end.

THE FIRST SEQUENTIAL AUTOMATA CONTROLLED BY A MODIFIABLE CONTROL MECHANISM

And yet the first solutions to this problem had already reached Europe by the end of the fifteenth century with the appearance of the first automatic carillons controlled by interchangeable mobile drums with pegs, like the one built in 1487 by Barthélemy de Koecke at Alost in what is now Belgium, or the Speeltrommel Carillon built in 1666 at Utrecht in the Low Countries.

Thus this development clearly marks the earliest appearance of sequential automata controlled by a modifiable control unit.

THE FIRST PROGRAMMABLE SEQUENTIAL AUTOMATA

These solutions, however, continued to be quite inadequate, because each additional sequence involved the construction of yet a further control unit. But with the great advance in clockwork and high-precision technology in Europe, far more elaborate solutions were developed in the eighteenth century when the Frenchman Jacques de Vaucanson, the Austrian Friedrich von Knaus, and yet others built their famous android or animal automata. Each one of these automata was equipped with a mechanism controlling the instructions needed to execute a given sequence of basic operations, thus prefiguring the modern concept of programming.

There were for instance the Digesting Duck, the Flute-player, the Pipe-player and the Tambourine-player, all by Vaucanson, as also the many replicas, often more sophisticated than the originals, which the same master created. There was also the Flageolet-player and the four Talking Machines by von Knaus.

THE CONCEPTS OF VAUCANSON'S AUTOMATA

With regard to the Digesting Duck (built c. 1738 but destroyed by a fire in c. 1879), there are many witnesses to confirm what Vaucanson said himself (as reported in Doyon and Liaigre), namely that he had aimed to produce a very faithful replica of the duck: "He stretches out his neck to go and take the grain from the hand, he swallows it, digests it, and excretes it, once digested, through the normal channels; all the gestures of a duck swallowing rapidly, speeding up the movement in his throat to pass the food into his stomach, are copied from nature." But it must be added that although Vaucanson had contrived to replicate the physical motions, he had to cheat a little in the simulation of digestion.

Nevertheless Vaucanson's automata were very interesting creations, and realised some novel concepts, as J. Perriault explains:

> While one might be tempted to think that it was a futile distraction, Vaucanson and his protector Louis XV, who must have promoted a policy of research, were nursing a grand design: to create a complete man, with a circulation of blood and lymph, a respiratory system, etc. The enterprise got nowhere, of course, but it left us with the employment of rubber in the form of tubing, which was to have formed the veins and arteries for the big model. While we still seem to be a long way from information technology, what was here achieved gives rise to a series of reflections that are to our purpose . . .

> Where information technology is concerned it is crucial to be able to replace a literary description of a given phenomenon with a meticulous and complete description that allows it to be implemented in an automaton. Vaucanson, like Euclid and many another, had been able to construct and regulate a working model that produced a well defined result, to wit, the simulation of reality.

> He had furthermore found in the technology of his day what he needed for his construction. Here, unlike a bead-counter, for instance, the mechanism is completely in charge, automated from start to finish. There is a control unit which directs the various mechanical elements which, in the eighteenth century, would have been connecting rods, cams, and pistons. Just as the arithmetical operation $a + b$ is rewritten in a sequence of transfers and counts, so the analysis of the duck's

waddle is to be found in the sequencing of these various elements. The control unit is a drum covered with various projections corresponding to the functions to be triggered.

Resuming in present-day terms the concepts which, as we see, go back a very long way, we can draw up a first balance sheet.

[As to what the eighteenth century contributed, we know that it was by now possible] to analyse complex problems and identify basic stages and working rules that would effectively achieve the end in view: the simulation of reality. There is therefore the capacity to create algorithms that had to be realised in terms of mechanics. To this end each organ of execution is endowed with an access position that identifies it by reference to a drum on which is mounted the ultimate triggering mechanism. These knobs ... represent an *encoding* of the rules of the algorithm. Today they would be called *instructions*. As for the drum, it is an implement that one would be quite wrong to consider a novelty: it is a *programme*.

This is not to say that our present-day understanding of these concepts has not evolved since then. Quite the contrary, but the concepts thus brought into play did possess a structure independent of the nature of the appliances in which they were set to work. Their organisation was such that, when applied to other devices and subject to new technology deriving from much more advanced theories, they were to give birth to the elements with which we are very familiar today.

JAQUET-DROZ'S AUTOMATA

Another notable advance in the field of android automata was made during the Enlightenment: around 1780 the Frenchman Pierre Kintzig made his famous Dulcimer-player which was, so it is said, modelled on Queen Marie-Antoinette. Some ten years earlier the Swiss inventors Pierre and Henri-Louis Jaquet-Droz had constructed their Musician, their Draughtsman, and their Writer. This latter can be visualised more clearly thanks to the description made of it by R. Carrera, D. Loiseau, and O. Roux:

The little fellow, a child of barely three, is seated on a Louis XV stool. In his right hand he holds a goose-quill while his left elbow leans on the little mahogany table. His eyes follow the letters traced out, his face is alert; his movements, although a bit jerky, are natural.

The mechanism that moves him is extremely complex – indeed it is the most complicated of all three androids. To make it Pierre Jaquet-Droz had to solve some very difficult problems; it was necessary, in particular, to house the complete mechanism in the child's body, and to control the wrist movements through the elbow and arm.

There are two sets of gearwheels alternately set in motion without pause, until the full stop is reached which brings everything to a halt.

The first mechanism, housed in the upper half of the body, turns a long cylinder on a vertical axis, comprising three sets of cams; each one has the task of activating levers that move the wrist in the three basic directions. So the pen can move in more than one dimension and use downstrokes and upstrokes to trace out the letters. With each turn of the cam one letter is formed. At this point the second mechanism is triggered to impart to the cam an up or down motion. The extent of the run is determined by interchangeable steel pegs, fixed around the circumference of a disc situated on the lower part of the movement. There are forty of them, each one occupying a sector of 9 degrees. Each peg assigns to the cam a clearly defined position that corresponds to a letter or a specific manoeuvre, such as moving to a new line, dipping pen into inkwell, etc.

Thus it is possible to prepare the device in such a way that the automaton writes any text comprising a maximum of forty letters or characters.

There are many other parts of the mechanism worthy of note, like the one that controls the head and the eyes, and the one that can regularly increase or reduce adjacent intervals. Another element moves the dot over the i in order to make it the full stop that ends the sentence, and then arrests the mechanism.

Such appliances were thus the earliest sequential automata with modifiable controls, structurally conceived so that certain command sequences could be employed at any given point in a sequence of operations: they may be regarded today as a true precursor of the modern concept of the *subroutine*. [62]

THE FIRST SEQUENTIAL AUTOMATA WITH FLEXIBLY MODIFIABLE PROGRAMMES

However advanced they may have been, these automata had very little in common with present-day computers: the control unit was in effect an integral part of the mechanism and thus was not independent of the material structure of the system. Furthermore, each modification of

---

62 A subroutine is a fixed sequence of instructions which may be called upon anywhere, and in any number of places, within the main programme. The subroutine is stored once and for all in a single part of the storage of the machine and, when its services are required at a particular point in the main sequence, control passes from the main sequence to the first instruction of the subroutine and, at the same time, a record is kept of whereabouts in the main sequence this occurred. When the instructions in the subroutine have been completed, control automatically returns to the instruction in the main sequence which comes after the recorded position at which the subroutine was invoked. [*Transl.*]

the sequences to be enacted required a delicate preliminary adjustment of the mechanisms affected.

To arrive at the computer as we understand it was to involve a radical change of direction in the field of artificial automata. The first step was to devise a machine whose functioning would be controlled by a modifiable control unit governed by a sequence of instructions recorded on a malleable input medium that was independent of the material structure of the internal mechanisms. A machine had to be invented in which the commands would obey the rules of a very precise code, following a method of communication previously established between operator and machine.

The first solutions to this problem were in fact discovered by French master-weavers in the eighteenth and early nineteenth centuries. It all began in 1725, when a weaver from Lyons, Basile Bouchon, invented a loom that accepted commands by means of a punched tape: it was a command mechanism similar to the embossed drums of Vaucanson's Duck, but the knobs were replaced by holes which corresponded to the functions to be triggered. Such a system is widely known, moreover, because it features in barrel organs in the shape of unfolding tapes variously perforated. This system will have been invented around 1727, in the days of Vaucanson and Bouchon.

In 1728 the weaving-loom system was improved by Falcon, Bouchon's assistant, who had the idea of replacing the flimsy tape by perforated plaques of cardboard chained together in an endless loop.

This stage, after a fashion, ushered in the era of communication between man and machine. And yet Falcon's system was not automatic: an operator had to be constantly on hand while work was in progress.

The next stage was achieved by Vaucanson in 1749 when he constructed the first loom whose functioning was completely automatic. He had the idea of replacing his predecessor's perforated cardboard by a perforated metal cylinder rotated hydraulically. So Vaucanson turned his loom into a programmable but cyclical automaton, one in which the commands were inscribed by means of perforations on a hydraulically-driven drum, and were regularly repeated.

JACQUARD'S LOOMS

So it was only at the beginning of the nineteenth century, thanks to Joseph-Marie Jacquard, that weaving-loom technology took its most crucial step forward, turning it into a fully operational and marketable piece of equipment. Collating Falcon's programming system based on punched cards with Vaucanson's principle of automation, he combined the use of the moving drum equipped with a sequence of punched cards and the concept of a swinging arm that lifted hooks. Thus it was that in 1804 the first fully automatic loom activated by a punched-card system was built: the system was to achieve a worldwide success.

THE FIRST STAGE ON THE ROAD TO MODERN PROGRAMMING TECHNIQUES

What evolved from Bouchon to Jacquard marked the birth of machines capable of modifiable programming, as also the advent of sequential automata that were programmable but non-cyclical.

In order to control the operational sequences needed for the patterns of a given piece of weaving, a flexible method had been conceived for the programming and execution of operations, based upon a language encoded by means of punched holes very precisely located on a hard cardboard input device, these punched cards following each other in an endless loop. This was a crucial discovery, the use of which was to be extended and made more accurate, adaptable, and universal with advances in the relevant technology, and thanks to the development of symbolic logic and contemporary mathematics.

8. The Development of Punched-tape Solutions

THE EMERGENCE OF THE NEED FOR ARTIFICIAL PROCESSING OF STATISTICAL DATA

One of the earliest practical applications of the punched-card system for artificial information processing was in a machine specially devised for the purpose in the United States in the late nineteenth century. R. Ligonnière gives a good account of the context for this crucial invention which was to incorporate the punched tape as a key element. It is an invention which in its day revolutionised the disciplines of statistics, accountancy, and business management, and it was to be the occasion for substantial advances in technology and industry – developments which were to reveal themselves as an essential component in the burgeoning computer industry:

> [A plan to hold a census of the population of the United States every ten years was approved in 1787 by the Constitutional Convention meeting in Philadelphia.]
>
> The first census, on 2 August 1790, limited itself to counting 3,893,637 people inhabiting the territories of the Federation. After 1800 the population was broken down under six headings: white males below the age of sixteen; white males above the age of sixteen; white females; free persons of black race; slaves; profession.

The questions asked by the census authority became more and more numerous, for it was soon realised that the information deriving from these constitutional censuses could serve as guidance in legislative decisions. From one census to the next the total numbers increased by an almost constant factor in the region of 34.6 per cent. In 1860 the population had grown to 31,440,000.

The point had been reached by 1850 where, to facilitate the extraction of data from the census forms, Congress was obliged to vote a law limiting the number of basic questions to one hundred. Nonetheless the census data became harder and harder to exploit, and more and more expensive too. Paradoxically no permanent organisation existed to oversee this major function. Periodically a "Census Bureau" was set up, for ever longer sessions. For each such period of activity this Bureau would recruit experts, on a temporary basis, from the Civil Service, the Military, and Academia to conceive, organise, and prepare for the following census.

In 1870 (40 million inhabitants) a timid first step was taken towards improving on the earlier methods. Colonel Charles W. Seaton suggested a mechanism with a rigid framework and a roller-fed tape, in order to facilitate the manual transcription between the sheets drawn up by the census officers and the central digest of records. A rather modest step forward!

The 1880 census was a nightmare: the population had risen to 50,262,000, and it was to take seven years to analyse the data and extract useful information from them.

By 1890 the system had simply ground to a halt: those who made economic and legislative decisions could no longer access the necessary data within a practicable time limit and the constitutional rules could no longer be adhered to. A radical overhaul was called for. [R. Ligonnière (1987)]

THE FIRST STATISTICAL PUNCHED-CARD MACHINE

So this story began in 1889 when the American Census Bureau, faced with the delicate problem of analysing the mass of statistical data relating to an ever growing number of souls within a reasonable interval, set up a prize for the invention of an artificial system of statistics that was accurate, reliable, and fast, in time for the next (1890) census.

The winner was the young technician Hermann Hollerith (1860–1929) who in 1884 had invented a statistical machine capable of counting units electrically by means of a punched-card system, whereby the passage of current through the perforation moved the needle of a counter up by one point. R. Moreau (1987) quotes E. Cheysson's description of this machine in action, in the issue of 8 July 1893 of *La Science moderne*:

The problem consisted of combining on a 12 × 6 cm rectangular card (as borrowed from Joseph-Marie Jacquard) the 210 boxes required to accept every possible entry, each one denoted by a brief conventional sign representing it.

It is a veritable algebraic or chemical notation in which a letter stands for a name: *m* for male; *w* for female; *rk* for Roman Catholic; *dt* for German; [other letters indicate illiterates, divorcees, and so on].

On the cards thus prepared holes are punched with the aid of a sort of pantograph, these holes corresponding to the data relating to each person in the census, so that the holes punched in each card give an exact definition of the individual to whom the card applies. All the individual's answers are transcribed into a precise language that the machine can understand, so the cards may be handed over to the machine.

All cards embodying the same item of data, for instance cards with the letters denoting "illiterate" (*An*), can be selected for special analysis, with no need for any painstaking process of hand-sorting. Here again, electricity simplifies the operation.

Next to the counting machine there is a compartmented sorting box, each compartment closed by a lightweight lid. In the case of the "illiterate" code, for instance, whenever the needle on the counting machine makes contact through a punched hole with the tube on the fixed tray which corresponds with the "illiterate" code, the lid is put into electrical contact with the tube and opens automatically. So the operator sees the entire sorting box shut all except for one compartment standing open, in which he simply has to place the card thus sorted automatically.

Now that we have seen the role of the machine and the card, we can easily visualise the complete operation. Once the forms for each household reach the central office, they are passed to specialised staff who complete them by translating them into conventional signs. And once this work has been checked, they pass to the workers who punch the cards, then to others who double-check the punching. The punched cards are then passed to the machine and the sorting box which counts them and groups them by series. The same card undergoes the successive sorting processes assigned to it by the section head. Thus by means of the counters it is possible to count the population by place of birth, by whether they are home-owners or not, or by their illnesses; then, using the sorting box, to reclassify them into groups by sex and by age in decades. Each one of these groups may be put through the counters which will give its numerical breakdown in terms of marital status, intellectual level, and status as worker or employer. They may subsequently be classified in the sorting box into various groups by profession. And, in a final run, each packet, already classified by age, sex and profession, may be classified by marital status, employment situation, and so on. The census chief is thus free to vary the combinations as he sees fit.

The system could be criticised for providing more information than could actually be published. A single census would produce enough material to fill a large library if the whole of it were put to use. So most of it has to remain in the state of manuscript documents, with only grand totals published, and summaries that embrace vast groupings if not the entire nation.

In a word, and it is a point in its favour, such an organisation only works with a centralised facility for analysing the data.

Hollerith's machine, then, comprised a hand-operated *punch*, being a sort of pantograph by which the individual data could be punched into the cards on a predetermined coding. It also comprised a sort of *tabulator*, itself made up of a reader which in turn comprised a plate with containers that held mercury, and a moveable reading-head fitted with as many needles as there were punched holes (the needles could not make contact with the mercury except through the punched holes). There was also a battery of dial-counters controlled by electromagnets, as also a group of 26 sorting boxes each one with a lid controlled also by electromagnets. Each card was placed by hand beneath the reading-head. When the head was lowered, the needles closed circuits and made the relevant counters advance by one unit. One of these circuits belonged to a sorting box which opened when the circuit was closed. The card was thrown into it and the lid shut by hand. So the operation involved simultaneously *counting* and *sorting*, the sorting preparing the cards for a further count in a different order. [See J. Favier and R. Thomelin (1965)]

The Hollerith machine therefore proved to be the true ancestor of those mechanographic machines known as *tabulators*, so called "because they allow for the display of results in tabular form". As for the *sorting box*, that was the ancestor of the *mechanographical sorters* which "were machines that carried out comparisons by solving sorting problems on objects that could be represented by the codes on mechanographical cards". Be it noted, lastly, that "in recognition of the cards employed by Hollerith, the mechanographical cards still in use not so long ago to enter data in a computer were still sometimes referred to as Hollerith cards, even though their format was different to the original ones." [R. Moreau (1987)]

BIRTH AND DEVELOPMENT OF MECHANOGRAPHY
FOR STATISTICS AND ACCOUNTING

Although very rudimentary, the Hollerith machine still required less than half the time needed by any other rival system. So it enjoyed a great success; it served for the 1890 and 1900 censuses and also, with a number of improvements, for the New York Central Railroad accounts from 1891, as also for many insurance companies.

In 1896 Hollerith left the federal administration to start up his own business in Washington, DC, the Tabulating Machine Company which, thanks to the considerable success it was soon to enjoy, was to contribute to familiarising the rest of the world with his system. In 1903 his firm was to market the first truly automatic sorters and tabulators.

Let it be noted in passing that he merged in 1911 with the Computing Scale Co. (then specialising in the production of automatic weighing scales), as also with the International Time Recording Co. (that specialised in clocks), and this all gave birth to the Computing Tabulating Recording Co. (CTR) whose innovation in 1914 would be to produce the Hollerith printer. We may also observe that in 1913 Thomas Watson (1874–1956) was recruited; he had previously worked for NCR and, after being Director General of CTR from 1914, became its President in 1924 – the same year he changed the name of the group to International Business Machines Corporation, better known as IBM.

Among the most significant technical achievements in the history of IBM mechanographics were:

1 the printing tabulator (1920)
2 the future standard-format of IBM cards with 80 columns, and the system of rectangular perforations (1928)
3 the tabulator with direct subtraction (1928)
4 the recapitulating perforator and the verifier (1930)
5 the invoicing tabulator (1931)
6 the alphabetic tabulator and the multiplier (1932)
7 the duplicator (1933)
8 the comparative classifier (1937)
9 the wholly electronic calculator (1946)

However, from 1905 the US Census Bureau, which had made use of Hollerith's successive machines since 1890, decided to build a customised appliance for the next (1910) census, and found in the engineer James Legrand Powers (1870–1915) the person who would henceforth be in charge of this task.

He bypassed the Hollerith patents to produce in 1907 a mechanical card-reader coupled to an electric punch, the reading of the cards being effected by little rods that could pass through the holes and press buttons. This system was of course purely mechanical, but it proved its efficiency in its turn, being even more reliable than the electrical reading method of Hollerith's first system.

In 1909, after obtaining a patent for this invention, Powers invented a semi-automatic tabulator with printing counters. But he in turn was tempted to strike out on his own and in 1911 he founded a specialist company, the Accounting and Tabulating Machines Co. (ATM), which set up factories successively in America, England, and France, and was soon in competition with Hollerith's group. Among his innovations we especially note the invention of alphanumeric printing, in 1925.

In 1927 Powers's patents were awarded to Remington Rand Inc., which originated that year out of a merger of Powers's ATM with the Remington Typewriter Co., the Dalton Adding Machine Co., the Rand Kardex Library Bureau (which specialised in filing systems), and a few other companies of lesser importance, with James H. Rand (1886–1968) at the head of the group.

In 1950 Remington Rand acquired EMC (Eckert-Mauchly Computer Corp.), founded two years earlier under the name the Electronic Control Company (ECC) by John P. Eckert and John W. Mauchly (inventors of the massive electronic analytical calculator known as ENIAC), which went on to assume the once familiar name of UNIVAC Division.

In 1955 Remington Rand in turn merged with Sperry Gyroscope Co. to become the Sperry-Rand group, later the Sperry Corporation. Lastly, in 1987 Sperry merged with the Burroughs Corporation. This latter had been founded in 1886 at Saint Louis, MO, by William Seward Burroughs (who invented the first commercial printing key-operated adding machine), as the American Arithmometer Co. In 1905 it moved to Detroit, then changed its name to Burroughs Adding Machine Co., and in 1952 to the Burroughs Corporation. Thus was born the group known today as Unisys.

But Powers's story did not end here. In Europe his patents were awarded to Samas, which became the Samas-Powers company, and in 1959 merged with the English company BTM which was exploiting IBM patents. Hence came International Computers and Tabulators Ltd (ICT) which subsequently bought up EMI's electronic calculator departments from the English company Ferranti before giving birth, by merger, to the group currently known as ICL (International Computer Limited).

Another important step was taken at the beginning of the twentieth century in the field of industrial mechanography: wishing to answer the needs of the insurance company who employed him, in 1919 the Norwegian engineer Fredrick Rosing Bull (1882–1925) invented an original tabulator embodying new characteristics, which he built in 1921. This was an electromechanical statistical machine with a card selector, a system patented in 1924, and first marketed in 1925 in Norway, Denmark, Finland, and Switzerland. On his death F. R. Bull bequeathed his patents to the Cancer Institute in Oslo, where he was being treated. The patents were bought back in 1928 by the Swiss firm Egli of Zurich, which up till then had been specialising in the production and marketing of direct-multiplication calculators like the *Millionnaire*, or direct-division ones like the *Madas*. Egli straightaway set about producing mechanographical punched-card machines. In 1931 the company, henceforth known as Egli-Bull, moved to Paris, while keeping offices in Zurich and Brussels.

Then in 1933, now under complete French control, Egli-Bull took the name of Compagnie des Machines Bull and went into mass-production, patenting new machines such as the tabulator with independent cycles (1934). In 1970 it merged with Honeywell until 1976 when it took the name CII-Honeywell Bull before finally, and after the making and breaking of associations with many other interests, becoming the Groupe Bull following nationalisation in 1982.

Among the technical innovations in the field of mechanography brought in by this company let us mention: the tabulator with three 10-digit accumulators (1931); K. A. Knutsen's printing mechanism which introduced continuously rotating print-wheels that allowed a printing speed of 120 lines per minute; the alphanumerical printing block (1934) with separate alphabetic and numerical print-wheels, and so on.

MECHANOGRAPHY: ANOTHER IMPORTANT STAGE IN THE TECHNICAL AND COMMERCIAL HISTORY OF THE COMPUTER

We can now see what an important stage the development of mechanography was in the commercial history of the computer, for it saw the birth and development of important businesses which later formed the club of the most important computer manufacturers – IBM, Unisys, Bull, ICL, etc.

Here also, indeed, were created the true conditions of a market in automatic data-processing machines. In fact, the preceding inventions have been at the origins of a whole dynasty of punched-card machines (tabulators, checkers, calculators, recapitulative punches, sorters, etc.) which we have seen undergoing a considerable development ever since, and which have emerged little by little from the simple domain of statistics to become applicable not only to all kinds of accountancy work but also to certain types of scientific calculations.

Without going into tedious detail let us make clear, in the words of J. Favier and R. Thomelin, that the originality of the punched-card machines [formerly] lay in their ability to read the documents directly, once these had been set up in the form of cards bearing the data encoded as perforations.

Two characteristics of their use in particular resulted from this: their speed of execution, which amongst other things allowed for the sifting of a huge mass of documents (applied to national statistics) without a prohibitive outlay; and automatic sorting.

Accountancy and statistical operations required *calculation, transcriptions*, and *sorting*.

[At the root of the system, at least at its advanced stage,] was to be found the transcription of each item of data and each group of data that corresponded to a fact of accountancy or statistics, on a document of standard format, the card, on which the numbers and letters were translated in accordance with a certain code by a series of perforations.

This once-for-all initial transcription offered a double advantage:

1 The cards provided the data on an independent material medium, and they could be sorted and classified in any order;

2 The perforations allowed machines capable of reading them to operate automatically.

Carried out by means of a punch, or perforator – a key-operated machine that could be operated at great speed – the once-for-all transcription did however present a serious danger in that a mis-keying at this stage introduced an error into every succeeding stage of the work. Therefore the keying was repeated by another operator at a second machine, the checker, which would only proceed if the second keystroke corresponded exactly with the first one. Thus one had the moral certainty that the data had been correctly entered, and that subsequent operations would *necessarily* give exact results, generally at a much greater speed.

Other machines then came into play, requiring no intervention other than feeding in the cards, switching on, and some measure of supervision.

These other machines were

1 two basic machines: the *sorter* which filed the cards in the required order; and the *tabulator*, a truly automatic counting machine, which executed simple arithmetical operations on groups of cards (totalling up, for example) and which printed on paper, in their definitive form, the accounts and statements;

2 various auxiliary machines, including: the *duplicator* which reproduced the data from one series of cards onto a different series; the *summary perforator*, a machine connected to the tabulator which punched the totals calculated by the tabulator onto a fresh set of

cards; the *calculator* which performed all the required mathematical operations on each card; the *merger* which completed the functions of the sorter, and in particular permitted the rapid merging of two packs of cards previously sorted in the same order; and lastly the *translator* which printed out the translation of the perforations in readable form on the card itself. [J. Favier and R. Thomelin (1965)]

Briefly, and in broad outline, therefore, this summarises the general principles of the mechanographical method which was developed, particularly between the two World Wars and up till the 1950s, to exploit the flexibility of punched-card programming for all the needs of statistics, as well as those of general accountancy, third-party accountancy, analytical accountancy, industrial accountancy, inventory and stock control, payroll, etc.

And so, thanks to an evolution which brought about an ever growing automation, reliability, and speed in the machines, mechanography has played a notable role in technical innovation. Indeed the overall view shows, as R. Ligonnière suggests, that mechanography has

1 played a most important social role, both in employment and social custom, by opening up workshops and offices to women (just as typewriters and accounting machines did in the second half of the nineteenth century);

2 constituted (along, of course, with calculators, typewriters, key-operated adding machines, cash registers, accounting machines, telephone switchboards, telegraphy, etc.) a powerful economic lever, as evident from the rise of companies like IBM, Remington-Rand, and Bull, who have made it their spearhead, as also from the growth of companies like NCR (cash registers), Burroughs (accounting and calculating machines), Bell (telephones), Ferranti (electrical engineering), etc., who have in turn caught up with them;

3 opened the way to new relations between states, society and individuals, by making it possible to create large databases on different sectors of the population and thereby enabling various kinds of social action;

4 revealed to the world that a demand and therefore a market existed for artificial data processing;

5 led to the realisation, partly thanks to initiatives by members of the scientific community, but chiefly as a result of its intrinsic structural limitations, that by a radical change of course another route could be envisaged which had unlimited potential for the mechanical resolution of a very wide category of solvable problems.

The punched card itself thus became the first universal medium to carry data readable by artificial means: "Conceived as a medium for data treated by mechanical or electrical methods . . . ; triumphing until 1945, it then survived until the threshold of the 1960s, gradually losing ground to the ever closer alliance between electronic calculators and magnetic media in the form of tapes, disks, and diskettes." [R. Ligonnière (1987)]

STRUCTURAL LIMITATIONS OF MECHANOGRAPHICAL MACHINES

But on this stretch of road that was to lead to the computer serious blockages were encountered, of a conceptual nature, in that mechanographical machines executed sequential operations only semi-automatically.

From the early 1930s this difficulty was perceived by scientists such as the British astronomer Leslie John Comrie (1893–1950) or the American astronomer Wallace John Eckert (1902–1950). Confronted at the time by mathematical problems which required the execution of repetitive operations on large amounts of data, these scientists, inspired by the work of the Spanish engineer Leonardo Torres y Quevedo, worked round the obstacle by connecting several punched-card machines in series, taking the output of one as the input for the next.

In this way they realised by electromagnetic means the concept of Charles Babbage's Difference Engine, mentioned earlier (pp. 173ff.), creating what were called *difference tabulators*. The appliances born of this were the first electromagnetic systems capable of reading punched cards, of executing chained sequences of operations, and of printing out the results, with no human intervention.

In spite of this, the path remained blocked, for two reasons at least:

1 The operations were executed not in the order that the logic of the problem dictated but under an order constrained by purely technical problems;

2 Except when they malfunctioned, these machines could not of themselves deviate from a fixed sequence of operations once they were set in motion; in other words, they were incapable of locating a given instruction in the programme so as to modify the sequence of operations by themselves.

So a complete change of direction was needed, going back towards Babbage, not towards his Difference Engine but rather towards his most ambitious ideas of where a solution was to be found. But it would be necessary to develop these ideas in accordance with a highly elaborate theory of abstract reasoning, starting from all that had been acquired in mathematics and symbolic logic, and benefiting at the same time from the entirety of progress in physics, in the science of artificial automata, and technology, in order finally to overcome the handicaps and see the gradual emergence of the concept of our present-day computers.

9. Charles Babbage, his Analytical Engine, and his Followers

In fact Hollerith was not the first to apply the system of programming using punched cards in a digital machine. The English mathematician Charles Babbage had done it far more successfully half a century earlier. This point is vital if we wish to emphasise the fact that this scientist was in many ways the grandfather of the modern computer.

Going much further than his "difference engine", this one was designed to enlarge considerably the scope of artificial calculation. Until his death, Babbage devoted himself body and soul to perfecting an extremely complex calculating machine which he called his Analytical Engine. To form an idea of the strong impression that this machine made on the inventor's contemporaries and immediate successors, it is sufficient to cite Maurice d'Ocagne's account of 1905:

> With the machine we are now going to discuss, we seem to be entering into a magical realm. In the mind of the inventor, it was intended to carry out any series of arithmetical operations on any numbers, in as great a quantity as desired, and to provide a printed solution accompanied by an algebraic explanation of the operations that had been carried out.
>
> At first sight, the mind retreats in horror at the very terms of the problem and dares not even entertain the possibility of there being a solution. It was nevertheless onto looking for just such a solution that Charles Babbage began to concentrate all his efforts in 1834 and, a few years later, he managed to overcome at least theoretically all of the difficulties raised by the question.

We should not smile at this. For that time, the mere idea of a machine of this sort was most impressive. Babbage's machine was designed theoretically between 1834 and 1836, and was equipped with a control unit with moveable sprockets on a cylinder that could be modified, and with 1,000 registers of 50 digits. It was structurally conceived to carry out automatically sequences of linked operations, of any arithmetical or algebraic variety, simultaneously on its one thousand 50-digit numbers. The power of the Analytical Engine's mechanism was thus supposed to cover the entire range of numerical and algebraic relations.

Babbage expressed his own astonishment at his discoveries and innovations in a letter that he sent in May 1835 to his friend, the statistician A. Quetelet: "For the last six months, I have been working on a machine that is more powerful than the first [the difference engine]. I am myself surprised at the power that can be conferred on this new mechanism. Just one year ago, I should never have imagined that such a result was possible."

But, apart from a few minor modifications, the general architecture of the Analytical Engine was fixed for good some five years later. To cite Maurice d'Ocagne again:

> As of 1842, General Menabrea, then a captain in the Piedmontese engineers, was able to give an initial description of the machine, based on Babbage's explanations, in the *Bibliothèque universelle de Genève* (XLI, p. 352). A curious fact should be noted, which General Menabrea mentioned to the Academy of Sciences in Paris when he presented the machine to them in 1884 (see CRAS, 2nd week 1884, p.179). His paper had been translated into English for *Taylor's Scientific Memoirs* (see vol. III, London, 1843). The translator, who had signed only with the initials L.A.L., had added notes that displayed a remarkable insight and great mathematical culture, and at the same time a profound knowledge of the subject. Intrigued, Captain Menabrea asked Babbage to reveal the secret behind the initials.
>
> How surprised he was when Babbage gave him the name of Lady Ada Lovelace, Lord Byron's only daughter. It would be hard to imagine a greater contrast between the poet and his daughter, who had applied herself to the exact and extremely arduous study of calculating machines.

This is indeed all the more strange when, through the magic of words and metaphors, this lady mathematician transformed the driest mathematical formulae and the most off-putting technical descriptions into the most exquisite algorithmic poems. As she put it in her text: "We can say that the Analytical Engine will weave algebraic patterns, just as Jacquard looms weave flowers and leaves".

Ada Augusta Byron, Countess Lovelace (1815–1852), was also the first weaver of coded instructions on punched cards. She had devised a certain number of programmes with the idea of one day introducing them into the machine of her friend and master. They were based on a language that was compatible with the Analytical Engine, which she had adapted from the very system of punched cards that had been invented by Basile Bouchon, and subsequently perfected by Joseph-Marie Jacquard.

"She was a woman of some ability, and she had a clear grasp of her subject. If the Analytical Engine had been built, she would have been perfectly qualified to be its chief mathematician." [M. V. Wilkes (1956) pp. 10–11] But figures of rhetoric were not enough and, with her great precision and extraordinary insight, Ada Lovelace brought up the epistemological problems that were to arise: "The Analytical Engine has no claim to originate anything. It can carry out anything it is told to do. It can follow an analysis, but it is incapable of imagining analytical relations or truths. Its role is to help us carry out that which we already master."

It must be said that, in its inventor's own view and according to his own terminology, Babbage's machine was equipped with:

1 an *input/output* unit;

2 a unit for setting the machine in motion (for which Babbage did not coin a term), which transferred the numbers from one section to another in order to place them in the correct sequence: it was the machine's *control unit*;

3 a store, which was a numerical memory capable of storing the intermediate or final results of the calculations that had been carried out: it was the machine's *memory*, able to receive the numbers used in the calculations and store the results;

4 a *mill*, which was designed to carry out the operations on the numbers that had been introduced into the Analytical Engine: this was the machine's *arithmetic unit*, in which numbers were combined according to the required rules – in other words it was the *processing unit* whose job it was to carry out the calculations by employing the data that had been introduced into the machine and transforming it in order to produce the desired results;

5 finally, a *printing device* to provide the results.

Thus Babbage's machine was, at least on paper, a true predecessor of modern computers.

In the history of automatic and sequential digital calculation, the great synthesis had thus been achieved by this inventor of genius, who succeeded in applying not only the progress that had already been made in the mechanisation of arithmetical calculation (contributions of Leibniz, Leupold, Stanhope, Thomas of Colmar, etc.), but also the discoveries of the great constructors of mechanical automata during the Enlightenment (Von Knaus, Vaucanson, Jaquet-Droz, Kintzig, etc.), as well as the conceptions of Joseph-Marie Jacquard, the period's greatest programming expert.

Approximately one century ahead of his time, and at a period when such conceptions were difficult to understand, Babbage thus clearly defined the main units and functions of analytical calculators (see pp. 201ff.).

Since Babbage's machine required no human intervention in the carrying-out of its sequences of operations, it thus also synthesised the concept of an automatic sequential digital calculator with a non-cyclical automaton governed by a flexible programming system and equipped with a modifiable control unit, independent of the material structure of the corresponding internal mechanisms.

Even more importantly, Babbage defined, for the first time in history, a true precursor of today's universal computers: multi-purpose analytical machines that are not specialised for solving only certain categories of problems, but are conceived to deal with a vast range of computable problems.

However, we must now examine the fundamental difference between Babbage's machine and modern computers. To return to the Analytical Engine, M. V. Wilkes pointed out (*loc. cit.* pp. 6ff.) that:

> For the store, Babbage proposed to use columns of wheels, each wheel being capable of resting in any one of ten positions and, therefore, of storing a single decimal digit. When numbers were to be transferred from the store to another part of the machine, a set of racks would be engaged with teeth cut in the wheels, and each rack would be driven until the wheel concerned reached its zero position. The motion would be transmitted by rods and links to the arithmetic unit, where, by means of another rack, it would be used to move one of the wheels of a register to the correct position. Babbage gave a lot of thought to the design of the arithmetic unit, particularly to the problem of how to effect the carry-over, or tens-transmission, as rapidly as possible . . .

> For the control, Babbage proposed to adapt the mechanism used in a Jacquard loom for the production of fabrics having complicated patterns . . . Before each passage of the shuttle a group of warp threads, selected according to the requirements of the pattern, must be lifted so that the shuttle may pass underneath. The various threads pass through wire eyes (healds) which are connected by strings to a set of hooks arranged in a framework, several threads being attached to each hook; any particular group of warp threads, therefore, can be lifted by lifting the corresponding hook. The lifting of the hooks is accomplished by means of a set of horizontal bars which rise just before the shuttle passes.

> The hooks which are to be lifted during any particular passage of the shuttle are specified by holes punched in a set of cards, one card to each passage. The cards are strung together with loops of string so as to form a continuous chain, and pass over a cylinder having a square cross-section. Just before the shuttle passes, the cylinder moves bodily to one side so that one of the cards is pressed against a matrix of horizontal rods, there being one rod associated with each hook. The faces of the cylinder have clearance holes drilled in them so that only those rods which come into contact with the card are moved, all those rods which are opposite holes punched in the card remaining at rest. The rods which move engage with the corresponding hooks and move them out of range of the rising bar. After the passage of the shuttle, the cylinder returns to its normal position and rotates so that a new card is presented to the matrix of rods. [. . .] (Wilkes, *loc. cit.* pp. 6–8).

One of the most striking features of Babbage's Analytical Engine [Wilkes subsequently continues] is the way in which what we should now call conditional operations were to be handled. Babbage introduces this subject by saying that one of the Italian mathematicians, Professor Mosotti, came to him one day with the following difficulty: "He remarked that he was now quite ready to admit the power of mechanism over numerical, and even over algebraical, relations to any extent. But he added that he had no conception how the machine could perform the act of judgment sometimes required during an analytical inquiry [i.e. numerical calculation, M.V.W.], when two or more different courses presented themselves, especially as the proper course to be adopted could not be known in many cases until all the previous portion had been gone through. I then inquired whether the solution of a numerical equation of any degree by the usual, but very tedious, proceeding of approximation would be a type of the difficulty to be explained. He at once admitted that it would be a very eminent one."

Babbage had in mind a method similar to the one now known as Horner's method. He showed that the decision as to which of two courses of action is to be taken depends on whether some calculated quantity comes out to be positive or negative. If it comes out negative it will be because a large number has been subtracted from a small one, and he considers, by way of example, what happens in a machine with a capacity for 20 decimals if 511 is subtracted from 423, thus:

$$00000 \ 00000 \ 00000 \ 00423$$
$$00000 \ 00000 \ 00000 \ 00511$$
$$\overline{99999 \ 99999 \ 99999 \ 99912}$$

The action of the tens-transmission causes 9s to appear successively in the decimal places to the left of those containing significant digits. The 9s are, as it were, propagated to the extreme left of the register and would go further if the register were extended. Babbage points out that the motion of the carrying lever which would cause a 9 to appear in the 21st place, if this place existed, may be used to set any desired action in train. Since this lever only moves if the result is negative, whether the action takes place or not is conditional on the sign of the number calculated. In his choice of the action to be initiated by a conditional operation Babbage again showed his characteristic soundness of judgment. He proposed that it should be used to advance or roll back the cards on the Jacquard mechanism to any specified extent (specified presumably by a number in a register or punched on one of the variable cards). If the cards are advanced, a section of the programme will be skipped; if they are rolled back, a section of it will be repeated.

In this way, a cycle of operations can be repeated until the answer at the end changes sign, when another cycle of operations can begin.

In the example given by Babbage the number whose sign changes is one which occurs naturally in the calculation. Lady Lovelace quotes a more complicated example in which the number which changes sign is one introduced specially for the purpose. The example is the calculation of Bernoulli numbers from a recurrence formula – a very ambitious piece of programming indeed. In the calculation of the nth Bernoulli number B_n a certain cycle of operations has to be performed $(n + 1)$ times. It is arranged that at the beginning of the calculation of B_n a certain storage register contains the number n, and operations are included in the programme which cause unity to be subtracted from the number in this register each time the cycle is performed. [. . .] A conditional operation, therefore, inserted at the end of the cycle, will cause the machine to go ahead with the next part of the calculation after $(n + 1)$ repetitions of the cycle, instead of repeating the cycle again. The allocation of a storage register to hold a number which is used solely for counting is a prominent feature of the technique used in preparing programmes for modern automatic calculating machines. (Wilkes, *loc. cit.* pp. 11–13).

In other words, Babbage thought that his machine should have the possibilities of what is now called *conditional branching*: a given programme can be put into action at certain moments according to a defined procedure (according to the value of a parameter, for example), the machine thus being able to carry out a repetitive calculation automatically without having to reprogramme the basic operation each time.

Another possibility that Babbage's machine had was to branch off to a short sequence of ancillary instructions, which it would carry out before returning to the main programme. This is what is now called the use of subroutines.

It is now easy to see just how audacious were Babbage's advances and concepts for the period in which he lived.

WHY DID BABBAGE NOT DISCOVER THE COMPUTER?

Apart from the question as to how Babbage arrived at such a powerful synthesis and at ideas that were quite so revolutionary (answered on pp. 245ff.), what we can now wonder is, if he reached such a degree of abstraction and generality, what stopped him from discovering the modern computer? For, even though this great genius almost found the answer, he did not in fact invent the computer. There are at least four reasons for this:

1 Despite the fact that the state of technology and knowledge

associated with machine tools was well established and in full development, they were not yet sufficiently tried and tested in Babbage's day to come up to the height of his conceptions;

2 Babbage never really managed to disassociate the description of his machine from its own mathematical structure – in other words, he never worked on a clear distinction between the physical form of this machine and the corresponding structural research, which should have been carried out using a formalised hypothetico-deductive theory (see pp. 227ff.);

3 Mathematics and logic were not yet sufficiently well developed to allow him to conceive of the essential notion of a programme symbolically expressing an algorithm to solve a class of problems (see pp. 244ff.);

4 Finally, from a purely technical point of view, the memory store of his Analytical Engine was far too limited to allow the physical recording of all of the instructions in a given programme (the idea of a stored programme was in fact of little use in a machine such as his, which had a purely mechanical, and hence extremely slow, internal memory (see pp. 302ff.).

BABBAGE'S ENGINE: A TRUE CHALLENGE FOR TECHNICAL KNOW-HOW

That said, if we return to the concretisation of the ideas contained in the Analytical Engine, it is easy to imagine the technical problems that Babbage faced. In 1914, the Spanish engineer Leonardo Torres y Quevedo (1852–1936), who had always openly appreciated his illustrious predecessor's work at its true value, paid homage to him in the following terms:

The large number of mechanisms to be considered, the multiple connections that it was necessary to make between them, the need to have apparatus that could at any moment alter these connections, the difficulty in combining everything without the different mechanisms disturbing one another, and without friction impeding the movements, plus many other practical difficulties that we could also cite, made the problem almost unsolvable.

It was necessary to have Babbage's mechanical genius to deal with it and, even though he lavished his intelligence on the solution, as well as his and his country's money, the results he obtained were not encouraging.

When he undertook the Analytical Engine project, Babbage's theoretical and practical experience was exceptional ... In order to carry out his scheme, he devised a system of mechanical notations that represented a huge amount of work.

He studied a large number of solutions. This project was in fact his life's work, and he laboured on it constantly for thirty years. Yet, despite his uncontested and uncontestable abilities, despite his intelligence, his commitment and his perseverance, he failed. His plans and models are conserved in the museum in Kensington, but it is to be feared that they will be of no use to anybody.

However, as Maurice d'Ocagne says:

With the subsidies he received thanks to the generosity of Queen Victoria, which he swelled by using his personal wealth, Babbage managed to produce some of the parts that were to make up his machine.

Unfortunately, death stopped him in his tracks when he had hardly begun to assemble the parts, and they remained scattered in the hands of his son, a British army general, who donated them to the South Kensington Museum [now the Science Museum], where they can still be seen.

What is more, General [Henry P.] Babbage assembled a large number of documents concerning his father's projected machine.

Will there one day be an especially brilliant mechanician who with the help of this description will complete Babbage's interrupted work?

This question was asked at the British Association for the Advancement of the Sciences, which entrusted a study to a commission made up of the best qualified savants [they were, according to a report of the time: Cayley, Farr, Pole, Glaisher, Fuller, Kennedy, Clifford, and Merrifield].

This commission reported that the section that had already been assembled was only a small part of the mill, yet it was enough for mechanical addition and subtraction to be executed using Babbage's procedure. The unassembled parts included the handles and gunmetal wheels, mounted on steel axles, and a greater number of wheels made of a tin and zinc alloy, produced in moulds that still existed. Finally, that there was a large quantity of drawings that clearly demonstrated the essential movements of the parts of the machine, but which needed to be completed with construction drawings.

However, the commission decided that what still needed to be done would require so much time and money (some £25,000 at least [at the time, of course]) that the Association was incapable of carrying out Babbage's scheme.

We should add that the conclusion of the experts was that the Analytical Engine was "an extraordinary monument to inventive genius, but it would forever remain a simple theoretical possibility." "Can we not hope,"

M. d'Ocagne complained (writing in 1905), "that the principles concocted by this brilliant English mechanician may one day receive a less broad yet still interesting application, such as one that would allow the calculation of those precious determinants in the resolution of systems of first-order equations with several unknowns?"

Today, this question seems so elementary that it might bring a smile to our lips. However, we should repress it, for Maurice d'Ocagne knew what he was talking about, and, as a great specialist in "calculation simplified by mechanical and graphical procedures", he had studied every possible aspect of the subject as it then existed. And, at that time, calculating a simple determinant appeared so difficult that mechanising it, let alone automating it, looked like an insane undertaking.

At the end of the twentieth century, the question that was then raised concerning the possible construction of a model of Babbage's machine obviously has lost its point. Unless it were an historical one, for the concept has since not only been put into action by mechanical, electromechanical and electronic means, but also superseded by the Analytical Engine's computing offspring.

HOMAGE TO A GREAT SAVANT, WHO DEVOTED HIS LIFE TO THE DEVELOPMENT OF SCIENCE

Nevertheless, with hindsight, we can now see the absolutely vital role that Charles Babbage played in the history of computing.

This quite exceptional genius had a sad fate. He devoted his entire life to making a physical reality of a revolutionary structure produced by his own imagination, but without success. Less than a century later, this structure was to plant seeds in other minds and prove exceptionally fertile. However, near the end of his life, he prophetically declared: "If I could live just a few more years, the Analytical Engine would exist and its example spread across the entire planet." Then he added, even more pathetically, to those who would later take up the cudgels: "If any man, who is not rebuffed by my example, one day produces a machine containing within it all of the principles of mathematical analysis, then I have no fear for my memory, for he alone will be able to appreciate fully the nature of my efforts and the value of the results I have obtained."

We could surely not pay better homage to that extraordinary inventor than by reproducing the words of his testament.

THE FIRST PERSON TO CONTINUE
BABBAGE'S WORK

After the inventor's death, Henry P. Babbage tried to continue his father's work. The project started in 1880 and ended in 1910 with the assemblage of part of the Engine, the processing unit or "mill" and the printer. But then, because of a lack of financial means and technical know-how in the field of mechanics, work was stopped for good.[63]

PERCY LUDGATE: THE SECOND MAN TO
CONTINUE BABBAGE'S WORK

As R. Ligonnière (1987) explains, Babbage's ideas were nevertheless not abandoned:

> About sixty years after Babbage had established his plans in 1840, and at the same time that his son Henry was abandoning the work, an Irishman announced a new project for an analytical machine. Percy E. Ludgate (1883–1922), from Dublin, was more of an accountant than a mathematician, although he did have a solid mathematical background. Was it his education, or his natural leanings that gave him such pragmatism and sense of proportion? By exploiting the progress that had been made in calculators over the previous fifty years, Ludgate also showed that one can happily combine imagination with reason. He worked for six years on his projected calculating machine. At the beginning, he was unaware of Babbage's concepts. When he heard about them, he carefully studied the available texts, recognised his predecessor's genius, but set about proposing a less ambitious and more mechanically advanced solution. Ludgate could work on his project only in his spare time, yet in February 1909 he published a fourteen-page paper explaining his ideas. Unfortunately, the detailed plans he drew up have been lost. Here are his project's main characteristics:
> - replacement of Babbage's various punched cards and readers by just one roll of perforated paper containing the instructions and elements of the problem to be solved. This would mean that each sort of programme would appear on a different roll of paper;
> - giving the machine two numeric keyboards with ten keys. The first

---

63 On the second centenary of Charles Babbage's birth, in 1991, the Science Museum in London did, however, build a complete working model of the Analytical Engine, relying entirely on mechanics, following the inventor's drawings and making use only of such parts as would have been available to him in his day. The engine stands six feet high and can be watched reliably performing various mathematical operations – which lays to rest the charge that nineteenth-century mechanical technology (to which Babbage himself had made a notable contribution) was not up to it.

keyboard would be used to enter data or instructions into the machine, without having first to make a strip of perforated paper, which would greatly increase flexibility for short, relatively simple calculations. The second keyboard would allow the operator to communicate with the machine's control unit. It prefigures the "control desks" or "consoles" which were to follow with the early relay machines thirty years later. When directed from the keyboard, the machine could punch a record of the operations onto a roll of paper so as to reproduce them on a later occasion;

- specialisation of routine ancillary operations – specialised sub-programmes would be given to particular cylinders, for example, a dividing cylinder, a logarithmic cylinder, and so on;
- simplification of arithmetical operations by means of rules and tables based on logarithms. This particularly applied to multiplication, which Ludgate intended to carry out directly, rather than by using repeated addition as Babbage had planned;
- delivery of results, either by printing digits on paper, or by punching a paper strip that could later be used in another programme.

These various ideas considerably simplified the Analytical Engine, as Babbage had imagined it. What is more, as Ludgate limited himself to a capacity of 192 variables with 20 digits, the result was a calculator that was theoretically transportable, thanks to its greatly reduced dimensions: 66 cm long, by 61 cm wide, by 51 cm high.

An electric motor was to give the main axis of the machine a speed of three rotations per second.

The performances that Ludgate predicted were 3 seconds for addition and subtraction, and 10 seconds for multiplying two 20-digit factors. Division, which was extremely slow and ran on an ancillary cylinder, could last up to 90 seconds.

The inventor had gone out of his way to allow his machine to be used by operators who, while they needed instruction, were not necessarily mathematicians. Use of the calculator would be facilitated by free access to a library of perforated strips, filed and identified according to their functions.

In 1914, Ludgate contributed a second paper to a collective work published to celebrate the tricentenary of John Napier's invention of logarithms. The author made a brief but enthusiastic summary of Babbage's work before mentioning his own research, pointing out that he had not yet been able to start constructing the machine . . .

However, he did still define important elements of technical progress and had produced an intelligent compromise between mathematical constraints and mechanical complexity. If his analytical machine had been built in the 1920s, the technical conception of modern computers would probably have been found more rapidly as a result.

TORRES Y QUEVEDO: THE GREAT FOLLOWER
OF BABBAGE'S WORK

Thus, Babbage's invention long remained a challenge for technology, and some even considered it to be a pure piece of fantasy dreamt up by an eccentric.

This attitude in fact persisted until 1914, when Torres y Quevedo started studying the problem and came up with a few answers.

But before his breakthrough, he too had had a rather pessimistic outlook (see p. 195): "Perhaps another will triumph one day where Babbage failed, but this does not seem easy, and it would to my thinking be foolhardy to follow in his footsteps until we are in possession of new mechanical principles that will make the favourable outcome of the journey seem probable."

However, he was later to declare: "The difficulties of a purely mechanical solution seemed to me to be unsurmountable, at least with the means at my disposal. To be more precise, while I had been considering just a mechanical solution to the problem of calculators, I shared the general view. I thought it impossible to solve the problem of mechanical calculation – with all the generality that presupposes – in the terms in which I have here examined it. But it was studying how to generate movement at a distance which put me on the track of these new machines and studies of automation."

Thus it was that Torres, by establishing the advantages of electromagnetic relays, showed the practical superiority of machines based on electromechanical technology.

TORRES'S ELECTROMECHANICAL DEMONSTRATION

He used some of his predecessor's mechanical procedures, which he combined with electrical resources, especially electromagnets, along with a set of detailed plans for an appropriate set of components which could carry out elementary arithmetic, thus showing that the Analytical Engine could be constructed using electromechanical means.

In 1953, his son Gonzalo wrote:

> My father did not construct an analytical machine, but he established its principles – which are similar in certain ways to modern machines [he is referring to the early electronic computers that had

just been produced] – in the paper referred to [Torres y Quevedo (1953)] and he constructed a demonstration model ... which calculated the value of the formula $a = (p \times q) - B$ for small numbers.

It was an elementary analytical machine and was exhibited in Paris in 1914. It had a tiny memory, not at all comparable to the ones being discussed in this symposium. He also constructed an absolutely automatic electromechanical arithmometer ... made up of a typewriter and various other equipment: a recorder, totaliser, multiplier, comparator and co-ordinator. A number was typed onto the keyboard, for example 12,347, which was the dividend of a division. Then the division sign (a colon) was entered, followed by the divisor, for example 60. The machine then began to work and when it had finished, it automatically wrote the "equals" sign followed by the result, 205 remainder 47.

The machine had a unit which at each moment compared the divisor with what remained of the dividend and, when the divisor became smaller than the dividend, it stopped removing the divisor and the carriage moved on one position.

However, despite the essential contributions of this brilliant successor of Babbage, the Analytical Engine was not truly completed in the 1920s. It was necessary to wait until the Second World War for the need for such a calculator to be felt once more. It was at that moment, because of significant technological progress, linked to far more significant advances in symbolic logic and mathematics, that Babbage's dream at last came true in a far more conducive environment.

10. Developments in Electromechanical Calculating Machines

FROM BABBAGE'S MACHINE TO THE DEVELOPMENT OF ANALYTICAL CALCULATION

Thus it was that, during the 1930s and 1940s, the need to perform large calculations on a machine became urgent. In other words, because of the striking failings of classical numeric calculators, the structural limitations of the analogue method, and the fact that the mechanographic processing of information had been blocked by increasingly complex scientific problems, the need for more suitable machines was increasingly felt in the interwar years. Accordingly, in the 1930s, certain researchers began to re-explore the paths Charles Babbage had traced out one century before: how to produce a real analytical calculator.

WHAT IS AN ANALYTICAL CALCULATOR?

Before proceeding, we should first clarify what these terms mean.

It will be noticed that this expression is as old as Babbage's engine itself. And yet, it still appears new because historians of the computer have never made proper use of it. The word "analytical" derives from the Greek *analusis*, from the verb *analuein*, which means "to untie", or "release" a part from the whole, including the most elementary parts. It is thus identical with the analytic method in mathematics, which consists in "reducing a given problem to a second problem, the second to a third, and so on, until we arrive at a problem we can solve" (J. M. Duhamel, *Méthodes dans les sciences de raisonnement*, I, p. 51).

Now, this is the precise meaning of numerical analysis, which is a basic part of automatic digital calculation. Its aim is to study practical ways to solve mathematical problems numerically and to refine them by means of appropriate techniques. Thus, in Condillac's words: "to analyse is to decompose, to compare and to understand relations . . . Analysis is thus the total decomposition of an object, and the distribution of the parts in an order such that generation becomes easy" ("Art de Penser", *Oeuvres philosophiques*, I, p. 410). This method is therefore the opposite of the synthetic method, which consists in deducing from a problem which we can solve the solution of the preceding one, which we then relate to the original problem, and then the solution of the one preceding this, and so on until we find the solution of the original problem itself (see J. M. Duhamel, *op. cit.*).

Now, *analytical* was precisely the term Babbage chose for his engine, which was thus the first analytical calculator to be conceived by a human mind. And, after Babbage, Torres y Quevedo continued to use the same term in his *Essays for the Automatic*, when studying "the general principles applied to analytical engines". In other words, even if Babbage did not exactly express it in these terms, he had invented what may be called *analytical calculation*. This consists in:

1 a preliminary investigation of any given problem of a solvable category to reduce it to a second problem, then to a third, and so on, in order to determine the mathematical rules and procedures needed to solve it;

2 the application of these rules and procedures by means of a programme of instructions in an appropriate machine;

3 finally, the problem is solved by the machine itself which, controlled by the programme, automatically carries out sequences of instructions, rigorously in the order of the various elementary steps defined by the procedure.

In terms of automata, the general definition of calculators based on this technique is as follows: an analytical calculator is a digital, sequential and programmable automaton:

- which is governed by a flexible programming system;
- which is equipped with a modifiable control unit, independent of the material structure of the machine's internal apparatus;
- and which necessitates no human intervention in order to carry out completely the sequences of programmed instructions to be executed.

This general definition (in modern terms) indeed corresponds to Babbage's machine. It also allows us to isolate the true concept, as a sort of algebraic structure independent of whatever technology is being used. Since it also clearly displays the potency of Charles Babbage's synthesis, it will also now be easier to see the difference between this machine and the belfry automata, astronomical clocks, automatic chiming bells or Vaucanson's and Jaquet-Droz's androids. What is more, it allows us to have a better understanding of the basic difference between the Analytical Engine and rigid, inflexible calculators in the classic categories of accounting machines, mechanographical and analogue calculators or those related to the difference engine (see pp. 169ff.).

THE INCONTESTABLE INFLUENCE OF BABBAGE ON RESEARCH BETWEEN THE WARS

To return to the history of analytical calculation, even though people have sometimes abbreviated it (to magnify their own glory or else, in a not always disinterested manner, to highlight a particular local or national contribution), it must be said that all of the research carried out at that time was inevitably influenced in one way or another by Charles Babbage's work. Ideas such as his are certainly not found overnight. It can be observed that most of the treatises and articles dealing with digital calculation by means of artificial or mechanical means that were published in the nineteenth century and the early twentieth century contain the very ideas that constituted Babbage's Analytical Engine. This applies equally to instruction manuals for researchers, or sources of information or inspiration concerning theories of automatic calculation that had already been more or less developed.

Among such publications, we can cite:

- the work of Maurice d'Ocagne (previously extremely well known, and republished several times after 1893);
- that of L. Jacob published in 1911;
- the important article *Numerisches Rechnen* published by R. Mehmke

of Stuttgart in 1902 in the *Encyklopädie der mathematischen Wissenschaften*, then translated into French by M. d'Ocagne in 1909 in volume one of the *Encyclopédie des sciences mathématiques*;

- the equally well known publications of Leonardo Torres y Quevedo, which appeared in 1913 and 1920, not only in the periodical of the Real Academia de Ciencias Exactas, Físicas y Naturales of Madrid, but also in the *Bulletin de la société d'encouragement pour l'industrie nationale* in Paris;

- the collective work, *Modern Instruments and Methods of Calculation*, edited by E. M. Horsburgh and simultaneously published in Edinburgh and London in 1914, to celebrate the tricentenary of Napier's discoveries, which was also extremely well known and dealt thoroughly with the subject including several descriptions of Babbage's Analytical Engine;

- all of the articles on this subject published in the *Comptes rendus* of the Academy of Sciences in Paris;

- all articles published on this subject in R. Taylor's *Scientific Memoirs;*

- without of course forgetting Babbage's own publications, and L. F. Menabrea's paper that was translated into English by Ada Lovelace.

Even Edgar Allan Poe (1809–1849) alluded to the impact of this subject in his *Tales of the Grotesque and Arabesque*, which he published in 1840. As for the English logician, philosopher and economist W. S. Jevons (1835–1882), who produced the first mechanical device for logical inference, his praise of this invention certainly did not pass by unnoticed: "It is to Babbage's profound genius that we owe the greatest progress in mechanical calculation, since the principles of the calculation of differences were embodied in his machine ... In his project, the Analytical Engine, Babbage then showed that a mechanical machine is, at least in theory, capable of competing with the work of even the most practised mathematicians in all the branches of their science. The mind seems thus able to imbue matter with some of its highest attributes, and to create its own rival in the wheels and levers of an insensible machine." (*Philosophical Transactions*, 160/1870, pp. 497–518.)

There was no more room for doubt. Researchers in Germany, France, Great Britain, America, and elsewhere who tried during the 1930s to produce an analytical calculator were all directly or indirectly aware of Babbage's work.

COUFFIGNAL: A SUCCESSOR TO BABBAGE
AND TORRES Y QUEVEDO

First among these researchers was a French engineer, as is shown in an interesting article published by M. d'Ocagne in the *Larousse mensuel illustré* in June 1937, and which begins with a summary of the situation which is now familiar to us:

> Torres, who was haunted by the failure of such an attractive project, started working on the problem himself, and found an electromechanical solution to it, which he was not to have time to complete. He nevertheless constructed a partial model, which we were able to see in 1920 in Paris, at the exhibition organised to celebrate the centenary of Thomas of Colmar's invention of the arithmometer. Since then, a purely mechanical solution to Babbage's problem, simpler than the inventor's, has been found by Couffignal, who is today a professor at the Brest school for mechanical engineers. It is now in the process of being constructed.

Louis Couffignal (1902–1966), a disciple of Maurice d'Ocagne, and obviously influenced by the work of Babbage and Torres y Quevedo, was indeed one of the first scientists to study this problem attentively in the interwar years. He began work in 1930, and in an initial paper published that year there is already mention of the general problem of a "universal machine" with a mechanical sequencing of calculations.

But his research took a decisive turn when, in 1933, he published his first important paper on "calculating machines, their principle and development", in which he carried out a detailed review and even an original classification of all the past or present calculators. This work then led, in 1937, to a complete theoretical solution of Babbage's problem, including definite improvements on the original concepts, thus making Couffignal a true innovator in this field.

Subsequently, in his PhD thesis, presented in 1938, he clearly defined the characteristic functions of the necessary mechanisms and even had the idea (already foreshadowed in his 1936 publication) of using binary numbers. That very year, he produced plans for the construction of an analytical calculator, based on electromechanical technology, and using the binary system instead of the decimal one (see pp. 86ff.).

In 1939, the Ministry of Defence asked the Centre National de la Recherche Scientifique to set up a mechanical calculation laboratory in order to study and construct calculators. Louis Couffignal was put in charge. But circumstances soon became unfavourable, and during the war he was cut off from any communication with similar laboratories among the Allies.

To sum up, Couffignal was on the point of making Babbage's dream come true when the Second World War suddenly broke out, France was occupied by the Nazis, and his projects were put on hold. However, after the war, in 1950, Couffignal did manage to produce a machine of the same sort for the Institut Blaise-Pascal, with the technical help of the Logabax company. It was a multi-purpose binary analytical calculator, but now based on electronic technology. Despite its impressive nature, and the success he enjoyed at the Cybernetic Congress held in Paris that year, Couffignal's calculator was no longer really as spectacular as similar machines had been just before and after the war. The concept had been overtaken by the first true computers, which had started to appear a few years before.

ZUSE: THE GERMAN PIONEER OF
ANALYTICAL CALCULATION

Another computer pioneer was the German engineer Konrad Zuse (b. 1910). From 1936 to 1939, he set about planning and constructing binary mechanical analytical calculators. His research then led to the successive construction of two calculators, Z1 and Z2 (from the initial of his surname), designed to solve algebraic equations. But these two machines remained at an experimental stage.

In 1941, Zuse actually constructed a calculator of a similar type, though with electromechanical equipment, based on electromagnetic relays and structurally designed for a broader range of applications. This was the Z3, which never worked properly and was unreliable. The Z4 was then constructed in 1944. This was also an electromechanical calculator which was, at least in theory, capable of carrying out any sequence of operations. But, because of the war, these machines remained unknown outside Nazi Germany. What is more, the first three were destroyed by Allied bombing. As for the fourth, it never became truly operational because of the defeat of Germany in 1945.

However, in 1948, Zuse started to construct new machines, before setting up his own company, Zuse Kg, in 1950, which specialised in the construction of electronic computers. Hitler had apparently had no interest in Zuse's work, so his ideas remained officially unrecognised, at least during the war, and he probably received no state subsidies. He seems to have tried everything to attract the interest of German universities. But, according to him, such attempts were in vain, for nobody believed in the future of the sort of fundamental research he wanted to carry out.

Zuse tells how a constructor of calculators, to whom he had applied for a subsidy, declared: "that it was wonderful that a young man should spend so much time and effort developing new ideas in the field of calculation".

The industrialist concluded by "sending him his best wishes for the most favourable possible results in his research, but, unfortunately, every possible technique for making calculators had already been discovered and there were no new ideas to be found in this field." (As reported in Zuse (1970) cited by R. Ligonnière in *Ordinateurs*, 20 September 1982, p. 44.) With hindsight, we can see just how astute the man was and how full of common sense, even though he didn't realise it!

We shall probably never know whether the Nazis really understood what was at stake in terms of analytical calculation, nor whether they had any plans to build a computer. On the other hand, what we do know is how much in Germany's scientific life the Reichsführer and his abominable Gestapo destroyed and sterilised. This is, of course, extremely fortunate for us, otherwise we might still be living through the thousand barbaric years of the Reich! We can tremble at the thought of what might have happened if Zuse and his fellow computer enthusiasts had received the full support of their nation.

STIBITZ: THE AMERICAN PIONEER
OF ANALYTICAL CALCULATION

In the United States, the first real attempts in this field were made by George Robert Stibitz (b. 1904), a physicist and engineer attached to the Bell Telephone Laboratories in New Jersey. His early inspiration came from a field that was conceptually limited, but where technically no mistake could be tolerated – betting on horse races. As J. Tricot explains:

Historically, it was the legalisation of the tote in Great Britain in 1929, which led to the construction of the first truly reliable totalisator. It should be remembered that, as opposed to simple betting, the tote implies adding up the bets that are made on a particular racecourse, or even those made outside it in the nearby towns. It is thus necessary that the number of bets made on each horse be displayed in each place where bets can be made, and at the racecourse itself, otherwise there is a risk of fraud (such as in G. R. Hill's film *The Sting*, with Paul Newman and Robert Redford).

The telephone (with electromagnetic relays, like doorbells, before the arrival of electronic systems) was of course used, coupled to total-isators with steel ball-bearings that fell into glass tubes (a technique reminiscent of the one used in marshalling yards, where each ball-bearing stands for a wagon, which is still used in France because of its reliability). But in January 1930, the British company Thompson Houston started using a purely electric totalisator, without ball-bearings, which could be transported from one racecourse to another, and which was capable of recording up to 12,000 bets on six horses per minute! It had in fact been invented in the United States (in Baltimore in 1928), and was the first to be produced by the American Totalisator Co, which set up a working model on the famous Arlington Park racecourse in 1933.

These totalisators could not calculate very rapidly. But they were reliable, and the use of telephones gave some people ideas, especially a remarkable engineer in Bell Telephone Laboratories called George Stibitz. Thanks to telephone relays, he conceived [in 1937] the first binary circuits (0 = closed, 1 = open), which allowed multiplication, division, binary/decimal conversion in both directions,[64] and other more complicated operations to be carried out (using i, the square root of -1, or j as it is called by electricians to describe the behaviour of alternating currents). There was an immediate and logically useful application in the planning of direct telephone networks. (J. Tricot in SV,741/ June 1979, pp. 114ff.)

So it was that Stibitz was able to construct a binary totalisator, using telephone relays, which was in fact the first true electromechanical binary calculator. As Moreau puts it:

Stibitz built his totalisator at home one weekend, using a few cast-off relays, two light bulbs and the pieces of a tobacco pot. The relays were cabled together in such a way that the two light bulbs lit up if the result was 1, and remained off if the result was 0. This was during the summer of 1937. This first machine, that allowed binary-coded elements to be processed, thus inaugurated the era of [automatic digital data processing] machines whose arithmetical and logical units, as well as their memories, were constructed from elements that could adopt one or other of two stable physical states.

In his spare time, Stibitz also produced multipliers and dividers. He then declared to his employers that he was in the position to construct an [analytical] computer that would cost roughly $50,000 [or £30,000 in 1980]. He was told that nobody would be willing to spend such an amount just to do sums. But Stibitz managed to convince Bell Labs of the interest in producing a calculator capable of carrying out operations on complex numbers. An engineer called

---

64 As R. Moreau explains in *Ainsi naquit l'informatique*, Stibitz "picked up on the ideas of a great Irish logician, George Boole (1815–1864), who reduced logic to the simplest algebraic systems, known today as Boolean algebra. This algebra uses two digits – 1 designating all objects as a group, and 0 emptiness as an entity – and two operations: *or* and *and*, otherwise known as Boolean addition and multiplication. It enjoyed no success when it was first published, but assumed great practical interest when attempts were made to devise a calculator. If the data were encoded in the binary system, they may be represented by chains of open (0) or closed (1) contacts, and Boolean algebra may be used to make a model of the calculation. Whence the success of this type of coding." (See pp. 251ff.)

Sam B. Williams, who taught Stibitz about number theory, was then attached to him and both of them presented their Complex Calculator at the end of 1939. It was later called Bell Labs Relay Computer, Model I [or else the Bell Telephone Labs Computer, Model I], and it was certainly the world's first [analytical] binary computer. Data were entered by means of a teleprinter. The memory and calculation hardware consisted of 400 telephone relays. [R. Moreau (1987)]

COMPUTING CONCEPTS INAUGURATED
BY BELL TELEPHONE LABS

In this way, Stibitz and the Bell Labs inaugurated a set of concepts that were later to be used in computing. Here are a few (the Bell Telephone Labs first used their calculator for their own purposes):

Two other groups of Bell Labs, who also had big calculations to do, connected a teleprinter to the Complex Calculator. This was undoubtedly the first example of what is now called a data communication system. The machine was shared in the following way: the first teleprinter to ask for access was admitted, that is to say the Complex Calculator put itself at the disposal of the first teleprinter that contacted it. When it had finished working, the one of the other two teleprinters that had been waiting the longest could then start its calculation, and so on. (This is the system that is often called FIFO, or "first in first out".)

But things went even further. In September 1940, Stibitz connected a teleprinter from Dartmouth College in New Hampshire, where the American Mathematical Society was holding its congress,[65] to the Bell machine that was in New York, by using the telegraph network. The participants could send a calculation to the Complex Calculator and receive the answer in under a minute. It was certainly the birth of remote job entry and the beginning of *telematics*.

As of 1939 the extremely important role that telecommunications was later to play in the world of computers could already be felt. By 1942, Stibitz had also invented floating-point arithmetic, which allowed his machine to divide very large numbers by 10, by 100, by 1,000, etc. and then to remember the necessary operations to indicate the correct result [see Chapter 27, previous volume]. It is used by

---

65 Two leading figures in the history of information technology attended this congress: John W. Mauchly and Norbert Wiener. The former was shortly to become one of the begetters of ENIAC, the first entirely electronic multi-purpose analytical calculator, and was a little later to be one of the originators of the UNIVAC I, the first commercial computer in history; and the latter, whose book, published in 1948, was to play a key role in the "entire field of the theory of command and communication, in machines and animals", was to coin the word *cybernetics* in its current meaning, as defined in the book's title.

practically all modern computers." [R. Moreau (1987)]

Thus George Stibitz, the great American pioneer of computing, became one of the first people in history to produce a physical machine that was operational and reliable, and structurally conceived as an analytical calculator, even though it could only be put to very specific uses (see definition, p. 202).

However, the possibilities of his machine were extremely limited. Because of the special nature of its calculation structure, it could carry out only the four basic operations on complex numbers.

Thus Stibitz [who was then working on essential questions such as modifying programmes according to predetermined sequences, and the detection of mistakes in the calculations, and thus received financing because of the war effort] produced a new computer [structurally conceived to carry out more varied tasks] that could "at least evaluate polynomial expressions" by carrying out instructions that were communicated to it by punched paper strips taken from a teleprinter. This was the Bell Labs Relay Computer Model II, also called the Relay Interpolator, which was built and put into operation in 1942 [and continued to be used until 1961]. To carry out its calculations, the Model II had a very limited form of conditional instruction, which means that it cannot be considered to be a multi-purpose analytical computer. The same goes for its successor, the Bell Labs Relay Computer Model III [also known as the Ballistic Computer, which was begun in 1942 and completed in 1944, and was designed for fire control in gunnery. Capable of 100 per cent checking of its results, and of automatically consulting numerical tables pre-recorded on punched tapes, it was used by the US Navy's research laboratories until 1958]. The first real multi-purpose [analytical] computer in this series was in fact the Bell Labs Relay Computer Model V. It was produced in 1946, used about 9,000 relays, weighed approximately ten tonnes, and took up a surface area of 105 square metres. The addition of two 7-digit decimal numbers took 300 milliseconds, multiplying them one second and dividing them 2.2 seconds. [R. Moreau (1987)]

NEEDS CREATED BY THE SECOND WORLD WAR

Going back to the beginning of the 1940s, the development of analytical calculation only really got off the ground after war broke out, with the scientific and military needs thereby engendered. In particular there were:

1 the paramount importance and urgency of achieving effective crypto-graphic analysis of the secret communications used by the Germans and Japanese. It was essential to find a simple, reliable, fast, and

effective method of cracking the codes used in enemy communications, and the Allies could find no way of breaking the cipher by either an intuitive approach or non-automated analytical methods;

2 the need to perform more accurate calculations than were possible with analogue computers, for example to achieve far more accurate gun aiming (see pp. 157ff.);

3 an ever growing requirement to solve simulation problems;

4 finally, the pressing need at the time to resolve complex problems such as those associated with the use of radar to intercept enemy aircraft.

WHEN BABBAGE'S DREAM BECAME REALITY

Note to start with that, with the exception of the Bell Labs Relay Computer, Model V (completed in 1946), each machine that we have just mentioned was structurally conceived to carry out only one kind of calculation. Zuse's early calculators, for instance, – even supposing that they had ever worked correctly – were only capable of solving algebraic equations. And even Stibitz's first calculator could only handle arithmetic (albeit with complex numbers).

So, little by little, the question arose of designing a far more generalised machine, a machine capable of solving a vast range of problems, instead of building a separate machine for a specific task. This was precisely the question about universal calculation posed a century earlier by Charles Babbage. Asking this question again during the difficult days of World War II was what led to the conception and construction of multi-purpose analytical calculators, in other words machines structurally devised to allow a broad range of operations to be performed.

In 1937, Howard Hathaway Aiken (1900–1973), Professor of Physics at Harvard, was faced with a set of complex mathematical problems whose solution would have required an inordinate amount of time by human hand alone. The following year he initiated a project aimed at constructing a machine that could carry out ordered sequences of operations on mathematical problems of any kind. The machine was to run without any human intervention until it had worked through all its operating instructions. In the meantime, it seemed obvious to him that the artificial devices employed in the punched-card machines used in mechanographical calculations or statistics could be physically borrowed to build his machine. There is little doubt that Aiken's original inspiration for his computer came from Babbage (indeed Aiken himself referred to it in the historical introduction he drafted for the article he wrote to this effect on 4 November 1937; (see Aiken and Hopper).

Research, funded by the university itself, got under way. It very soon became apparent, however, that far greater investment was needed if the work were to come to fruition. It was then that Aiken made contact with Thomas Watson, President of IBM, who enthusiastically agreed to foot the bill for the project. Overnight Aiken acquired the skills, assistance, experience, and technical expertise of the Corporation's core business machine enterprise. In particular, he started working with three of its best engineers and technicians: C. D. Lake, F. E. Hamilton and B. M. Durface. So one fine day in January 1943 Babbage's dream finally came true in electromechanical form. What had emerged, in fact, was the machine that IBM called the ASCC (IBM Automatic Sequence Controlled Calculator).

The machine that Aiken designed, best known as the *Harvard Mark I*, was actually built in the IBM laboratories at Endicott. It remained there for some time while testing was carried out.

This, then, was the first multi-purpose analytical calculator in history that was entirely completed, operational and reliable. The calculator was subsequently dismantled, shipped to Cambridge (MA) and then reassembled in the basement of the Cruft laboratory at Harvard University. It became fully operational only in May 1944.

THE HARVARD-IBM MACHINE

The operation of the Harvard Mark I was essentially mechanical. It did, however, make use of electromagnetically-controlled relays and gears.

Its moving parts were driven and synchronised by gears connected to a drive mechanism system that ran for practically the entire length of the machine. The calculator was 16 metres long, 2.6 metres high and 0.6 metres deep. It weighed 5 tonnes and had about 750,000 different components, including 1,000 ball bearings and a little over 850 kilometres of electrical wiring with 175,000 connections and three million solder joints.

From one end to the other, this impressive machine was made up of:

1 two panels each fitted with several rotating switches (similar to the control knobs on electric cookers). Each switch had ten settings from 0 to 9. Numerical data and the constants of the problems to be processed were input manually. The entire system allowed 60 constants, each with 23 digits, to be input at once for use by the machine in the course of a calculation. Nonetheless, without human intervention the machine itself could not alter their values;

2 a series of panels on which were grouped 72 registers that provided number storage. Each register could hold a 23-digit number and was equipped with a mechanism for transferring 10s as well as a look-ahead carry system. This was the accumulator group for the

machine's *memory*. It also served as a buffer for intermediate results that would be used later on, and as stand-alone addition and subtraction machine;

3 a certain number of specially designed panels used to carry out multiplications and divisions;

4 a number of other panels dedicated to the calculation of sin x, log x and 10^x. This enabled the values of other trigonometric, logarithmic, and exponential functions to be calculated when required;

5 other panels bearing devices to read punched tapes carrying pre-calculated values of mathematical functions;

6 a *control unit* that read instruction lines from holes punched in 24 positions across the width of the tapes. These contained the programmes that the machine was to carry out. Guided by cylinders, the punched tapes were fed over a reading device with 24 pins which created electrical contacts through the holes. Depending on the arrangement of the holes punched in the tape, the system controlled the order of operations programmed for calculation. Each set of 24 holes on the punched tape could be taken as three groups of 8 holes, corresponding to an instruction such as, "Take the number on device I, then the corresponding number on device II and execute the operation using device III".

Machine input facilities comprised two punched-tape readers and a perforator. Output facilities consisted of two electric typewriters, driven by the control unit. These provided hard copy of the calculations performed.

It should also be mentioned that the calculator included a synchronising clock, beating every 3/200 of a second. For its day, the Mark I's performance was remarkable: 0.3 seconds for addition; approximately 6 seconds for multiplication; an average of 11.4 seconds for division; 1 minute for a sine calculation; 1.12 minutes to determine a decimal exponential; and 1.84 minutes to calculate a decimal logarithm.

Nonetheless, despite his knowledge of Babbage's work, Aiken did not include any provision for conditional branching. He was to remedy this later on by incorporating an auxiliary control unit into the machine.

One last note on this calculator. While the reliability of results was eminently superior to that of the ENIAC (see below), its speed was far slower than its electronic successor. With regard to breakdown rates, the technicians who personally worked on the Mark I when it was running 24 hours a day, noted that it broke down about once a week. At best it took a few minutes to repair, the average was about 20 minutes, and only in the worst cases did the work take a matter of hours (see J. Pérès, L. Brillouin and L. Couffignal).

During the final months of World War II, the Harvard Mark I was used exclusively by the US Navy to solve problems in ballistics. (Naturally it was later used to solve a far wider range of problems during its service life, which lasted until 1959 when it was decommissioned.)

AIKEN'S OTHER COMPUTERS

In the meantime, Howard Aiken started his own computation laboratory. He built machines not just for the US Navy, but also for the US Air Force and so on. These calculators were to become known as the *Mark II, Mark III*, and *Mark IV*. As R. Moreau explains:

> The initial idea for the Mark II goes back to 1944, even though the calculator was not actually finished until 1947. The machine, which was decimal, worked with ten decimal digits. Among other interesting features, it was the first machine to use *BCD* encoding.[66]
>
> Another interesting feature of the Mark II was that it used built-in *floating-point* arithmetic. It was therefore only the second machine to do this, after the Bell Labs one.
>
> Mark II was followed by Mark III (1949) which abandoned electromechanical technology in favour of electronics. As we will see later, electronic technology was already well established by 1949. It should be noted that the Mark III contained one of the first *magnetic memory drums*.
>
> Then came the Mark IV (1952), which had one of the very first *ferrite core memories*. [R. Moreau (1987)]

LIMITATIONS OF ELECTROMECHANICAL CALCULATION

So, in the history of automatic numerical calculation, the Harvard Mark I was the first completed multi-purpose analytical calculator developed using electromechanical means. The technology in question, however, would never have produced anything but bulky, costly machines with poor performance. It was just too limited to allow progress to take place, especially with regard to analytical calculation and artificial data processing. Here the key lay in the development of electronics.

66 "What is the actual problem? Let us take the number 27. We have to code the number in binary in order to store it in the calculator's registers if these are made of elements that can be in one or the other of two stable states (relays, diodes, etc.). Two types of coding are therefore possible. From the Mark II onwards, in the decimal machines, each decimal figure is represented by a code with at least 4 bits [in other words a minimum of 4 binary digits]. The most commonly used code at the start was the Binary Coded Decimal (*BCD*). In this, 27 is written as 0010 0111, where 0010 codes the number 2 in BCD and the 0111 codes the number 7. The calculator then carries out the operations as though it were dealing in decimals. In binary machines, each decimal number is converted into its binary equivalent. 27 becomes 11011 and the machine works with numbers thus converted to base 2. The Mark II was thus a decimal machine using BCD." [R. Moreau (1987)]

11. The Electronic Revolution

THE ELECTRONIC REVOLUTION

Any history of electronics must go back to the early groundwork. Looking back over the years, we can pick out several German names, among them those of Heinrich Geissler (1815–1879), Julius Plücker (1801–1868), and Johann Hittorf (1824–1914). And we should not overlook the contributions made by an Englishman, William Crookes (1832–1919); a Frenchman, Jean Perrin (1870–1942); another Englishman, Joseph Thomson (1856–1940); and the Irishman George Stoney (1826–1911). Their researches led to the development of the physics of *electrons*.

In 1897 the German Karl Ferdinand Braun (1850–1918) invented the cathode-ray oscillograph. Aside from this, the real history of electronics began in 1883 when the American inventor, Thomas A. Edison (1847–1931), made the discovery that electric current can pass inside a vacuum tube from a hot electrode to a cold one (but not the reverse) and that, in heating a wire by passing an electric current through it, free electrons are expelled.

The English physicist John A. Fleming (1849–1945) took the next step. He was trying to find a means of detecting electromagnetic waves when, in 1904, he struck on a way of putting to use Edison's so-called "thermionic" effect, produced inside the vacuum tube. What Fleming did was to place a small, positively charged metal plate inside the tube. This meant that the free electrons expelled by the heated metal filament all precipitate onto the plate, thus generating electric current. This invention was called *Fleming's valve*, now better known as the *diode*.

This was followed in 1907 by the *triode*, the idea of American Lee de Forest (1873–1961). He thought of inserting a third electrode into the tube, between the plate and the filament whose effect was to amplify the incoming current.

Together, these opened the way to the development of electronic tubes, which made radio receivers possible and gave birth to the radio industry.

BIRTH AND DEVELOPMENT OF ELECTRONIC CALCULATION

But it gradually became clear that there were other applications for these vacuum tubes. They could be used to *switch* electrical signals. And it was this that opened the way to the development of *electronic calculation*, especially after it was discovered that instead of using electromagnetic relays, it was possible to carry out arithmetical calculations using electronic tubes. By establishing in advance a code for the data circulating in the machine, it became possible to develop *numeric* electronic calculators.

In the meantime the English scientists W. H. Eccles and F. W. Jordan came up with the famous *flip-flop* device. This was a dual triode with a special characteristic: any pulse coming into either of the inputs flipped each of the two triodes into an opposite state. In other words, the flip-flop was a bistable electronic device.[67] Perfected in 1919, this discovery was to open the way for the *first binary electronic circuits*.

This meant that by constructing and activating appropriate circuits, it was already possible to carry out arithmetical calculations by using the bistable system of two vacuum tubes, in exactly the same way as with the electromagnetic relays. By encoding the data code using *binary code* (a set of rules that allows data to be represented bi-uniquely by reducing numbers to a combination of two digits, usually 0 and 1) technology gave birth to *binary calculators*. Over the years, the ease of the binary system became increasingly obvious. This led – as we have seen in Chapter 4 above – to the gradual adoption and development of the technique of electronic binary calculation.

In order to achieve this, however, a further difficulty had to be overcome. The initial data were supplied to the machine in decimal form. It was therefore necessary to introduce an input device that converted decimal to binary and an output device that converted binary to decimal. This requirement was made even more imperative by the fact that the machines in question (analytical calculators and then electronic computers) carried out increasingly complex sequential calculations which in turn involved an ever greater number of intermediate stages. Quite aside from the use of the binary system, electronics made the operations of these machines far easier thanks to the use of electronic circuits and ever greater technological advances. This meant that the machines' performances improved considerably and they became more reliable, safer, faster, and smaller.

As far as artificial calculation is concerned, therefore, electronics permitted the introduction of simple, elegant and extremely effective solutions to centuries-old problems, be they elementary or complex. Over and above this, it allowed calculation to go far beyond the very limited capacity both of normal calculators and of analytical calculators based on Babbage's structure. On the other hand, technology itself gradually forced engineers and technicians to completely revise the way they conceived and built

---

67 In fact, by taking a special type of multivibrator and replacing its shunted condensers by counter-batteries, a device is created with stable equilibrium positions. In position one, one of the two tubes is at zero, while the other is at maximum voltage. In the other position, it is the opposite, with the first tube at maximum and the other at zero. This means that the electric pulse makes the system flip between one state and the other.

the machines. For instance, it obliged them to radically modify the basic nature of the units and their components.

There is no doubt that the advent and growth of electronic calculation was one of the defining events of the twentieth century. It produced totally new techniques in which electricity was no longer merely a source of energy, nor simply a propulsive force that facilitated certain tasks. Used directly, electricity became an extremely effective means of automatic data processing and, with the development of the theory of information, further became a medium of communication. And it is precisely this brand-new function that enabled the creation and development of *electronics*, a discipline that stands distinct and separate from electricity and electrical engineering.

Electronics meets a whole range of requirements and has numerous possible uses. Because of this, it opened up the way for all manner of concepts and undertakings. It allowed mankind to glimpse the broadest possible horizons where solutions to unimaginable problems, or to questions long held insoluble, were soon to become everyday realities. Among them are radar, radio, television and telecommunications, supersonic aircraft, rockets, etc.

Because it benefits from several centuries of progress in the fields of mathematics and symbolic logic as well as in areas such as automation, programming, and artificial numerical data-processing techniques, this revolutionary technology also and above all provided the possibility of achieving ever more daring concepts and projects. Among them, clearly, are such things as the current generation of electronic computers, whose genesis was largely facilitated by it.

But there has been a progressive feedback in return. Today, in fact, it is the requirements thrown up by information technology (in the broadest sense of the term) that provide one of the main driving forces behind progress in the field of electronics.

THE EARLIEST APPLICATION OF ELECTRONICS
TO AUTOMATIC ARTIFICIAL CALCULATION

Coming back to the early days, it was in 1939, in the United States, that the first attempts to harness electronic technology to the application in question were undertaken. This happened at the Iowa State College where John Vincent Atanasoff (born in 1903) and Clifford Berry (1916–1963) built (but never finished) a small binary numeric calculator. Known as the ABC (Atanasoff-Berry Computer), it was conceived for the purpose of resolving systems of linear equations. Its interesting feature was that (at least in part) it utilised electronic components. But the physical and logical structures of the machine were fairly rudimentary and it was never an analytical calculator in the true sense of that term (in that it was not programmable). Furthermore, it never worked properly.

Another attempt along the same lines was made at more or less the same time but was apparently quite unconnected with the first. In 1941, at the Berlin Technisches Universität in Germany, an engineer called Helmut Schreyer tried (albeit unsuccessfully) to produce an electronic version of one of the calculators that Konrad Zuse had aimed to perfect using electromechanical means.

Similar attempts along the same lines were made in 1942. J. Desch, H. E. Kniess and H. Mumma developed the "four operations" experimental electronic calculator in NCR's labs at Dayton, Ohio. And again, there was the experimental electronic multiplier designed at around the same date by B. E. Phelps in IBM's Endicott laboratories.

THE BRITISH COLOSSUS: THE FIRST COMPLETELY ELECTRONIC CRYPTO-ANALYTICAL CALCULATORS

In fact, the first major breakthrough in this field was made in Great Britain, just before the end of World War II. It happened at Bletchley Park, home to the British Code and Cipher School. Here a team led by M. H. A. Newman and T. H. Flowers built powerful cipher-breaking machines under the name of *Colossus* [see R. Ligonnière in InPr, 24/June-July 1984, pp. 97ff.; see also I. J. Good (1979); S. H. Lavington (1980); N. Metropolis, J. Howlett and G. C. Rota; D. Michie; B. Randell (1973)].

These machines were produced from 1943 onward, under conditions of strictest military secrecy. The Colossus machines were totally electronic binary analytical calculators that were structurally designed to solve logical problems, to give plain-English read-outs of encrypted texts, and to reconstruct the processes and the key or keys of the German cipher systems. The aim was simply to halt the advantage the Nazi troops had acquired thanks to the use of cryptographic machines, including the famous *Enigma* machine (see Fig. 5.1).

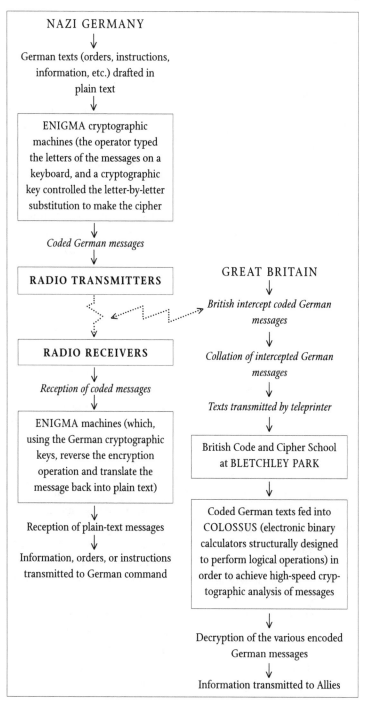

FIG. 5.1. *Cipher war by radio and electronic calculator waged at the end of World War II*
(Table adapted by the author from R. Ligonnière, *Préhistoire et histoire des ordinateurs*)

Among many other interesting features, it is worth noting that these machines were:

- capable of performing binary logic calculations;
- capable of conditional branching;
- capable of automatically printing (by means of electric typewriters) the processed results;
- capable of storing programmes already written for the purpose of executing pre-selected functions; etc.

We should also mention that, "during the conflict, the British decoded 75,000 to 80,000 of the messages they intercepted. The entire progress of military operations was influenced by the use of the information thus obtained. After the war a number of countries continued to use encryption machines similar to *Enigma*. The British classified all information which might reveal how decryption was possible using [analytical] electronic calculators." (R. Ligonnière, in *Ordinateurs*, 19 November 1984, p. 42).

THE AMERICAN ENIAC: THE FIRST COMPLETELY ELECTRONIC, MULTI-PURPOSE, ANALYTICAL CALCULATOR

Meanwhile in the United States two physicists, John Presper Eckert (born 1919) and John William Mauchly (1907–1980), took the history of electronic calculation another major step forward. They were doubtless inspired (at least in part) by the ideas of Atanasoff and Berry. Working together with the mathematician Hermann Heine Goldstine (born 1913), they designed the *Electronic Numerical Integrator and Computer*, better known by its acronym ENIAC.

Designed in 1943, the machine was completed in 1945. It was built at the Moore School of Electrical Engineering, part of the University of Pennsylvania. Electronic circuits provided memory storage and calculation functions and the machine itself was structurally designed to perform a vast category of operations. The ENIAC was, therefore, truly the first completely electronic, operational, multi-purpose, analytical calculator in history.

The machine itself was impressive. Weighing 30 tonnes, it covered an area of 72 square metres, arranged in a U-shape 6 metres wide by 12 metres long. It was made up of forty or so metal cabinets, each 3 metres high, 60 cm wide and 30 cm deep. There was a multitude of switches and display lights. Completely electronic, the ENIAC contained 18,000 vacuum tubes and consumed 200 kilowatts of power when in operation. Clearly, its designers were obliged to install a ventilation shaft and a cooling system in the room containing thousands of electronic tubes. In addition, the machine had more than 10,000 condensers, 6,000 switches of various types and 1,500 relays. It incorporated 50,000 resistors and its assembly involved something like 500,000 solder joints.

It included a reader capable of scanning punched cards at the rate of 120 cards per minute for input of data into the calculator. The results of the calculations were either output from a card perforator that worked at the rate of 100 cards per minute, or, once the accumulators halted, read directly from numbers formed by lighting up tiny display lamps on the front panel of the accumulators. Depending on the type of data processed, input could also be done by means of three mobile panels that could be connected to various definite parts of the machine selected according to the task that the calculator was to perform. These panels acted as "function tables", receiving instructions for the programme to be followed.

Each control panel, which contained 1,456 dial switches, could store up to 104 items of information in 14 numerical positions. Their setting, however, had to be done manually. The operator had to input, one by one, each of the digits that made up an item of numerical data or the encoding of an instruction. And since it was necessary to enter all the data and instructions into the three panels at the same time, the operator had to set some 4,368 switches before starting the machine. To carry out a particular procedure not included in the instructions of the initial programme, it was necessary to give the machine absolutely every detail, as well as making all necessary adjustments, before starting it up.

In addition to these painstaking preparations, it also took quite a long time to change a programme and to prepare for the next one. This was because it was necessary to change the instructions, the connector panels, and the positions of the switches all at the same time. This produced a considerable number of errors. Worse still, the machine often broke down. Even when it did work, it did not always calculate accurately. The inventors of the ENIAC themselves admitted that, after taking account of human error, the machine only got the correct result 20 times out of 100.

But none of this detracts from the fact that the machine constituted a first-rank demonstration that electronic tubes could be used to carry out a broad range of calculations. Nor should we forget that, for its day, the ENIAC's performance was truly remarkable since it reached hitherto unequalled speeds. An addition took 200 microseconds (200 millionths of a second), a multiplication (of ten digits by ten digits) took 2.8 milliseconds and a division (with a ten-digit quotient) took 6 milliseconds. Hence the slogan in praise of the calculator: "The ENIAC is able to calculate the trajectory of a large-calibre naval shell in less time than the shell takes to reach its target!"

Nor should it be overlooked that the ENIAC worked one thousand times faster than the Harvard Mark I calculator, which took $1/3$ of a second for calculations that took the ENIAC only $^1/_{3000}$ of a second. So against this quality, which was quite exceptional for its time, the only drawbacks were its lack of reliability, its breakdowns and its uncertain operation. Quite naturally, advances in technology were soon to correct all of these defects.

STRUCTURAL LIMITATIONS OF PREVIOUS ANALYTICAL CALCULATORS

Neither the Harvard Mark I, nor the ENIAC, nor any of the previous calculators were really computers, in the true sense of the term. Structurally speaking, these calculators were still closely related to Babbage's Analytical Engine. The point is that none of these machines, in fact, was self-controlling. In other words, all of their programme systems were external: the device storing their instructions was completely separate from their internal memory, which was intended solely for storing data received and processed.

Moreover, their programme instructions were executed independently of the results of their calculations, so nothing that happened during the processing could change the way in which they worked. Once the process was under way, the calculators were not able to locate a given programme instruction automatically. In turn this meant that, unaided, they could not modify the sequence of execution of any programme. It was therefore always necessary to have a human operator on hand to make all the decisions that might be required.

Clearly, the solution was to design a programmable and multi-purpose sequential numeric automaton in which the programme itself would be stored in the internal memory. This would mean it could process instructions in the same way as it processed data. This is the *stored programme* solution. On the surface, it seems very simple. Nevertheless, its implementation meant attaining one of the ultimate mathematical abstractions of the day.

12. Fully Programmed Machines

From the mid-1940s a number of engineers anticipated the essential importance of stored programmes. Some of them quite rightly emphasised the huge advantages to be had from storing instructions in the machine's internal memory. They could see that this would provide a far more flexible working tool.

Robert R. Seeber put this idea to Howard Aiken in 1945. Seeber was a Harvard mathematics graduate who, until then, had been working in the Mark I team. Aiken, however, was violently opposed to the idea. More and more differences separated the two men until Seeber was forced to quit the Harvard team and went to work for IBM full-time. It should be noted that in the same year, thanks to its own levels of expertise and technological advances, IBM was already planning to build a more powerful analytical calculator, with better performance and wider scope than Aiken's machine.

So in August 1947 Wallace J. Eckert, an astronomer from Columbia University, led a team that included Frank E. Hamilton and R. R. Seeber. Together they created a gigantic machine that was christened SSEC (*Selective Sequence Electronic Calculator*). On 27 January 1948 Thomas Watson, President of IBM, publicly unveiled the calculator. He dedicated it "to world science". A plaque on it read, "This machine will provide the assistance required by scientists working in research institutes, government, and industry. It will enable them to explore the product of human thought to the farthest limits of time, space, and physical conditions."

In fact the SSEC was mainly used on astronomical problems. It calculated a table of lunar ephemerides that was far more accurate than anything previously produced. In particular, it enabled the tabulation of 12-hourly variations in latitude, longitude, and parallax of the Moon's horizon. Twenty years later these were the basic elements used for the Apollo project when determining the conditions which would enable the first men to land on the Moon, as planned for 20 July 1969 (see R. Ligonnière in *Ordinateurs*, 24 September 1984, pp. 38ff.).

Technically speaking, the machine was really a compromise. On the one hand, some of its designers wanted to utilise electronic components because of their proven performance in the ENIAC. On the other, a number of team members wanted to stick with electromagnetic relays in view of their great reliability as had been proven by the experience with Mark I. In the end it was decided to use vacuum tubes for the control unit, the arithmetic unit, and for a 160-character memory, while electromagnetic relays were used everywhere else, for instance in the electromagnetic 3,000-character-capacity random-access memory.

It should be noted in passing that the SSEC also used punched-tape readers with an overall capacity of 300,000 characters and table readers with 100,000-character capacity, etc.

Leaving aside the mass of technical detail that has little relevance here, it should be mentioned that various elements used in this machine were of undoubted use to IBM as a company which was now fully committed to a race for what would shortly become the computer market. The SSEC brought IBM the fruits of solid technical experience from which it would profit before long. And for some while, IBM continued to make use of this knowledge in the way it built analytical calculators and computers.

For the days when it was built, the SSEC was exceptionally fast. In fact it could perform 60,000 multiplications of two 14-digit numbers in a little over 20 minutes, that is to say the same time that a human calculator took to do just one. After several months of programming, the machine took 103 hours to process a problem of physics involving nine million elementary arithmetic operations. Using a desktop calculator, the same calculation would have taken nearly a century to complete.

An advertisement of the time proclaimed that the SSEC was 250 times faster than the Harvard machine, although more trustworthy comparisons set the ratio at about 100. If the SSEC was slower than the ENIAC's totally electronic technology, its memory capacity and its reliability worked in its favour. Quite a few clients preferred to commission major algebraic calculations from the SSEC, rather than the ENIAC which was prone to breakdown and to multiple sources of errors.

The IBM machine worked with four-address instructions (of which one was used to store the result and another for the following instruction), the instructions being extracted from memory in pairs. Furthermore, it could automatically modify the sequence of instructions, based on the sign of a result, or determine the address of the next instruction to be used.

Moreover, it was furnished with the capability of *conditional branching*, and of executing short sequences of auxiliary instructions before returning to the main programme. In modern terms, it was able to run *subroutines* [see B. Randell (1973) pp. 219ff.].

Best of all, it had a *stored programme*. There is no doubt that IBM's SSEC was the first operational multi-purpose analytical calculator in history that was designed with the ability to both process and execute ready-stored instructions which could later, therefore, be processed in the same way as the data (see C. J. Bashe, L. R. Johnson, J. H. Palmer and E. W. Pugh).

IBM'S SSEC: THE FIRST "NEAR-COMPUTER" IN HISTORY

In short, this machine was capable of automatically locating a given instruction in a programme in order to modify the sequence being followed. It had no need of a human operator to make the requisite decisions. This made IBM's Selective Sequence Electronic Calculator the first "near-computer" in history.

Why "near"?

First of all because, although it undoubtedly marks a major stage in the history of analytical calculators, the SSEC nonetheless remained a transitional machine. Another reason is that it really only continued along the line laid down by the Harvard Mark I calculator. Furthermore, quite apart from a serious *lack of synchronisation* in the steps of a calculation, even in its day it was completely *anachronistic*, as a result of the hybrid technology that mingled electromechanical components with a variety of electronic ones.

These shortcomings, however, were fairly minor compared to its far more serious failings. Even for its creators and operators, it was extraordinarily *complex*, in the way it was conceived, designed, built, and operated. The SSEC also embodied serious *inconsistencies of logic*. This meant that, despite all its advantages and its performance, the difficulties in programming the SSEC often made clients prefer to commission the ENIAC or even the Mark I for their calculations rather than this complex and anachronistic machine.

This is why after the SSEC, from 1945 onward, all machines built by IBM were conceived and developed around *computer architecture*, based on the theory laid down by the mathematician John von Neumann (see pp. 286ff.).

FROM A PURELY THEORETICAL IDEA TO
A FULLY JUSTIFIED CONCEPT

In this connection, people have often asked whether or not the designers of the IBM SSEC took the idea of storing the programme in their analytical calculator's internal memory from the celebrated report that von Neumann had published in the same year. In fact, the question has no importance whatever and the whole debate on it, one may even say, is a red herring. After all, it is perfectly plausible that the idea of a stored programme could equally well have been brought out by W. J. Eckert, R. R. Seeber and F. E. Hamilton without any outside influence whatsoever.

But, and this is the most important point to stress, the IBM team only developed the idea in an empirical fashion. All they were really aiming at was a strictly technical way of applying their minds to product improve-ment. In other words, for this team of engineers and mathematicians, the stored-programme concept always remained at the purely theoretical stage. It could not be otherwise, since the necessary structure had, in fact, not yet been made concrete for lack of a true theoretical justification. This meant that they were unable fully to capitalise on the potential offered by the crucial discovery of the stored programme. The people who built IBM's SSEC were still effectively locked into the by-then-outdated ideas originated by Charles Babbage.

This is why, contrary to what is often stated, IBM's SSEC cannot be classed as one of the computer family. It was, in fact, an electronic-electromechanical version of Babbage's Analytical Engine. True, the IBM team made definite technical improvements to it, but they were locked into the external programme and the whole series of drawbacks that naturally came with that.

Under the circumstances, one cannot but recall a famous remark made by Andrew D. Booth, a British pioneer of information technology, when speaking about Eccles and Jordan's celebrated discovery of the *flip-flop* in 1919. "At any time from that moment on a modern computer could have been built ... Remember that the Dane, Valdemar Poulsen, had already invented the magnetic recorder twenty-five years earlier and the basic suggestion of magnetic memory was advanced by an Englishman, Oberlin Smith, in 1888."

Although Booth's own contributions were and remain substantial, he certainly never understood – and indeed many of our own contemporaries, not all of whom are beginners, still fail to understand – one fundamental point. It is not enough just to have all the necessary ingredients, or even the right technology, to perfect an artificial device as revolutionary as the computer. In addition you must possess the mathematical aptitude and the scientific expertise to put forward a full and rigorous justification, theoretical and formal, for the underlying concept. It is precisely this that distinguishes simple "technical tinkering" from the work of true professionals. The latter results in the practical implementation of a concept that has been properly evaluated and reflected on at length, as a physical system complete with a well-established logical and mathe-matical structure.

So, to return to the designers of IBM's SSEC. Despite having the revolutionary idea of implementing the stored-programme concept, they then shut this off within purely technical constraints, thus making of their machine a device which was logically inconsistent. No matter how ingenious an idea is, no matter how fertile the mind that produces it, no matter how sound or inventive the technician who implements it, that idea will never attain its full scope or bear the fruits it deserves unless it is underpinned by rigorous theoretical studies. In fact, "the purpose of theory [. . .] is to classify and synthesise results obtained and to set them out in rational form. This permits not only the interpretation of that which is already known, but also, as far as possible, the foreseeing of that which is still unknown". (L. de Broglie, *Sur les sentiers de la science*, p. 14) "The purpose of mathematical theories is not to reveal the true nature of things to us; that would be an unreasonable pretension. Their sole aim is to co-ordinate the laws of physics which are revealed to us by experience but which, without the help of mathematicians, we would never be able to describe." (H. Poincaré, *Leçons sur la théorie mathématique de la lumière*, Preface, I)

Let us not forget that Ancient Greek mechanicians certainly possessed all the technical elements needed to build a calculating machine. But they never did it. This was above all because they did not have the necessary number-system (i.e. a numeric notation as perfect as that which came out of India). But, even more importantly, they lacked the appropriate mathematical knowledge to produce the requisite theoretical justification, which in turn was needed before the corresponding physical machine could be built. Without the necessary intellectual tools and in the absence of any advanced mechanical theory, they found themselves not only quite unable to devise such a concept, but also and above all they were mentally incapable of even perceiving its possibility.

Let there be no doubt that this is one of the reasons that explain why any given invention is produced at a particular moment in history rather than at any earlier time.

THE CONCEPT OF "MACHINE": THE ROAD TRAVELLED SO FAR

Nonetheless, the fundamental truth that has been illustrated above was not particularly new in the immediate post-war period. It had indeed already been formulated forcefully, and with conviction, more than a century earlier. What a road has been travelled since the French mathematician Gaspard Monge (1746–1818), teaching at the École Polytechnique in 1808, gave this definition of machines that today is considered classic to the case in point: "[Machines] are engines destined to bend the forces of nature, of weight, wind, water, and horsepower to the work that man desires" [see L. Couffignal (1963)].

Following the invention of the steam engine and later the development of systems of material bodies in movement, a tendency emerged in Europe that progressively separated the description of what a machine was from its theory. In other words, as early as the end of the eighteenth century, studies increasingly led towards an ever clearer separation between physical devices and their corresponding mathematical structures. In 1864 Haton de La Goupillère determined that the "invention" of a machine could be only the practical application of a fully justified scientific theory to an assembly of specific mechanisms. And in 1877 F. Reuleaux proposed a formalised deductive theory covering the classification of different mechanical components.

Nonetheless the idea remained of according the term "machine" to any mechanism that operated under material forces. This did not, however, prevent Koenigs from rightly stating, as early as 1905, that "mechanisation is truly an artificial zoology in which the creator is man himself, guided by a wide-ranging and mysterious intuition" (see L. Couffignal, *ibid.*).

In fact, the first major step towards modern concepts was taken in 1893 when Maurice d'Ocagne discovered the famous collection of calculating machines in the Conservatoire des Arts et Métiers as well as the equally important collection belonging to General Sebert (now owned by IBM). Since he could not relate these machines to any contemporary mechanical theory, d'Ocagne had the highly original idea of placing them into categories for which he devised his own hierarchy. To achieve this, he borrowed his classification criteria from biology. From 1905, the date the new edition of his *Calcul simplifié* ("Calculation Made Simple") appeared, he always referred to "the comparative anatomy of calculating machines". This stripped mechanical calculators of the uniqueness that they had previously enjoyed and which conferred on each one its peculiar bizarreness or curiosity. Without question, d'Ocagne's approach paved the way for an axiomatic theory of mechanical calculating machines. After that, the study of machines was viewed as a discipline that could be rational, objective and, therefore, scientific.

L. Couffignal has pointed out that, "to Reuleaux's way of thinking in the study of this class of machines, the steam in a steam engine was not part of the machine." Bearing this in mind, let us take a look at the example of a rocket where the fluid is the only internal moving element. How can we assert that the fluid is not part of the machine itself? So, if we were to stick to the simple, classic idea of how to define a machine, just as in the case of the fluid, machines in this class would be equally impossible to categorise. As difficult as defining the ether that fills the space the rockets explore.

The only way to overcome the problem, therefore, was to devise a true mathematical theory, complete with axioms, criteria, definitions, deductions, demonstrations, theorems, corollaries, and conclusions (see pp. 244ff.). This is because "just as comparative anatomy, guided by the principle of identifying a plan, sees through picturesque variety to identify homologous organs, so, by detecting formal analogies, axiomatics reveals unsuspected correspondences between different areas of the same science. It even discovers relationships between sciences that seem foreign to each other" [R. Blanché (1965), p. 69].

In the twentieth century, logicians and mathematicians have constructed a mathematically rigorous and very general axiomatic theory of abstract mathematical entities designated by the term *machine*. In the meantime, and especially from the end of the 1930s, an idea emerged that man's role in this consisted above all in endowing his machines with purpose. Subsequently, cybernetics defined the relevant basic concepts and formulated a very general definition for artificial devices: *a machine is a physical device perfected by man the aim of which is to replace man in the execution of an action*.

In any case, what is a machine if not an assemblage of physical units arranged in a manner that obliges mechanical forces to act *towards a predetermined purpose*? Otherwise stated, it is the idea of a physical device applied to certain operations and which logically brings with it the idea of a purpose, just as the idea of cause implies the idea of effect.

In short, according to L. Couffignal and M. P. Schutzenberger's formulation of how we have finally come to think of the machine. "[A machine is] an assembly of inanimate entities, organised so as to replace a human being in the execution of a series of operations defined by man himself [for the purpose of achieving a definite goal]."

So it is that, over several generations, man has designed and then built the following types of machine. Their level of complexity varies in accordance with the purpose they are designed to achieve:

1 *mechanised tools* whose primary function is to allow us to bring an external source of energy to bear directly on specific materials;

2 *energy tools* which use a certain quantity of energy within their own operation, while still acting on matter;

3 *mechanised vehicles* which are used to transport objects or persons and also act upon matter;

4 *motorised machines* that use energy as raw material by applying it to engines, generators, transporters, etc.;

5 *transformation machines* that change the physical, chemical, or biological nature of matter by the action exerted on its constituent parts;

6 *information-transforming machines*, drawing energy from an external source, that are initiated by predetermined physical actions, expressed in the form of "data". These machines apply a certain number of transformations to the data in order to generate "results"

(this category of machines includes classic calculators, analogue computers, analytical calculators and digital computers, amongst many others).

FROM MECHANISATION TO THE ZOOLOGY OF ARTIFICIAL AUTOMATA

Until the twentieth century revolution of *artificial automata* had taken place, it was impossible for the information revolution to happen.

We owe the first, vital contributions in this and many other fields to the Spaniard Leonardo Torres y Quevedo. Thanks to the general principles he drew up, he was, from the technical point of view, the first pioneer in the sector of analytical calculators. In addition to this, however, he was also a pioneer in a far broader field and can be considered the father of modern theories on *automation*. It was he who first established, in a systematic, broad-ranging and very detailed manner, the theoretical possibility of building artificial automata. He demonstrated astounding insight with regard to a fair number of modern machines that go by the same name (see p. 256).

This is what Torres himself had to say about some of the results of his revolutionary theories involving concepts that were remarkably advanced considering that he was writing in 1914:

These automata will have senses: thermometers, compasses, dynamometers, manometers . . . , equipment that will be sensitive to the circumstances which are intended to affect their operation.

The impression received by each of these machines will generally be translated into movement, for example the movement of a needle over a graduated circle.

The automata will have limbs: machines or devices able to carry out the tasks they have been given. They can be controlled by very simple means, even if the task itself is complicated. This can already be seen in certain timepieces in which a mechanism similar to an alarm triggers figures that perform various movements.

These automata will have all the energy they need: batteries, currents of water, compressed-air tanks which provide the motors with the energy they need to run and operate the machines designated to carry out the desired tasks.

Furthermore – and this is the main aim of automata – they will be able to discern. At all times, taking account of the impressions they receive, or even of impressions they received earlier, they will be able to control the desired task. It is necessary that automata imitate life by adjusting their actions to fit their impressions and by adapting their behaviour to fit the circumstances.

In theory, the construction of devices that play the same role as our senses presents no difficulty. Every day we invent new devices that measure and record the various elements present in the phenomena of the physical world. Things that cannot be measured today, will be measured tomorrow, or at any rate ... there is no reason to think it will not be so.

We can make the same observation about the machines controlled by the automaton and that carry out the work. Surely no-one can say where the progress of mechanics will end. No-one can refute the idea that we may one day be able to invent a machine to do any task we choose.

This is not the case if we ask ourselves whether it will ever be possible to build an automaton that, in order to decide how to behave, weighs up the circumstances in which it finds itself. It is considered possible, I believe, in a few very simple cases. The automation of purely manual mechanical tasks, such as a labourer may perform, is considered a possibility, but tasks requiring greater mental input may never be carried out by mechanical means. But this distinction simply does not hold water because, except in the case of reflex movements that do not concern us here, every human action involves the use of *mental faculties* ...

From the purely theoretical point of view ... , it is already possible to build an automaton whose every action depends on a greater or lesser number of determined circumstances, according to rules that one can impose arbitrarily when we build the automaton. The rules must clearly be sufficient for the automaton always to be able to determine, without any uncertainty, what it must do next.

This problem could be solved in a thousand different ways. In order to make myself more easily understood, however, I shall abandon pure abstraction and instead illustrate an electromechanical method that could, in my opinion, provide a general solution to the problem.

[The principle in question is "an electromechanical method. By means of a bank of spaced switches, it is possible to arrange that any given combination of circumstances makes one of the switches move so as to activate an electromagnet whose armature, when it moves, in turn triggers the action that corresponds to the set of circumstances in question." (G. Torres-Quevedo)]

It is also possible [Torres y Quevedo adds] to increase at will the number of switches and the number of contacts connected to each of them. In other words, you can increase indefinitely the number of specific factors that an automaton must take into consideration before acting. You could make its life as complicated as you wished.

From the theoretical point of view, all of this can be achieved without any difficulty whatsoever. There is no fundamental difference between the simplest of machines and the most complex automaton. Both come down to a material system subject to the laws of physics. This follows from the way they are made. When the laws of physics involved are complicated, however, it is necessary to go through a lengthy process of reasoning to deduce from the laws themselves what manoeuvres are required. It will seem as though the machine that performs these manoeuvres has also done the reasoning itself. It is this last illusion that generally throws people looking at the problem off track.

Without a doubt it was theories developed from such fundamental bases which led to the development of computers – which, after all, are only automata of a very specific kind. It also led to the advent of robotics and the science of other artificial automata, such as we know them today in the fields of cybernetics and artificial intelligence.[68]

13. On the Road to the Computer: Recapitulation

THE LEAD-UP TO THE APPEARANCE OF COMPUTERS

In order to gain a better understanding of what lay behind the scientific discovery of the concept of the computer, we must now rapidly recapitulate the knowledge that had been acquired over several centuries in the field

---

68 It is therefore easy to see how important Torres's contribution was with regard to the way thinking developed about machines and above all about artificial automata.

All of a sudden it becomes clear just how far we have come since René Descartes's *Discourse on Method* (Part 5), the ideas in which were indeed developed by Torres y Quevedo.

He [Descartes] freely admits that one can consider an animal's body as if it were a machine which, having been made by God, is infinitely better ordered and contains movements far more admirable than any that could ever be invented by man.

He adds that "if we had machines which had the organs and external appearance of a monkey or any other dumb animal, we would have no way of knowing that they were not really these animals as nature made them."

But Descartes denied that anyone, even God Almighty, could ever build automata capable of imitating human actions guided by reason.

He believes that it is metaphysically impossible, for instance, for an automaton to use words or any other sign "to respond to the sense of everything that is said in its presence, even as the simplest of men can do."

He freely admits that an automaton could speak, but he could not conceive that it could talk sense . . .

[In fact,] Descartes . . . was thrown off track by the idea that the automaton, in order to talk sense, would itself be obliged to do the reasoning, whereas in this case, as indeed in all others, *it is in fact its constructor who does the reasoning for it.*

Let us add, however, that the reality of discoveries in today's cybernetics and information technology is outstripping even Torres y Quevedo's scientific fictions (see pp. 302ff. and 369ff.).

of artificial automata and automatic calculation which allowed the era of computing to begin.

This history is, of course, directly related to the development of what can be termed calculation in the general sense of the word, that is to say the application of a set of well-defined arithmetical, mathematical, or logical rules and procedures to the elements of an appropriate set, in order to obtain a certain category of mathematical entities constituting the result. But, above all, it concerns the permanent attempt to simplify methods of calculation, with a view to replacing them by economical thought processes, whose performance demands no reflection or active contribution from the human calculator.

So, this history is in fact linked to the ongoing attempt to automate human thought. Hence the idea gradually emerged to replace the old ways of calculating, which require reflection, skill and intelligence when used, by algorithms for calculation.

This is also the origin of those countless arithmetical or mathematical "tricks" (the most famous being Euclid's algorithm to determine the highest common factor) which can be found throughout the history of these disciplines since the eras of Sumer, Thebes, Babylon, Athens, Bagdad, Cordoba and up to our contemporary civilisation. Such techniques, of course, derive from extremely varied mathematical properties, a generalised study of which finally produced, by a kind of ultimate abstraction, a general theory of algorithms. It is important to remember that this theory is now used to analyse and study any finite succession of elementary operations carried out step by step, in a strictly rigorous sequence, so as to solve a particular class of problems.

Another factor that greatly contributed to this movement was a fundamental need, which scientists and all users of numerical or mathematical calculation had felt for a long time, arising out of the fact that their profound reflections were constantly interrupted by the necessity to perform long and repetitive calculations. This need was obviously expressed in the whole-hearted desire to have specially constructed machines to carry out such intellectual tasks. Thus it was that various sorts of calculators were developed and constructed, specialised for the four basic arithmetical functions. Then a variety of artificial devices was conceived and produced which could receive data and transform it, in certain ways that are inherent to their own structure and functioning, in order to obtain the required results.

But, in the end, the greatest advantage to be gained from such developments was the fact that it opened the way to digital computers. Such machines transform information. The data received and processed are represented by a variable and discontinuous physical phenomenon, expressible in a one-to-one way in a discrete form; that is to say, by means of figures and symbols that answer to the extremely precise rules of a code. In order to arrive at the idea of the computer, it was necessary first to approach the notion of a machine which "accomplishes man's work while checking its own operations and even correcting its own mistakes" (Gardellini, quoted by P. Foulquié).

The fundamental advances in this direction were:

1 progress in techniques of automation, from the early productions of Greek and Arab mechanicians to those of twentieth-century Western experts. This was initially linked to the desire to imitate men and animals, then gradually drew away from its original context to be applied to extremely varied applications, with a view to eliminating human intervention in a series of operations;

2 the modelling of certain processes of human intelligence, based on observations of material reality at different levels, which then received a logical and mathematical basis, in the most general and abstract terms;

3 systems originally devised as assemblages of units equipped with commands, produced mechanically for the first time by European inventors during the Enlightenment (Vaucanson, Von Knaus, Jaquet-Droz, etc.) which mathematics later provided with a more formal and far more rigorous status.

Thus it was that, during this approach to artificial automatic calculation, several automatic calculators were produced that were specialised for arithmetical operations or for accountancy. However, they possessed a fundamental flaw. The person carrying out the series of operations had to intervene many times. As well as entering the initial data, he had to note down by hand all of the intermediate results and then enter them again, step by step, into the machine.

This was accordingly not the right approach. The required automatisation was more of a sequential sort: it was necessary to retain certain concepts that had been applied, for example, in astronomical clocks, belfry automata and the automatic chiming of bells. Hence arrived the idea of automatic and sequential calculators which, with a minimum of human intervention, were able to carry out sequences of linked operations, according to a process that had been set up beforehand by means of a control unit.

But it was still necessary to avoid unmodifiable fixed-instruction calculators which, like Babbage's Analytical Engine, had their commands integrated into their internal mechanisms. This lack of adaptability meant that these machines could carry out just one sort of operation (the one for which they had been structurally conceived). What is more, any modification of the sequences to be performed required a preliminary adjustment of the commands.

It was also necessary to avoid the path leading to those modifiable-instruction calculators, whose commands were not independent of the material structure of the internal apparatus, such as the classic electro-mechanic calculators of the first half of the twentieth century (for example the *Monroe*, *Mercedes-Euklid* or the *Supermetal*), or the multi-register automatic accounting machines with intermediate figure displays (such as the *Logabax*) in the 1930s and 1940s, or else the non-programmable "four operations", "economic" or "scientific" pocket calculators.

The way forward was shown by programmable calculators, in which the instructions given for the sequences of operations are governed by a flexible programming system, itself based on a "language" which provided the means of communication between man and the machine.

Another blind alley was the variety of programmable calculators with semi-automatic processing, which require one or more human interventions during the carrying-out of a complete programme, and in which it is impossible to modify automatically a sequence once the process has begun. It was thus necessary to circumvent the tabulating machines and punched-card calculators of the beginning of the century, which had reached a dead end. Operations were executed under restrictions that were purely technical (irrespective of the logic of the problem). In other words, it was necessary to find a machine capable of locating a particular instruction in a programme in order to modify the sequence of operations itself.

The solution was thus the path towards programmable calculators, with fully automatic processing. Such a machine could solve a problem completely automatically, so long as it was provided with the appropriate programme. This programme would be worked out in advance by means of a preliminary *analysis* of the problem's characteristics, performed in such a way that the corresponding instructions could give the internal apparatus precise orders as to how the different elementary steps should be executed and the problem thus solved.

This is how we arrived at the concept of analytical calculators, which are none other than programmable automatic and sequential digital machines. Such calculators were equipped with the following items:

1 an *input/output* unit, allowing contact to be maintained with the exterior by means of a pre-established language. The input unit converts the data introduced into the machine into coded information, while the output provides the results, that is to say, information that can be used in a variety of ways;

2 a unit to *store* the programme's instructions, so that they can be used by the machine;

3 a *control unit*, to carry out and supervise the execution of the sequences fixed by the programme that has been given to the machine;

4 a *memory*, in order to store intermediate and final results;

5 a *processing unit*, which arranges the data that have been entered, transforms the way that they are represented, and then automatically transfers them to the machine's memory.

This is the direction which was taken for the first time by the English mathematician Charles Babbage.

But it was necessary to turn off this road once again, for the analytical calculators equipped with the same structure as his machine required external programming systems. It was from the need to surmount this structural limitation that the idea of stored programmes gradually emerged. The fundamental consequence of this was the invention of machines that could process their instructions in the same way as the data to be processed, in other words *computers*.

It is now clear how extraordinarily complex the history of computing was, revealing an approach which, even more than the universal history of numbers, now looks like the staggering of a drunk (see Fig. 5.2A–D).

But, in order to crown all of this research and make the best possible use of this fundamental concept, it was first necessary to arrive at a real mathematical theory, the essential idea of a universal machine, whose logical structure would be far more general and abstract than the multi-purpose analytical calculators that had existed until then.

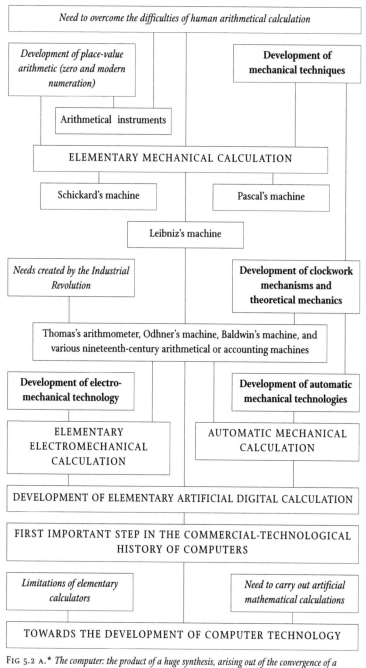

FIG 5.2 A.* *The computer: the product of a huge synthesis, arising out of the convergence of a multitude of needs and of many individual initiatives, some from a long time ago*

* In this figure, and the following three, the texts in italics indicate needs, structural limitations, or scientific developments (mathematics, logic, etc.) while the texts in bold Roman refer to technologies (mechanics, etc.) which became associated with analytical calculation before the arrival of the computer.

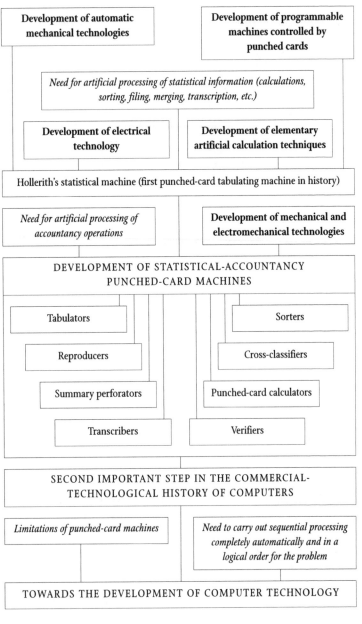

FIG. 5.2 B.

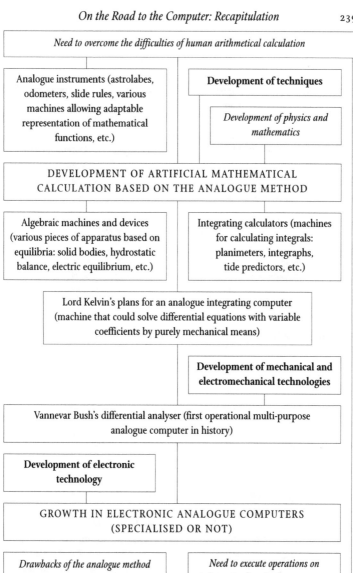

Need to overcome the difficulties of human arithmetical calculation

Analogue instruments (astrolabes, odometers, slide rules, various machines allowing adaptable representation of mathematical functions, etc.)

Development of techniques

Development of physics and mathematics

DEVELOPMENT OF ARTIFICIAL MATHEMATICAL CALCULATION BASED ON THE ANALOGUE METHOD

Algebraic machines and devices (various pieces of apparatus based on equilibria: solid bodies, hydrostatic balance, electric equilibrium, etc.)

Integrating calculators (machines for calculating integrals: planimeters, integraphs, tide predictors, etc.)

Lord Kelvin's plans for an analogue integrating computer (machine that could solve differential equations with variable coefficients by purely mechanical means)

Development of mechanical and electromechanical technologies

Vannevar Bush's differential analyser (first operational multi-purpose analogue computer in history)

Development of electronic technology

GROWTH IN ELECTRONIC ANALOGUE COMPUTERS (SPECIALISED OR NOT)

Drawbacks of the analogue method

Need to execute operations on information kept in the memory, to provide results as precise as required, to solve a considerable variety of problems, etc.

Development of elementary artificial calculation

DEVELOPMENT OF ARTIFICIAL MATHEMATICAL CALCULATION BASED ON THE DIGITAL METHOD

TOWARDS THE DEVELOPMENT OF COMPUTER TECHNOLOGY

Fig. 5.2 c.

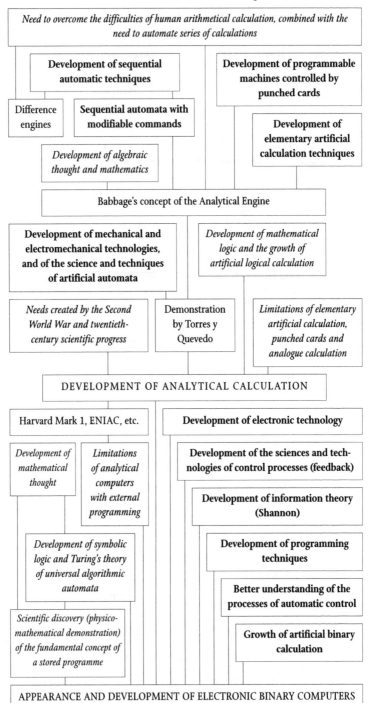

Need to overcome the difficulties of human arithmetical calculation, combined with the need to automate series of calculations

Development of sequential automatic techniques

Development of programmable machines controlled by punched cards

Difference engines

Sequential automata with modifiable commands

Development of elementary artificial calculation techniques

Development of algebraic thought and mathematics

Babbage's concept of the Analytical Engine

Development of mechanical and electromechanical technologies, and of the science and techniques of artificial automata

Development of mathematical logic and the growth of artificial logical calculation

Needs created by the Second World War and twentieth-century scientific progress

Demonstration by Torres y Quevedo

Limitations of elementary artificial calculation, punched cards and analogue calculation

DEVELOPMENT OF ANALYTICAL CALCULATION

Harvard Mark 1, ENIAC, etc.

Development of electronic technology

Development of mathematical thought

Limitations of analytical computers with external programming

Development of the sciences and technologies of control processes (feedback)

Development of information theory (Shannon)

Development of symbolic logic and Turing's theory of universal algorithmic automata

Development of programming techniques

Better understanding of the processes of automatic control

Scientific discovery (physico-mathematical demonstration) of the fundamental concept of a stored programme

Growth of artificial binary calculation

APPEARANCE AND DEVELOPMENT OF ELECTRONIC BINARY COMPUTERS

Fig. 5.2 d.

WHAT LAY BEHIND THE SCIENTIFIC DISCOVERY OF
THE STORED-PROGRAMME CONCEPT

We must here emphasise the considerable difficulty which the pioneers of computing had in reaching the scientific discovery (that is to say the indisputable and rigorous theoretical proof) of the concept of a stored programme. The apparent simplicity of this idea today makes us forget that it was far from obvious just half a century ago, and in fact conceals questions of quite incredible complexity. Here the technological aspects of the problem are presented retrospectively by R. Moreau (1987):

> Storing a programme [in the internal memory of an analytical computer] involved a [radical] change in the scale of the processing. [Going back to the time of the ENIAC,] It must not be forgotten that until then programmes were read directly by the processing unit on punched tapes or piles of cards placed at the input of the machine. As in the days of Jacquard or Babbage, it was thus necessary to introduce by hand the necessary punched tapes or cards to carry out a given part of a programme. Recording programmes directly into a direct access memory completely did away with this disadvantage and accelerated processing.

> Then, as soon as the first [analytical] calculators appeared, the idea of *instruction* became clear. It is known that all algorithmic processing comes down to a series of logical or arithmetical operations [see pp. 302ff.]. Each of these operations is indicated to the machine by [what might be termed] a *machine instruction*. Such an instruction, which we shall call *elementary* – in order to distinguish it from those used in highly developed programming languages – is communicated to the machine by means of a digital code. Generally, the first number indicates the operation to be executed, the second number the position in the memory where the operands (quantities to be operated on) are to be found, finally the third number gives the precise place in the memory where the results are to be placed. These three numbers are placed side by side in the instruction. The instruction thus appears as a single number in the form of data. So the increase in the size of the memory created the possibility of also placing instructions in the memory.

> This solution had several advantages. The first was a speeding-up of calculation. [In an electronic analytical machine] the transfer between the memory and the calculation unit is done purely electrically, whereas reading at the input unit of the machine, by means of punched cards for example, uses far slower mechanical methods.

Another essential advantage results from the fact that instructions could now be treated as data. It thus became possible to use the same part of the programme several times to process different data situated in different locations in the memory. It was no longer necessary to reintroduce into the machine the series of instructions, indicating locations in the memory, each time the operation was executed, as had been the case before. All that was needed now was to modify in each elementary instruction the address of the position in the memory where the value of the operands was indicated, thus modifying the number corresponding to that location. Hence the neologisms to *address*, meaning "to assign a memory address to an object", and *addressing* to denote the action of assigning an address.

A further advantage of the placing of the instructions in the memory was that it now became possible automatically to skip some parts of a programme. It is easy to understand the flexibility that was thus given to the machine. Take for example the addition, by means of a [non-programmable] calculator, of the first 100 integers, without using the formula $1 + 2 + 3 + \cdots + n = n\ (n + 1)/2$. It would be necessary constantly to give the instruction for addition by hand and enter the new data, whereas this would be performed automatically by a machine with an already stored programme . . .

A computer is [thus] a universal [analytical] calculator with stored programmes. It can by itself, without human intervention – and that is its essential distinction from the other machines derived from Babbage's engine – direct the execution of any algorithmic processing that it is given [see pp. 302ff.].

In a computer, the control unit must be considerably developed in order to be able to supervise the execution of the processing programme, to execute the necessary changes of address, to fetch the sequences of executable instructions, etc.

It is composed of a set of devices which co-ordinate the workings of the memories, the calculation unit and, to a certain degree, the input/output apparatus. The role of this unit can be defined by opposition to the arithmetical and logical unit. It executes a programme corresponding to a particular problem. It checks that any given programme is correctly executed and thus has a general role.

From a theoretical point of view this time, the elaboration (we are wrong to say "invention") of a new type of machine in fact comes down to the perfecting of a characteristic structure, which itself derives from a well-determined mathematical theory, that a human must apply to a set of inanimate entities, if he wants the artificial apparatus that thus comes into being to carry out for him all of the operations which have been fixed as objectives. But, "when it comes to a physical theory, three questions are raised:

1 Is the theory intrinsically logical, that is to say, does it contradict itself somewhere?

2 Is it in accordance with the facts?

3 Is it in better accordance with the facts than pre-existing theories?" (J. Hadamard, in BSFP, 22/1922, p. 96.)

This implies that the theory in question must be complete, non-redundant and non-contradictory if it is actually to produce the envisaged artificial machine and if the machine is to be consistent and compatible.

It is thus necessary to devise a formalised deductive science by means of a series of axioms, from which the entire sequence of propositions will be deduced. This means that we must make an abstraction of experience and physical reality in order to arrive at such a series of axioms, the only conditions being that the axioms must be:

1 *independent*, in other words, in no case can any one be deduced from the others;

2 *coherent*, that is to say, forming a complete set;

3 *compatible*, in the sense that they must never contradict one another.

So, in order scientifically to define the concept of the computer, it was necessary to:

1 establish mathematically the existence of such "universal" machines;

2 demonstrate, again mathematically, that by retaining the elements and main functions of a multi-purpose analytical computer, but adding corresponding programmes to its internal memory, we have in fact produced a physical example of such a universal machine.

In other words, for the theory of such machines, it meant elaborating a *hypothetico-deductive system*, the only law of which is logical coherence; any application to empirical objects is abstracted. It meanwhile remains independent of any subjective interpretation.

We can now understand why the possibility of this fundamental discovery came into view only after the requisite theoretical progress had been made, as a result of various sorts of pressure. Thus, by bringing together delicate mathematical questions, and in particular formal symbolic logic, the problems that had to be solved before finalising the structure required by stored programmes far surpassed the simple schemes of engineers and technicians. They were rather in the field of theorists and mathematicians of a broad culture and a solid scientific background.

14. The Contribution of the Mathematical Logicians

THE THOUGHT MECHANISMS THAT ALLOWED THE BREAKTHROUGH AND THE EXPLOSION OF COMPUTING

Indeed, it was not by perfecting the abacus that the computer was invented! As in the history of the candle and electricity, this highly structured presentation of the story shows that the appearance of computers did not come about as the final step of a uniform and linear evolution, which would have created only an extremely rudimentary calculating machine. We must once more emphasise the complexity, the non-linearity, the non-independence and the characteristic interactions of modern fundamental science's various fields, which now of course include computing. For, as with all the major innovations of our era (the steam engine, the car, the railway, aviation, photography, cinematography, the telephone, the television, video, telecommunications, etc.) this event was the cumulation of an infinitely complex and varied evolution, whose fruit matured only in the middle of the twentieth century, though it was formed by an extraordinary convergence of different and extremely ancient approaches (see Fig. 5.2 A–D).

We must also note that, if technical experience and technological progress certainly played an important part in this story, they were not the sole determining factors, being necessary but not sufficient. In fact, of all the thought mechanisms that allowed this breakthrough and the explosion of automatic digital calculation which finally led to the birth and growth of computer science and technology, the development of mathematical thought and the basic concepts of modern logic are certainly the first and most fundamental. Without wishing to complicate the following presentation with highly abstract and abstruse considerations, we are going to return once more to these fields, in order to bring home the vital importance of such intellectual mechanisms.

FIRST FUNDAMENTAL MECHANISM: ZERO AND THE PLACE-VALUE SYSTEM

At the risk of repeating ourselves, we must first point out that this development would never have occurred without the place-value system devised by the Indians. Among the numerous fundamental consequences of this discovery is the possibility to define and unify a large number of apparently distinct concepts, and by consequence to build general theories on its foundations, which were previously inconceivable. In this extremely vital innovation, it is impossible to exaggerate the essential importance of zero, which did not simply allow the empty positions in the place-value system to be expressed; nor did it just provide a word, a figure or symbol. It above all created a notion, understood at once as a numerical concept and an arithmetical operator, and as a real number inversely equal to mathematical infinity and simultaneously a member of the sets of integers, of fractions, of real numbers, of complex numbers and so on, from one generalisation to the next.

This fundamental concept is thus positioned at the meeting point of all the branches of modern mathematics: arithmetic of integers and fractions, the algebras of scalar or vector quantities, tensor calculus, matrix calculus, numerical analysis, infinitesimal analysis, differential and integral calculus, set theory, algebraic and topological structures, etc. And it naturally plays just such a fundamental role in all the other scientific disciplines, from physics, astronomy, astrophysics, statistics, economics or econometrics to biology, chemistry, genetics, linguistics, computing, robotics, and cybernetics.

To sum up, the vital discovery of zero gave the human mind an extraordinarily powerful potential. No other human creation has exercised such an influence on the development of mankind's intelligence.

SECOND FUNDAMENTAL MECHANISM:
THE DEVELOPMENT OF ALGEBRAIC THOUGHT

Another fundamental tool, without which the computer would have been impossible to conceive, is the fruit of several centuries' development of symbolic thought. This is especially true for the movement from ordinary arithmetic to algebra and from the representation of specific numbers to a literal notation for unknowns and undetermined constants. This marked one of the first steps from individual to collective reasoning (see Chapters 2 and 3 above). For it was by just such a thought process that the English mathematician Charles Babbage – whose famous Analytical Engine was the true ancestor of our modern computers – arrived at such a powerful synthesis and such revolutionary ideas (see pp. 189 ff.).

And there is in fact a striking analogy between the histories of artificial calculation and algebra. In the case of artificial calculation, the lack of any general theory meant that this field was long limited to producing machines with extremely specific functions, based on procedures that had come together by chance in order to carry out a precise range of arithmetical or mathematical operations. Just as the discovery of literal notation created modern algebra, so did the concepts applied by Babbage in his Analytical Engine open an utterly new era in the history of artificial calculation.

In brief, the step forward made by Babbage, when he discovered analytical calculation, revolutionised digital calculators based on individual structures, just as much as modern algebra changed ordinary arithmetic's notations and operations.

But this revolution was, of course, not enough to lead at once to the computer. Another step forward had to be made, which we shall examine later. It is nevertheless true that the symbolism used in Babbage's "engine" did correspond to a first stage of a generalised conception. Even though it was far inferior to the true universality of the calculating capacities of a computer, the very idea of generality which was conferred on the Analytical Engine can in fact be explained only by the fact that Babbage attained a higher degree of abstraction than had previously existed.

To arrive there, Babbage had had mentally to detach the specific, strictly individual property in each of the concepts he had fully synthesised, in order to examine it in isolation and then combine it with the other properties that he had abstracted from their contexts in the same way. Indeed, what possible relationship could be perceived, in the as-yet-ungeneralised concept of mechanical studies at the time, between a belfry automaton, a clock or a carillon, and automata such as those produced by Vaucanson, Von Knaus and Jaquet-Droz? What did automatic systems such as Jacquard looms have in common with highly specialised mechanical calculators, which were not automatic and above all not really technically developed? Apparently nothing. Only Babbage was capable of separating what had not yet been separated, and what was in fact inseparable in reality.

It is true that "real abstraction begins with awareness of similarities and differences. To abstract is to distinguish the common point between various objects and different characters of each object." (A. Burloud, *La Pensée conceptuelle*, p. 163). And abstraction leads naturally to generalisation. "It is natural for our [modern generalising] mind to form general propositions from the knowledge of particular details." (Descartes, *2e Rép*, IX, 111.)

As abstraction and generalisation are closely linked, Babbage accordingly produced a sort of "algebrisation" of the fundamental concepts of mechanical calculation. This led him, thanks to his obsession with the difficulties of human calculation and his realisation that existing calculators were very inadequate, little by little to a desire to leave behind the great variety of specific data, and so arrive at a much larger construct that approached a universal view. It was thus because his mathematical reasoning had been extremely abstract (according, of course, to the degree of generality that one could attain at the time) that Babbage, being an excellent mathematician, strove to conceive a machine that would rise far above any specific operation, and be able to solve all sorts of problems contained within a vast category, regardless of their individual characteristics (see pp. 190ff.).

Basing himself on the unlimited number of operations that he could carry out on his machine, on the also unlimited number of variable cards he could use to command his Analytical Engine, as well as on the ease with which conditional branching was constructed and operated from elementary sequences, and finally on the fact that both the processor and input/output were completely automatic, he declared: "It seems that all of the conditions that allow a finite machine to carry out an unlimited number of calculations have been fulfilled by the Analytical Engine ... I have converted infinite space, which was required by the conditions of the problem, into infinite time." This quotation speaks so well for itself that there is no need for any comment.

THIRD FUNDAMENTAL MECHANISM: THE DEVELOPMENT OF LOGICAL THOUGHT

Another intellectual mechanism that is indispensable to the development of theories of reasoning, and thus to the growth of computing and cybernetic concepts, naturally lies in how logic has evolved over several centuries. Without going into all of the details related to this discipline, it is nevertheless necessary to mention a few essential milestones and a few fundamental ideas.

Firstly, the etymology of the noun *logic* is the Greek expression *logikê technê* ("the art or science of discourse"), itself derived from *logikos* ("related to *logos*, i.e. what concerns words, reason and thought"). Thus logic concerns all of the intellectual processes of the thinking individual. No knowledge would be possible without it – neither spontaneous everyday thought, nor deeper scientific thought. As opposed to psychology (which studies the realities of psychic existence or, in other words, intellectual functions) the aim of logic is to study the very products of such functions, or mental activity. Whereas the psychologist operates from a genetic viewpoint, trying to retrace the birth of functions or mental structures from the objective observation of psychic realities, the logician looks at these structures and studies their foundations, their validity and the operations they bring into play.

In the same way, it is important not to confuse rational order and logical order, even though the Latin root of the former is the same as the Greek root of the latter. The former term derives from the Latin *ratio* (or "reason") and the latter from the Greek *logos*, as we have seen. Rational order considers things in themselves, while logical order studies the construction of propositions, the form and the order of language, which is for us the instrument of thought, and the means to communicate it (see A. Cournot, *Enchaînement des idées fondamentales*, I, p. 64).

Logic has also been explained as "the art of correctly conducting one's reason in the knowledge of things, both for instructing oneself and for instructing others". This art consists "in the reflections that mankind has formed concerning the four main operations of the mind: to conceive, to judge, to reason and to order". Such was the notion developed in 1662, in the introduction to the *Logique de Port-Royal*, which recycled certain scholastic concepts from the Middle Ages while taking into account the ideas of Bacon and Descartes (see Arnauld and Nicole).

This interpretation of logic in fact goes back to Aristotle (384–322 BCE) whose main objective was to identify the general laws of thought and to determine the processes we must follow if we are to establish the truth in various fields of knowledge. His *Organon* – which set out to compare grammar to the basic mechanisms of logic and to unite them together in language – had provided the principles and fundamental elements of classical formal logic (theories of the syllogism, of deduction and of induction). But, in real terms, logic is a science, because it belongs to the field of reasoned knowledge and because it is related to a particular object. Thus, in terms of classical philosophy, logic is the science of the forms and the laws of thought, and its aim is "to determine which intellectual operations that lead towards knowledge of the truth are valid and which are not" (A. Lalande, *Vocabulaire technique et critique de la philosophie*).

Now, "the truth can be seen as the agreement between thought and itself and as the agreement between thought and things" (E. Bloncourt, in GLE, 6, p. 815). From this viewpoint, we can say that classical philosophical logic is a normative science whose objective is knowledge of the truth. When seen in this way, logic includes formal logic, which is independent of any concrete demonstration, and which (as William Whewell put it) teaches us to dispose our reasoning in such a way that its truth or falsehood can be seen from its form. This sort of logic accordingly "abstracts from the material truth or falsehood of stated propositions, that is to say from their conformity with reality: its business is not to know, for example, whether the French are or are not honest, or the Bretons are or are not French. It considers only the formal truth of a statement, that is to say the validity of its form and whether the propositions are coherent with one another: 'The syllogism draws its force of conclusion from what the premises say, not from their worth' (Pradines, *Psychologie* III, p. 129). Its role thus consists only in determining whether the conclusion must be accepted if the first two propositions are accepted" [P. Foulquié (1960)]. Formal classical logic is thus the logical relationship "which, abstracting the matter of thought, that is to say the object being dealt with, studies only its forms in order to determine which forms of reasoning are correct and which are not" [P. Foulquié (1982)].

Now, as E. Bloncourt explains (*op. cit.*):

A person who thinks is a person who judges and reasons; but judgment and reason are operations which can be executed only thanks to concepts, or general abstract ideas which form their fundamental elements. In formal logic, the theory of the concept is thus of capital importance.

The logician [in the classical sense of the term] considers the concept from the point of view of its extension and intension. The *intension* is the set of qualities that characterise the object of the concept. The *extension* is the set of entities that have the same qualities. Extension is thus the inverse reasoning of intension. For example, the extension of the concept "vertebrate" is larger than that of the concept "dog", but the intension of the concept "dog" is larger than that of the concept "vertebrate".

Judgment is a linking of concepts. But the logician considers a judgment as formulated by means of terms, or verbal expressions of concepts, which together form a proposition containing a subject and a predicate, linked by a copula, the verb "to be", which must be given no meaningful existence. The logician considers propositions from the viewpoint of their quality (they can be affirmative or negative) and of their quantity (they can be universal or particular). Some examples: "the weather is fine" (affirmative proposition); "the weather is not fine" (negative proposition); "all men are mortal" (universal proposition); "some men are wise" (particular proposition).

The relationships expressed in the propositions are predicative relationships of conformity, inherency, inclusion, or exclusion between the subject and the predicate.

A chain of judgments constitutes reasoning. Thanks to reason, which plays a vital role according to Aristotle, we pass from principles to consequences, whose necessity and universality are uncontestable. Reason allows us to make inferences, or reasoning corresponding to the sort of knowledge qualified as "discursive", as opposed to knowledge that results from an intuition.

There is immediate reasoning, which can pass from the data to the conclusion without any intermediary, and mediate reasoning, which employs a intermediary whose conclusion is justified by its conformity to the principle of identity or non-contradiction.

The typical case of mediate reasoning is the syllogism. According to this form of reasoning, once two items have been postulated, a third necessarily arises from them. A syllogism consists of three propositions, of which the first two are the premises, and the third the conclusion. The first of these premises, which contains the term with the greater extension, is called the major proposition; the second, which contains the term with the lesser extension, is called the minor proposition. There is also a middle term that connects the two premises and makes the conclusion valid.

For example: "All men are mortal (major); and Socrates is a man (minor); therefore Socrates is mortal (conclusion)." Since the conclusion was implied by the propositions, syllogistic reasoning must necessarily formulate it. This conclusion is true, and its truth is uncontestable.

But syllogistic reasoning, which is a formal deduction, lacks the fecundity of mathematical deduction, which is a demonstration that enriches knowledge. Formal logic, whose worth is far less than the logic of the scientist, applied to a given field, reduced science during the whole of Antiquity to a *tabula logica*, to combinations of concepts, according only to the rules of the syllogism. During the Middle Ages, scholastic philosophy continued the speculative discussions of formal logic, and it was only with Bacon and Descartes that a new form of logic was born, which corresponded to a new conception of science.

Since then, logic has undergone other transformations, among which we must cite: symbolic logic, intuitionalist logic, polyvalent logics and dialectical logic, which may be far distant from Aristotle and scholastic philosophy, but are nevertheless their more or less direct descendants.

Thus has logic evolved, on one side towards a more general philosophical logic, whose aim is to study intellectual processes that conform to the laws of rational thought, and on the other towards mathematical logic, based on far more abstract conceptions.

When seen in this light, philosophical logic has finally come to include: methodology (which consists in a form of logic relative to particular sciences and allows unrefined knowledge to be transcended and scientific knowledge to be organised); epistemology (whose aim is the critical study of the principles, rules, hypotheses, methods, and conclusions of various sciences, in order to establish their validity and their objective range); and the theory of knowledge (whose objective is to determine the conditions, worth and limits of human knowledge).

FOURTH FUNDAMENTAL MECHANISM: FROM
CLASSICAL LOGIC TO ALGEBRAIC AND BINARY LOGIC

But, in order to arrive at the logical automata that are computers, algebraic logic (or the algebra of logic) had first to become fully developed. This story thus began in Antiquity, when Greek philosophers such as Socrates and Aristotle defined the syllogism (the rule of the sort: "If . . . , then . . ." which we have just seen). They thus elaborated the basis of mathematical logic and deductive reasoning.

"When three terms are related to one another such that the minor proposition is wholly contained within the middle term, and the middle term is contained or not contained in the whole of the major proposition, then there is necessarily a perfect syllogism between the two extremes." Such is Aristotle's explanation of the mental process by which a third proposition is drawn from and implicitly included in two given propositions. As we can see, a syllogism is thus quite simply a formal operation. The word in fact derives from the Greek *sullogismos* (from *sullegein*, "to bring together") which literally means "counting, calculation, reasoning". But, at this early stage, there was still a long way to travel before such "operations" finally joined the abstraction of modern-day calculation and reasoning. What was required, (to paraphrase George Boole) was that their validity as calculations no longer depended on the interpretation of the symbols themselves, but simply on the laws governing the ways they were combined. In other words, it was necessary that no interpretative system affected the relationships these laws permitted.

It was thus necessary to wait for the decisive step taken by the German mathematician and philosopher Gottfried Wilhelm Leibniz (1646–1716) who undertook one of the first thorough studies of binary arithmetic (calculations performed in a positional number-system with base 2) and at the same time was the first to formulate the basis of modern symbolic thought (see Chapters 3 and 4 above).

This is what is explained by Georges Bouligand, who begins by establishing a brief parallel between the contributions of Descartes and of Leibniz:

> Descartes's ideal was for mankind to learn to decide for itself, freely and sanely, in any field; hence his desire for a universal method which would, in whatever subject, join up with the method of the general science of mathematical operations, and most importantly algebra, which was becoming the driving force of geometry, mechanics, and physics.
>
> Leibniz had the same objective, but his ambition was to organise wider fields of formal operations. He even envisaged applying

Descartes's universal method to the activity of research. With this in mind, he proposed a permanent symbolism, to be applied in the first instance to logic, the doctrine of proof. But whether it was a matter of reasoning or of discovery, he wanted to reduce the combinations of ideas to the combinations of their symbols, and mental operations that had been used since Antiquity ("how to operate in order to . . .") to mathematical operations. Hence the care he took with his notation, particularly in his *Rules of Infinitesimal Calculus*, (then, and still, an incomparable tool in the physics of continuous media). The scope of this work was extended, and the power of symbolism doubled, through the use of binary numbers, which reduced to the figures 0 and 1, repeated over and over again, the notation of any number expressing the result of a *measure* (*length, duration*), experimental or theoretical. Its benefit was thus not merely the encounter between the determinant and the subtle operational mechanism which it represents (see F. Gillot).

With hindsight, it is now possible to measure the true long-term influence that Leibniz had on the foundation and even the evolution of modern symbolic thought. But to arrive at that point, it was necessary first to make a further intermediary step which was just as vital. For this, we must pay tribute to the British mathematician George Boole (1815–1864) who was one of the first instigators of modern mathematical logic. For it was he who, in a certain way, enabled Leibniz's programme to be realised using such a symbolism as he had dreamt of. This is W. S. Jevons's explanation:

Until the nineteenth century, logic had remained substantially the same as when Aristotle had created it, two thousand years before. If the science of quantity had remained in the same stationary condition since the time of Pythagoras and Euclid, it is certain that we would not have had Pascal's arithmetical machine, nor Babbage's Analytical Engine [nor, indeed, the computer]. I would take the liberty of describing the logical machine presented here[69] as the result and the revelation of a deep transformation in logic, as well as an extension of this field of knowledge that had been produced by a series of nineteenth-century British authors, among whom we can make special mention of Jeremy Bentham, George Bentham, Augustus de Morgan, Archbishop Thomson, William Hamilton [plus Venn and Lewis Carroll] and that eminent member of the Royal Society, George Boole.

The result of their efforts was to cause a breach in the citadel of Aristotelian logic and provide us with a system of logical deduction,

69 The *Logical Piano* built in 1869 as a "mechanical device for logical inference", which was the subject of his article under this heading published in 1870 in PTRSL 160, pp. 497ff.

almost infinitely more general and more effective than anything the ancient authors had discovered. The old syllogism was hardly likely to be transformed into a mechanical process, because of its rather naïve and coarse nature. It is only when a system is founded on the fundamental laws of thought that we can arrive at a deductive procedure which a machine making simple uniform movements can carry out.

This step forward in the logical doctrine was due to George Boole more than to any of the other logicians we have just cited. In his *Mathematical Analysis of Logic* (1847) and in his even more remarkable work on *The Laws of Thought* (London, 1854) he posed for the first time the generalised problem of logic: given certain premisses or conditions, determine the predicates of a class of objects subject to these conditions. Such was the general problem, which ancient logic had solved only in a few isolated cases: the nineteen modes of the syllogism, the sorites, the dilemmas, the disjunctive syllogism and a few other forms.

Boole has shown incontestably that it is possible, by means of mathematical symbols, to deduce not only the conclusions of all the ancient modes of reasoning, but also an unlimited number of other conclusions. Finally, Boole has shown by the same method that it is possible to deduce from a series of premisses or conditions, no matter how extensive or complex they may be, a conclusion which can be calculated (see F. Gillot).

Boole's other fundamental contribution was to have embedded the procedures of propositional logic within the operations of a true algebra. This is what is known as Boolean algebra. In his calculus of propositions, he adopted a highly abstract conception of algebra, which was completely independent of any notion of number or magnitude. His operands are only logical propositions, with no further explanation of their nature and without the use of any interpretative system. His logic thus deals only with the relationships permitted by the laws governing combinations.

Using only the criterion of the truth or non-truth of propositions, and denoting all true propositions by 1 and all false propositions by 0, and using x and y to denote arbitrary propositions, he represented relationships and operations such as negation, conjunction, disjunction, etc. He thus was able to devise procedures for calculating with propositions, which were similar to the procedures of ordinary arithmetic and algebra.

"A time will come", W. S. Jevons proceeded astutely, "when the obvious results of the late Dr Boole's admirable research will be appreciated at their true worth ... There is no doubt but that Boole's life marked a turning point in the science of human reasoning ... We must thus recognise that Boole discovered a true general form of logic and essentially gave to that science the form it will keep for ever" (see F. Gillot).

This was, indeed, an advanced generalisation of the fundamental notions of elementary logic, which had been reached by the confluence of:

1 the generalisable idea of number (see Chapter 27, previous volume);

2 the use of letters to denote quantities, and the calculus of algebra, respectively generalisations of ordinary numerical notation and of calculation using numerals (see Chapters 1, 2, and 3 above);

3 finally, the idea of a number-system with arbitrary base *m*, itself the most general idea that can be of a positional number-system with zero and with the same structure as our decimal system (see Chapters 1 and 4 above).

15. The Advent of Set Theory

FIFTH FUNDAMENTAL MECHANISM: THE DEVELOPMENT OF ARTIFICIAL LOGICAL CALCULATION

It was from this point that the idea gradually germinated of mechanising, or even automating, the very operations of a logic that Boole had transformed into a genuine form of algebraic calculus. As Alain put it "if a counting machine is possible, then a reasoning machine is also possible. And algebra is already a reasoning machine: you turn the handle and effortlessly produce a result which thought would have arrived at only after infinite pains" (*Propos*, p. 732, La Pléiade).

This was precisely W. S. Jevons's point, in the article cited above, in his justification of the existence of his Logical Piano. This was the first logical machine ever produced, and its purpose was to solve complex problems of a syllogistic sort more rapidly than the human brain, with no intervention except pressure on keys similar to those of the piano, which corresponded to operations and propositions expressed as logical equations.

After presenting several landmarks in the history of mechanical calculation, and emphasising the possibilities of Babbage's Analytical Engine, which was well known at the time (1870), Jevons added this comment, to explain the reasons why logicians (and he in particular) had set out to find concrete answers to this fundamental need:

> It is extremely remarkable that when one turns towards the related science [to the calculus and to mathematics], which is logic, we find no mechanical assistance or tool. Yet, work with logic abounds with metaphorical expressions that imply that our ability to reason requires just such an assistance, even in the course of the most abstract processes of thought.

In the fifteenth century, and even before, the logical labours of the greatest of the logicians were commonly termed the *Organum*, or "instrument", and, for several centuries, logic itself was defined as being the *ars instrumentalis dirigens mentem nostram in cognitionem omnium intelligibilium* ["the instrumental art directing our mind to knowledge of all intelligible things"].

It is indeed rare for an invention to be made without its having been anticipated at some time or another. But, to this day, I know of absolutely no previous attempt to invent or construct a machine to carry out the operations of logical inference. And it is only, I believe, in Swift's satirical writings that we find any allusion to a reasoning machine.

The only reason I can see to explain the total incapacity of logicians to conceive a real logical instrument is the great imperfection in the doctrines that they accept . . .

The logical machine that I am going to describe is not a simple model designed to make the fixed forms of the syllogism concrete. It is an analytical machine, lacking in any complications, which can carry out a complete analysis of any logical problem which we may pose to it.

By expressing the premises or given information of an argument on a keyboard that represents the terms, conjunctions, verb, and end of a proposition, the machine is led to undertake a comparison of the premises in such a way that it can deliver an appropriate logically deduced answer. It is thus loaded with a certain quantity of information, which is then reconstituted in the desired logical form.

The true process of logical inference has thus been reduced to a purely mechanical act. We therefore arrive at a machine that makes concrete [George Boole's] *Laws of Thought* and which realises the vague notion of an *organum*, or logical instrument, that has plagued the minds of logicians for centuries (see F. Gillot).

The beginning and the end of the history of logical machines are summed up briefly below.

Chronological landmarks

12th century. Arab magicians and soothsayers used (at least since Abū'l 'Abbas As Sibti, a native of Ceuta) "circular tables of the universe" (*zā'irjat al 'alam*), which were sorts of "machines to think out events" made up of concentric circles containing the Arabic letters and their numerical values, thus allowing any possible combination with a view to establishing symbolic and mystical correspondences. These were the first *combinatorial instruments*.

1275. In his *Ars compendiosa inveniendi veritatem*, Ramon Lull suggested a scheme of "syllogistic mechanics", a sort of combinatorial instrument that had clearly been inspired by the Arabs' *zā'irjat*, which consisted in his own words of "a universal method destined to prove the truths of faith".

1666. In his *Dissertatio de arte combinatoria*, the German philosopher Gottfried Leibniz, inspired by Lull's ideas, projected a universal system of notation based on the designation of all the elementary ideas that constituted the foundation of human knowledge, by symbols of an algebraic nature. The aim was to arrive at an "alphabet of thought", allowing all possible combinations to be made. It was his attempt to construct a *lingua characteristica universalis*, later supplemented by a method of mathematical reasoning, that anticipated the modern concept of logical calculus.

1810. The English scientist and logician Charles Stanhope constructed his *Demonstrator*. It was the first machine to offer (manual) help to humankind in solving problems of Aristotelian logic.

1847–1854. George Boole published his *Mathematical Analysis of Logic* (in which he defined his logical or binary algebra), and then his *Laws of Thought*.

1869. W. S. Jevons produced his *Logical Piano*, the first machine made to solve syllogistic logical problems, and "capable of providing an answer to any question given to it concerning the possible combinations which form any given class".

1881. The American logician Allan Marquand constructed an even more efficient logical machine that had a greater capacity than Jevons's.

1886. The American logician Charles Peirce sensed the isomorphism that exists between Boolean algebra and electrical switching circuits (as shown in a letter dated 30 December 1886 addressed to his former pupil Allan Marquand).

1903. The Italian logician Annibale Pastore constructed a reasoning machine based on the rules of the syllogism, which performed physical rotations and was made up of an arrangement of three pulleys (the first representing the subject, the second the predicate, and the third the term).

1910. The Spaniard Leonardo Torres y Quevedo constructed his *automatic chess player*, an entirely automatic electromechanical machine. Equipped with electric sensors that identified the positions of the pieces on the board and a retractable mechanical arm to move its own pieces, this automaton's programme always led to checkmate at the end of a game (the game had only three pieces – a king for the human adversary and a king and a rook for the machine). Able to identify cheating (by lighting a bulb at the first and second attempts, and blocking the game at the third), this chess player was thus the first logical automaton able to make certain decisions on its own.

1936. The American Benjamin Burack constructed an electrical logic machine able to solve complex syllogisms.

1943. In the greatest military secrecy, the British constructed the first generation of *Colossus*, powerful electronic computers, specially designed to solve logical problems with a view to deciphering the secret codes generated by the Nazis' cryptographic machines. These computers were structurally conceived to perform binary logical calculations and were endowed with conditional branching (see pp. 217ff.).

1946. The Americans Theodore A. Kalin and William Burkhart built a small logical truth calculator, electrical and programmable, where data could be automatically transferred from one part to another within the machine, and which had an automatic control unit supervising the sequences of logical operations.

From Boolean algebra to the theory of switching

The logicians who built the early logical machines, from Jevons and Marquand to Pastore and many others, never for a moment imagined that their inventions would one day have a practical use. Essentially, they were devices that applied in concrete form the elements of theoretical techniques and were thus destined only for university students.

But it is nonetheless true that these inventors opened up possibilities that had never been glimpsed before: the fact that a machine could reason for itself. Such possibilities were to turn out to be extremely productive, since they were to lead to the idea of automating calculations and processes, to the construction of computers and later to expert systems, once they had been combined with theories of reasoning.

As for Boole's theoretical ideas, it was necessary to await the twentieth century, with its progress in electric technology and a better understanding of electromagnetic and electronic phenomena, for a practical application of these fundamental discoveries to be made at last. Beginning with the work of the American physicist Claude E. Shannon who, in his 1937 thesis (*A Symbolic Analysis of Relays and Switching Circuits*) proved that the rules of Boolean algebra could be applied to electric circuits and that these circuits could perform the fundamental operations of the algebra. This vital discovery, once linked with the knowledge then available of two-state physical systems (sense of magnetisation, perforation, on and off current, etc.) which could be symbolised by the two figures 1 and 0, was thus a major step towards the technical realisation of data media to be used first by analytical calculators, and then by modern computers.

Thus, in order to arrive at the technical realisation of computers, it was necessary that logic should become detached from its purely philosophical context, that it should be abstracted from Aristotelian and scholastic ideas, with their highly specific language and ways of thinking that were not formally structured and were far too centred on physical reality, and finally that it should arrive at the logic of Boole.

But to reach even the conception of a theory of computing, it was also necessary that Boolean logic was in turn considerably generalised until it arrived at the stage of *symbolic logic*. This "replaces ordinary language, which is always imprecise, by a system of symbols with precise clearly determined meanings" and "is applied not only to relations of inclusion or inherency, but to every possible logical relation" (A. Cuvillier).

SIXTH FUNDAMENTAL MECHANISM: FROM CLASSICAL ALGEBRA TO SET THEORY

This leads us to another essential intellectual mechanism, the conception and language of set theory, which are now an integral part of mathematicians' and computer scientists' everyday lives.

Even child psychology has recourse to this approach to show, as in the work of Piaget, the fundamental role played by the theory of groups, the logic of classes and more generally the processes of set theory in the formation and development of rational thought in children.

These ideas have today developed into an extremely abstract theory, which even more thoroughly generalises calculations performed on whole numbers, magnitudes or propositions, and thus calculations performed on any class of mathematical entities.

In what may be termed the *calculus of sets*, a highly abstract concept of algebra has been adopted, which is entirely independent of the nature of the elements involved, and considers only the sets with no additional specifications.

Hence we have arrived at a superbly generalised algebra: the algebra of the elements of a set, in which we define relations such as inclusion and identity between corresponding subsets, as well as operations such as union, difference, intersection, etc. These are to sets what the relations and operations of inequality, equality, addition, subtraction, multiplication, etc. are to numbers and magnitudes. As for zero in such calculations, it is quite simply the empty set – the set which includes no elements and whose "number of elements" (or cardinality) is thus equal to zero.

It was Georg Cantor (1845–1918), a German mathematician of Russian origin, who made these essential contributions, the development of which (when added to the fundamental discoveries of Babbage, Boole and others) was to be indispensable in the creation of the concepts that concern us here. However, the algebra of sets is just a part of set theory, which includes among other things the study of relations between elements of parts of a set (relations of equivalence, of order, etc.) as well as notions of graphs and of correspondences (functions or mappings from one set to another, or to itself, finite or transfinite cardinalities, etc.).

As for general algebra (in the modern sense of the term), it is now used to study the different internal operations – called laws of composition – between elements of the same set (generalisations of ordinary arithmetical, algebraic and logical operations), and thus to define what are called algebraic structures (for example, the structures called group, ring, field, ideal, etc.). It can be observed that, in this context, Boole's algebra of logic is quite simply a special case of the algebra of the subsets of a set. For the fundamental operations of Boolean algebra are:

1 *logical addition* (the disjunction of two given propositions, that is to say, the proposition which is true if and only if at least one of the two is true);

2 *logical multiplication* (the conjunction of two given propositions, that is to say, the proposition which is true if and only if both of them are true);

3 and *logical negation* (the proposition which is opposite to a given true or false proposition: true if the given proposition is false, false if the given proposition is true).

Now, by considering all of the elements of a set (the set being considered to be the "totality", the "universe of discourse" or the reference set), it is possible to construct an algebra which can then be identified with Boolean algebra. To do so, it is sufficient to examine the logical propositions that can be formed from what is termed the characteristic function of a subset. This function allots a 1 or a 0 to an element in the reference set, according to whether it does or does not belong to the subset in question (1 thus corresponds to a true proposition, 0 to a false one). We then identify the operations of logical addition, multiplication and negation respectively with set operations which can be defined as follows for the parts or subsets of the reference set: *union* (which gives the subset of elements which belong to one or other or to both of two given subsets); *intersection* (which gives the subset of elements which belong to both of two given subsets); and *complementation* (which gives the "complement" of a given subset, that is to say the subset of elements of the reference set which do not belong to the given set).

This is what led to the study of switching networks in electric and electronic circuits, whose structures formed the basis for the construction of automatic calculators and binary computers, as well as for the corresponding binary languages, codes and symbolisms.

We should also note that one of the structures envisaged by such an algebra is the *lattice* which, on a much higher level, generalises Boole's binary algebra (which, by the way, is also called a *Boolean lattice*). But it also covers other kinds of algebra (such as free lattices, modular lattices etc.) which can be applied, for instance, to study the transmission of electrical signals in the form of long and short pulses. Following the formalisation of internal operations, the introduction of laws of external composition led to the study of the structure of *vector space*, a generalisation of classic vector calculus which is the basis of the theory of linear mappings of one vector space either into itself or into another vector space. The analytic representation of this was in turn a substantial generalisation of matrix calculus.

General topology, with its concept of *neighbourhood* of a given set (allowing an extremely precise definition of the notions of limit, convergence and continuity), considerably generalised mathematical analysis by introducing the structure called *topological space*. It is a branch of mathematics with many important consequences and applications, both in mathematics and in physics: point geometries, affine geometries, metric geometries, projective geometries, topological vector spaces, differentiation, measure theory, normed space, integration, the theory of probabilities, etc.

In turn, the recent *theory of categories* (created in the 1950s from homological algebra and algebraic topology) once again considerably generalised set theory and opened the way to even greater possibilities.

This list quite simply gives the names of some of the countless areas which modern mathematics has explored with new and fascinating techniques for whoever takes the trouble of mastering them. It is nevertheless amply sufficient to provide an understanding of how much these new concepts, marked by the growing importance attributed to set theory, have radically altered the very spirit of the science of mathematics. We have, in fact, arrived at "a science whose objective is to study the relations that can exist between certain defined abstract entities with the sole condition that their definitions do not create contradictions and [incidentally] that they can be used in other branches of the sciences" (J. Bouveresse, J. Itard and E. Sallé).

In other words, contemporary mathematics is characterised by an increasingly pronounced algebraicisation of its symbols and concepts, and by a thought process that no longer concerns itself with the physical reality or non-reality of things, but rather with constant concern for the unity and non-contradiction of the theories that compose it. It is in such terms that mathematical truth must be distinguished from physical reality.

In fact, as Théodule Ribot explains: "The history of the mathematical sciences is, in part, the invention and use of increasingly complex symbols . . . Words were first substituted for things that were considered to be countable; then particular signs, called numbers; later, with the invention of algebra, letters replaced numbers, or at least took their place in the problems to be solved; later still, the contemplation of geometric figures turned to contemplation of the equations that represented them; finally, for calculations of infinitesimal quantities, negative quantities and imaginary numbers, new symbols were employed" (*L'Évolution des idées générales*, p. 164).

In this progression towards the highest abstraction, symbols have also been used in a way that has become increasingly generalised; and these symbols are now constantly employed by computers, thanks to the abilities given to them that derive from the advance of pure mathematics. "For those unaware of its past, twentieth-century mathematics could seem to be in the grip of a strange contradiction. An irresistible current draws it towards a formalism and abstraction that is increasingly stripped of reality; yet a second vocation, which is no less imperative, pushes it to draw nearer and nearer to experience . . . We do not see this as the sign of a coming scission, but rather as the systole and the diastole that guarantee continued circulation" (François Le Lionnais, *50 ans de découvertes*, p. 182). It should rather be seen as a motive force, characteristic to fundamental research, which guarantees that it gets all the oxygen and red corpuscles it needs for its development and future existence.

Now, regarding the future of mathematics, Henri Poincaré declared one day: "Before, there were prophets of evil. They claimed that all the problems had been solved and, after them, we would just have to pick up the crumbs . . . But such pessimists have always been forced to recant . . . so much so, that now I don't think that there are any left." It is true that, for a long time, people said that mathematics had been "stabilised", that the edifice had reached its culmination, but in fact it is constantly evolving and being questioned.

In this context, Voltaire's jest takes on its full meaning: "Theories are like mice, they slip through ninety-nine holes, but the hundredth stops them." Yet, this irritating hundredth hole is an important force for progress, because of the crises it inevitably sets off, the profound reflections that result and the maturing of the concepts in question.

This is, in fact, what has allowed us to define and develop mathematical logic, the theories of deduction and of proof, and more generally the theories of abstract reasoning. And, in the same way, this process has allowed us to make great progress, particularly in the field of mathematical logic, and to establish the foundations and unity of contemporary mathematics. It is the origin of the gradual development of axiomatic constructions, with their hypothetico-deductive systems. Here follows a brief presentation of their basis and approach:

> At the beginning of the nineteenth century, the axiom was considered as a truth that was evident and necessary in itself, which did not need to be justified, and which would lend its character of inevitability and absolute truth to any deduction founded upon it. At that time, the postulate was distinguished from the axiom, in that it was not seen as self-evident; it was a mere hypothesis. Non-Euclidean geometries were created by denying the hypothesis of Euclid's postulate, and replacing it with other hypotheses. But, led on by this success, it was then noticed that further geometries could be created by denying some of Euclid's axioms that had until then been considered self-evident (non-Archimedean geometries, for example), and given that all axioms now seemed provisional, axiom and postulate merged together. All that was left now was a system of hypotheses, which no longer had to be self-evident, just compatible with one another, such that their consequences did not lead to contradictory statements. This is the criterion of internal consistency. The unconditional truth, deduced from self-evident axioms, thus gave way to the conditional truth of a hypothetico-deductive system (J. Ullmo, *La Pensée scientifique moderne*, pp. 190–1).

This is precisely what occurred with set theory. It was not due only to Cantor, who had in fact been preceded by certain mathematicians at the end of the Enlightenment (who had felt the need for a rigorous construction of the set of real numbers), and then in particular by mathematicians such as Bernhard Bolzano (1781–1848), Karl Weierstrass (1815–1897) and Richard Dedekind (1831–1916), who were the first developers of the theory of sets of points. However, as Nicolas Bourbaki put it: "It is to Cantor's genius that we owe set theory, as we know it today."

But this theory was certainly not defended by everyone, and was vigorously attacked by mathematicians such as Leopold Kronecker and Henri Poincaré, especially when it took its most serious blow with the discovery of its famous "paradoxes" (such as the one formulated in 1905 by Bertrand Russell of the set of sets that are not members of themselves, which produces the contradictory conclusion that this set is both a member of itself and not a member of itself). This sparked off the most serious crisis in the history of mathematics, with Cantor's detractors seizing on the opportunity presented by such antinomies to reject the entirety of set theory.

However, this crisis was surmounted thanks to a thorough study of the exact causes and precise nature of such contradictions, before they were completely eliminated by developing the necessary formalism. (This was especially so after the 1910 publication of *Principia Mathematica* by Bertrand Russell and Alfred North Whitehead, who among other things made a profound study of the nature of the axioms on which set theory had been founded, and with it the entirety of mathematics.) To borrow Alain Bouvier's metaphor "if you want to build a house, you have to check that its foundations are solid. But the example of the Tower of Pisa is there to remind us that sometimes the foundations are not enough to guarantee the balance of the edifice. To work correctly, it is necessary to establish 'pre-foundations' to support the functions and to study construction techniques in detail in order to make them as perfect as possible. The basis of all of contemporary mathematics is set theory; but set theory itself has not been set down for all eternity. We do not know if it is or is not contradictory."

It was on this very point that a further crisis concerning the foundations of mathematics was sparked off in connection with what has been called the *axiom of choice*. This leads us to say a few words about the German mathematician David Hilbert (1862–1942) and the Austrian-born American mathematician and logician Kurt Gödel (1906–1978).

It was Hilbert who proclaimed one day that "no-one shall ever expel us from the paradise which Cantor has created for us." He was one of the keenest defenders of Cantor's ideas. His early work on the "theory of invariants" was already revolutionary in the very simplicity of the presentation, which synthesised and generalised notions already established in the field, but the mass of calculations it contains makes it difficult to extract the general ideas. He then studied the theory of integral equations. But it was in 1899 that he published his most important work, *Grundlagen der Geometrie* ("Foundations of Geometry"), in which he developed the principles of elementary geometry based on a certain number of axioms.

Thus axiomatics was created by Cantor for sets, and by Hilbert for geometry, at the same time throwing off new ideas that many mathematicians were later to refine and improve in order to embark on the coherent construction of mathematics.

> [At the base of Hilbert's conception] there is the conviction that symbols and operations on symbols constitute the irreducible kernel of mathematics. The realisation of an axiomatic system can be established only by means of exhibiting such a realisation. Hence the necessity to have recourse to a proof of consistency, that is to say a proof which shows that no contradictions can be deduced from the axioms. In Hilbert's view, this was to become the central problem and was to be identified with the problem known as "the foundation of mathematics": how to give, for each branch of mathematics, a proof of the fact that the accepted procedures of proof never simultaneously produce, as theorems, a proposition and its negation. In order to carry

out his programme, Hilbert created what he called a "theory of proof". The principle is as follows: any mathematical theory can now be expressed as a rigorously formalised system, that is to say a set of formulae which differ from normal formulae only in that they contain certain logical symbols in addition to the ordinary ones. A proof is a series of symbolic formulae, each one being either an axiom, or being obtained from preceding formulae using a pre-established rule of inference. Hence the possibility to produce proofs that are themselves the subjects of mathematical study; once ordinary mathematics had thus been formalised, a metamathematics was created, in which the techniques of the former are considered to be only operations on written forms (J. Bouveresse, J. Itard and E. Sallé).

Thus a new discipline, *metamathematics*, was created, whose subject is the foundations of mathematics and the very nature of proof, and which instead of the formalised language of mathematics, uses an everyday language that is hardly suited to such formalisation. It is a branch of mathematics that not only allows us to "decide what proof is, and to build from scratch a theory of proof", but it also "gives mathematicians the rules for the manipulation of the abbreviated symbols of mathematics and their deductive criteria" (A. Bouvier).

Hilbert was therefore a pioneer in the axiomatisation of geometry and in the establishment of the notion of decidability as a criterion for a complete and non-contradictory hypothetico-deductive method. As for Kurt Gödel, he made an even more general contribution to the foundations of set theory in his notion of the undecidability of a proposition as regards a non-contradictory hypothetico-deductive system. In 1931, he sparked off a veritable revolution in the world of logic by establishing that a non-contradictory arithmetical system cannot be complete, because it will always contain an undecidable formula.[70] In other words, an arithmetical system cannot contain the formal proof of its own non-contradictory nature. This thus establishes the impossibility of a system formally proving itself to be non-contradictory without having recourse to more powerful methods than those contained within the system itself. [For example, Gödel himself showed in 1931 that the proposition in set theory known as the *axiom of choice*[71] is consistent with (does not contradict) the other axioms of set theory and so cannot be disproved in that system, and in

---

70 More precisely, Gödel's *incompleteness theorem* states that if a formal system is such that its axioms and rules of inference are sufficiently powerful to imply the whole of ordinary arithmetic, then it is possible to write a proposition using the terms of the system which can be proved not to be any one of the propositions deducible within the system. Such a proposition is therefore *undecidable* within the system. [*Transl.*]

71 The Axiom of Choice asserts, in plain language, that for any "reference set" there is a rule of selection, given once and for all and applicable to every possible subset of the reference set, which, when applied to any subset, selects an element of that subset: in other words, it is possible to choose an element from each one of any collection whatsoever of subsets of the reference set. [*Transl.*]

1963 P. J. Cohen proved that it cannot be proved either, so it is undecidable in the context of the other axioms. *Transl.*] It was thus possible to add either this axiom or its negation to the foundations of set theory. Since we now find further undecidable hypotheses, we can then construct several other set theories, depending on whether the system of axioms that has been adopted does or does not contain these affirmations or their negations.

The practical consequence of Gödel's theorem is as follows: whatever the worth or flexibility of a general theory, it will always contain certain propositions which may be considered to be true, but which can never be proved to be true.

We thus see how logic disturbed traditional conceptions – not only in respect of mathematical truth, but also in the understanding of mathematical reasoning – which set off a movement of permanent doubt as regards the possible unity of mathematics. But, in turn, this contact with mathematics has caused a profound conceptual change in logic, which in particular has paved the way for modern computer science.

SEVENTH FUNDAMENTAL MECHANISM: FROM PHILOSOPHICAL LOGIC TO MATHEMATICAL LOGIC

Let us return, then, to classical logic, which would seem to be, in Piaget's words "the study of true knowledge, envisaged in the most general terms". This is also what constitutes – though in a far more abstract way – the perspective of elementary mathematical logic (or the *first-order predicate calculus*) which, starting with a theory of quantification defined on elements of a given set, gives a theory of the propositions which are valid for any objects belonging to any world.

However, these relations concern only what is known as bivalent logic, which recognises only two truth values (*true* and *false*), being based on the principle of the excluded middle according to which there is no other possible alternative between two contradictory propositions ("*P* is true" and "*P* is not true"), so that if one is true then the other is necessarily false. In other words, if two propositions in this logic have identical terms, and differ only in their veracity or non-veracity, one must be true and the other false; they cannot be either both true or both false.

It should be remembered that the *principle of identity* ("what is, is; what is not, is not") from which the principles of the excluded middle and of contradiction derive, and the *principle of sufficient reason* (that is to say, in Leibniz's words "nothing ever happens unless there is a cause or at least a determining reason"), constitute what are called the classical rational principles. They make up the ruling principles of all rational activity (in the traditional sense), particularly in the search for the reason and the cause of things, and are thus the basis of classical knowledge.

However, the development of relativity has cast doubt over some of these first principles, and there arose an attempt to relativise theories of reason, for example by limiting the validity of the excluded middle, which used to be seen as a universal principle, but did not, in fact, answer to all of the situations that may arise in scientific or even daily life. Such is the case, for example, with propositions that cannot be declared true or false, and which accordingly are considered to be undecidable. Hence, alongside the undecidable, there has been a growing use of terms for the non-true, such as the doubtful, the impossible, the non-necessary, the undetermined, etc.

It is to solve these problems that contemporary logicians have developed *multi-valued logics*, based on the recognition of several truth values and rejecting the excluded middle, which asserted that more than two truth values could not exist. Among these, we can mention the trivalent logic of J. L. Destouches and P. Destouches-Février (with the true, the false and the impossible), H. Reichenbach's logic which also has three values (the true, the false and the undetermined), and that of J. Lukasiewicz (which uses the true, the false and the possible, or "contingent").

Philosophically, we have thus gone from classical rationalism, which considered contradiction to be an unsurmountable barrier, to relativistic rationalism which instead sees this situation as stimulating, and even a necessary condition for progress leading to the required synthesis. By using a similar technique, Louis de Broglie overcame the classical contradiction between the particle theory and the wave theory of light and thus arrived at a synthesis that allowed him to develop his wave mechanics: "What should not be forgotten", he explained, "is that 'contradiction' is not the correct term. The contradiction of: 'the phenomenon of light consists of a wave' is 'the phenomenon of light does not consist of a wave,' and any synthesis between these two antitheses is impossible. Thus the principle of contradiction retains all of its power. We can form a synthesis only from different propositions, such as 'light is made of waves' and 'light is made of particles.'"

This underlines how, in relativist rationalism (which can also be called modal rationalism), human thought is subjected to a constant evolution, where new acquisitions can have repercussions on even the most basic assumptions (as opposed to classical rationalism which claimed that reason was immutable). The example of Louis de Broglie demonstrates how reason can accommodate itself to a quantum indeterminism which violates the principle of sufficient reason (one of the essential classical principles).

But modern mathematicians and theoretical logicians have broadened this science and taken it much further, opening an infinite number of possible directions. This has led certain logicians, such as S. Kanger,

S. Krippe and J. Hintikka, to develop the mathematical theory of modal logics, which envisage logics with *n* values through the introduction of (*n* − 2) intermediate values between the true and the false (multi-valued logics among which classical logic with its principle of the excluded middle is the particular case *n* = 2), or even logics with an infinite number of values between the true and the false. (We should note in passing that these logics have led to important applications, in particular in the fields of epistemic logic, and of semantics and pragmatics in linguistics, as well as in physics where each theory requires its own modal logic and consequently marks the corresponding formal structure.)

Then, in other directions, just as we have been led to envisage a plurality of geometries and set theories in mathematics, so a plurality of theories of logic have been elaborated. As a result, the very nature of language and of the workings of the mind have been abstracted through the creation of hypothetico-deductive systems from sets of axioms chosen arbitrarily, but subject to the one obligation of satisfying the criterion of internal consistency.

Thus have we arrived at formalised mathematics and formalised logics, in which "the criterion of the coherence of a system of axioms, expressed as relations between symbols, has replaced the criterion of the self-evidence of the principles and theorems of a deductive theory" (L. Rougier, *Traité de la connaissance*, p. 34). This is exactly what describes the approach of contemporary logicians, whose thinking tends towards such coherence by means of a chain of thoughts that form a logical whole, that is to say a rationally organised set with "co-ordination in its composition" (see Lamouche, in *Logique de la simplicité*, p. 162).

We can therefore say that "there are no logical constants. In logic, as in any deductive theory, the symbols must be freed of all pre-established intuitive meaning: they thus become nothing more than signs that are combined according to the rules laid down by the primary propositions [i.e. the axioms] ... Logic thus becomes as arbitrary and empty [of all concrete reality] as mathematics. Everyone can construct his own logic – or logics – as he wants. All that is required is that the system should not be self-contradictory [in other words, that it is not logically incoherent] and that it should be explicit" [R. Blanché (1948), pp. 116–17].

It is now clear how the nineteenth and especially the twentieth centuries have progressively drawn logic away from its mother, philosophy, and tied it ever closer to mathematical science and thought. This phenomenon is all the more interesting when we see that contemporary logical philosophy now tends to restructure itself in order to concentrate on one or other of the parts of this new generalised logic, according to the philosophical or epistemological events which it gives rise to.

EIGHTH FUNDAMENTAL MECHANISM:
THE DEVELOPMENT OF SYMBOLIC CALCULUS

"Logic has become more mathematical and mathematics more logical. The consequence is that it is now wholly impossible to draw a line between the two; in fact, the two are one. They differ as the child differs from the man; logic is the youth of mathematics and mathematics the manhood of logic" (Bertrand Russell, p. 194).

This is precisely what has led to what used to be called logistics (from the Greek *logistikos*, "concerning calculations", from *logismos*, "calculation" and the *logistikê technê* "the art of calculation"). It is the art of organising a calculation into successive steps, rather as in the logistics of the sergeant major, or in the military art of the same name whose task is to organise movements of troops and supplies. It is a science closely related to what we now call symbolic calculus: "a discipline which, instead of starting from language [or from mental activity] like classical formal logic, establishes a system of symbols . . . modelled on the forms of thought and combines them, according to the rules of its axioms, while completely abstracting their meaning" [P. Foulquié (1982)].

The following summary will help us gain a better understanding of this phenomenon and of how far we have come since the logic of Aristotle: "Originally, logic was a reflection on effective operations of thought. It analysed our usual reasoning, as it is presented in verbal expression, in order to extract the rules that determined its validity. Now, in the same way as our geometry is no longer *geo-* nor even necessarily *metry*, so logic has broken the chain that tied it too closely to the *logos*. Finally, it has left behind the *logos*-reason and even the *logos*-language in order to involve itself only in *logos*-calculus. It thus abstracts logical meaning from its symbols, in order to deal only with how these are combined and how these combinations are transformed" [R. Blanché (1957), p.14)].

It is in this very trend towards the ultimate abstraction that various disciplines called inductive logic, dialectical logic, epistemic logic, logic of invention, logic of science, etc., which used to work separately and in a dispersed fashion, have now been unified in a far more general logic, whose subjects of study are the mechanisms and structures of rational thought.

As one of the countless possibilities thus offered to the development of rational knowledge, we have arrived at one of the most powerful levers that contemporary mathematics and logic have given to the birth and development of computer science and technology. For it was during all of these successive extensions and generalisations that twentieth-century logicians finally developed the theory of algorithms. This not only led to the scientific discovery and theoretical justification of the concept of stored programmes in the middle of the 1940s, but it also and most importantly provided the fundamental basis for the creation of today's computer languages and programming methods. This theory is still today an essential element in the conception and theoretical evolution of computing concepts.

The *algorithmic approach* is, in fact, a technique in the calculus of thought, a kind of high-powered generalisation of classical reasoning. It corresponds to an entirely formalised logic, based on a series of coherent axioms, whose operations are executed entirely on symbols that are totally independent of the nature of the reality they are supposed to represent. It is thus a logic in which the terms of the propositions are themselves formalised propositions, and in which the relations that exist between the propositions are expressed by means of algorithms, that is to say finite series of symbols and manipulative procedures (see pp. 274ff.).

Accordingly, *symbolic logic* is a science in which human thought is necessary, firstly to establish the system of axioms (that is to say the initial set of propositions), as well as the rules and procedures which govern the various relations between the propositions, secondly to check the consistency, the compatibility and the independence of the axioms thus posited. But thought is no longer necessary once the algorithm has been established, since it can be carried out by a specially designed machine, that is to say, a computer.

Among the most important and fertile branches of contemporary logic, we can also mention the following: the theory of models, the theory of recursive functions, combinatorial logic, cellular logic, the theory of automata and formal languages, connexionist logics, etc. which have played a decisive role, casting fresh light on the field of mathematical constructions, as well as on the development of cybernetics, artificial intelligence and modern linguistics.

We should also point out that the *theory of* (abstract) *automata* governs the sequences of operations required to solve, step by step, logical and arithmetical problems of a certain class. It also provides mathematical models whose structure can be applied to data-processing machines, as well as the methods and procedures of calculation to be used in the lowest level sequential automata.

We have now acquired a better understanding of the thought mechanisms that were essential to the foundation and development of computer science. In fact such spontaneous human activities as reasoning and deductive thought have greatly evolved. These intellectual processes have developed as a result of countless critical examinations of the knowledge and experience that mankind has acquired over the centuries, and later led to the creation of the various sciences, technologies and procedures that have provided reason not only with the means to avoid errors and inconsistencies, but also with powerful tools to allow it reliably to carry out the different operations necessitated by activities which may or may not be related to reality. Thus these modern developments allow procedures to be modelled, at least to a given point that is determined by the current state of science and technology. The foundations of these procedures are, therefore, mechanisms of thought whose objective is to establish the sort of "logical relation" that exists between the different elements that constitute reasoning.

In the history of philosophical and mathematical thought, therefore, these mechanisms were:

1 first *classical formal logic*, from Aristotle to the modern thinkers, with the mediaeval scholastic philosophers along the way;

2 then, in a far more generalised and highly developed abstract form, modern *symbolic logic*, on the road from Leibniz and Boole to the most advanced creations of contemporary logicians.

There is, of course, a radical difference between these two approaches, which contrasts the strictly specific nature of the former with the ultimate generality of the latter. We should not forget that "[classical] formal logic makes abstractions from the truth of propositions, that is to say from their conformity with reality. It deals only with formal truth, that is to say the legitimate nature of the form and the coherence of the reasoning. But it never questions the real truth of the rules that allow for this coherence, nor the veracity of its axioms. These rules are considered to be the law of every mind and of all thought; thus, their truth is a real truth and not simply a formal one" [P. Foulquié (1960)].

On the other hand, in symbolic logic "truth is purely formal. We do not affirm that *a* is true, but that if *a* is true, then *b* is also true. What is more, the conditioning of *a* upon *b* depends on rules and axioms that have been freely chosen. These axioms, standing as [symbolic] propositions . . . and propositional functions, are not [in classical logical terms] real propositions: they announce nothing that could be really true or false. They are simply the matrices or moulds of propositions, and they become true or false only from the matter which is poured into them, that is to say from the meaning given to the symbols, which themselves are meaningless" [P. Foulquié (1960); see also R. Blanché (1965), p. 63].

This means that the principal task of contemporary logicians has been to elaborate a sort of highly abstract logical ideography that allows for logical processing, in other words a symbolic calculus. "This has a double importance," as A. Cuvillier has explained, "in replacing the imprecise language of daily life with symbols whose meanings are exact; and in being able to address any possible logical relationship, no longer being limited to relations of inclusion or inherency."

In the field that concerns us here, this fact will become even clearer thanks to the following comparison concerning the development of logico-mathematical thought from the time of Babbage and Boole up to the age of the computer. Conceptually, to arrive at Babbage's ideas, it was first necessary to achieve the abstraction which would permit:

1 that the rules of ordinary arithmetic could be transferred to an *algebra of magnitudes* (a far more general form of algebra concerned only with magnitudes and no longer with the specific nature of the objects referred to);

2 that a level of deductive logic could be reached, which depended on the rules of classical formal logic.

To arrive at Boole's ideas, it was first necessary to achieve the abstraction that would permit:

1 that the rules of ordinary deductive reasoning could be transferred to an *algebra of propositions* (a logical algebra consisting in a considerable broadening of classical algebraic and logical thought, entirely independent of any notion of number or magnitude, but now based on given propositions, without being concerned by their specific nature or their different possible meanings);

2 that a level of logic could be reached that was far more generalised than Babbage's.

And to arrive at the basic ideas of the computer, it was first necessary to achieve the abstraction which would permit:

1 that the rules of the preceding algebras could be transferred to the *algebra of sets* (a far more generalised algebra than the others, the most general possible, based now on sets and not concerned by the specific nature – numbers, magnitudes, propositions, etc. – of the mathematical entities involved);

2 that a level of highly abstract *symbolic logic* could be reached, based on fixed methods of manipulation of symbols which are of variable complexity and belonging to an arbitrary set;

3 that an artificial device could be detached from its physical reality to give an abstract structure, according to a hypothetico-deductive theory governed by logical rules bearing on a very general domain.

It is precisely because these symbols have become the most general ones possible that we have been able to give computers their infinite range of possibilities and their universality in the field of calculation.

16. The Contribution of Alan Turing

The greatest theoretical advance leading to the birth of computer science actually took place shortly before the Second World War, in 1936, when the British mathematician and logician Alan Turing (1912–1954) published his famous article: "On computable numbers, with an application to the *Entscheidungsproblem*" ["The problem of decidability"], in which he rigorously defines the notion of algorithm and introduces the fundamental concept of what has been called since then a "Turing machine" [72] and of the "universal Turing machine" [73] or *universal algorithmic automaton.*

Alan Turing's discoveries were outstanding conceptual breakthroughs, and a decisive contribution to the logical foundations and to the development of computer science.

In order to understand the circumstances that led this man, barely twenty-four years old, to produce such fertile concepts and, soon afterwards, to provoke a veritable revolution in mathematics, it is worth outlining the early stages of his life before 1936.

This account is from M. Vajou:

> Alan Turing was a child of a strict elite of civil servants from the British Empire in India. He was sent away from his parents every autumn to the isolation of boarding school. Lack of physical affection left the boy socially rather inept and with a precocious inclination for the "abstract". Fascinated by symbolic notation, he stopped at every lamppost to read its serial number. An awkward adolescent, Turing did not fit in well with the cold, elitist British Public School system. He took refuge in his love of science. This was somewhat frowned upon, not only because Alan's chemistry experiments created noxious smells around the school, but also because this was a time when the only studies that were respected in these schools were the Classics.
>
> Moreover, he was nicknamed the "Math Brain" at Sherborne School, and he enjoyed the solitary pleasure of reading high-brow mathematical texts. He neglected the syllabus set by his schoolmasters, handing in extremely careless work. Even in mathematics he obtained

---

72 An abstract logical specification of an ideal machine, capable of computing the result of executing a given algorithm or class of algorithms suitably defined in terms of the structural elements of the "machine". [*Transl.*]

73 Since the specification of how a Turing machine operates is itself an algorithm of the type which the general specification of the Turing machine was designed to execute, it is possible to specify a single Turing machine capable of being programmed with the specification of *any* Turing machine; the former would then appear to function in exactly the same way as the latter. Such a Turing machine, capable of "emulating" any Turing machine, is what is referred to as a "universal Turing machine". [*Transl.*]

only mediocre marks. To his masters he must have seemed rather stupid.

It was Alan's friendship with a boy in the year above him that made him come out of his shell, his solitude, and his intellectual stagnation. Within a year, Turing rose to the top of the class in all subjects and (in 1930) even managed to obtain a scholarship to King's College, Cambridge.

At that time, Cambridge, along with Göttingen in Germany, was the world's leading mathematical research centre. Men such as Newman, Hardy and Russell were remarkable as both teachers and researchers, on the pulse of worldwide developments within their sphere of learning. A first-year student of mathematics at Cambridge did not enter into a normal educational institution; it was more of a creative intellectual club, the seniors inducting their acolytes in the principal mathematical challenges of the day.

This must have been a particularly congenial environment for Alan Turing,[74] and he soon became engrossed in the work of von Neumann, who had made use of the most recent mathematical developments to prove the logical coherence of the still-young theory of quantum mechanics.

To Alan Turing, Cambridge brought not only the empowering sensation of being at the heart of contemporary mathematical creativity, but also the climate of liberalism that was unique to King's College.

Under the inspiration of Keynes the economist and Forster the novelist, in a country that was still rather puritanical, King's became the influential centre of the "liberal-libertarian" school of thought which championed not only individual freedom to make one's own moral choices but also the need for a more rational social order. Alan became a firm supporter of these ideas.

Cambridge furnished Turing with the logical weapons for his work which was later to be published in "Computable Numbers", and enabled him to identify an immediate intellectual target: the problem of decidability.[75]

74 Turing went up to King's College in 1931; Von Neumann's *Mathematische Grundlagen der Quantenmechanik* was published in 1932, and Turing was studying it soon afterwards. [*Transl.*]

75 The question of "decidability" can be expressed in several contexts. As originally expressed in 1928 by David Hilbert, it took the form "is mathematics decidable?", in the sense that, given a formal, abstract and symbolic system defining mathematics, does there exist a definite (i.e. pre-defined) procedure guaranteed to be able to decide, in a finite number of steps, whether an arbitrary assertion stateable in the system is true, or is false? This is closely cognate to the question which was settled by Gödel's incompleteness theorem (also of 1928): provided arithmetic is consistent, mathematics is incomplete in that there exist propositions which cannot be proved. In Turing's terms, a "computable number" was one which could be obtained by the operation of some "Turing machine". By a method of proof similar to that used by Gödel, Turing showed that there must be uncomputable numbers, thereby settling, in his own way, Hilbert's *Entscheidungsproblem*. [*Transl.*]

Such circumstances led to Turing's rigorous definition of the notion of algorithm and made him the first to develop the concept of a universal algorithmic automaton.

WHAT IS AN ALGORITHM?

Intuitively speaking, an algorithm is a finite succession of elementary rules, governed by a precise and fixed prescription, allowing the step by step execution, in rigorous order, of certain types of executable operations in order to resolve problems of a certain class.

A very basic example of an algorithm is a cookery recipe. More to the point, there are such routine procedures as multiplication; determining the quotient in dividing one number by another; finding the highest common factor or the lowest common multiple of two whole numbers.

An algorithm is a method "always expressed as definable directions, to be regularly followed in a mental activity" (M. Bernès, in Lalande's *Vocabulaire*). Conversely, a method (such as those used for learning the piano or a language, for instance) is an algorithm which economically guarantees a determined result.

Basically, a method of reasoning is an algorithm. It is a process which can be carried out mechanically (in the true sense of "mechanically") and without a thought, but which reflects an action deployed according to a calculated plan determined in advance.

According to Valéry, an algorithm is "the intention of diminishing the amount of effort that is required each time an operation is performed, and substituting a uniform treatment (at times a method of automation) in place of the need to invent a solution to each single problem . . . In other words, it is the search for a system of operations that does the work of the mind better than the mind itself can" (*Variété* IV, p. 220).

In psychology, numerous algorithms are used in the form of procedures that are each applicable to a given type of psychological problem, and which guarantee a solution (or if they do not find a solution, then the result is the certainty that there is no solution) within a finite number of stages (however large the number, in theory). In certain circumstances, the algorithm becomes a simple "calculation"; in others an operational programme, and in yet others it represents a collection of rules – which, from the point of view of contemporary symbolic logic, is strictly the same thing. There are diagnostic algorithms, enabling the recognition of a class of psychological problems, and corresponding resolution algorithms, which find the solution to the former (see H. Piéron).

The same notion is used in educational psychology, notably in techniques of programmed learning, which, according to Piéron, work as follows: "Learning material is presented to the pupil broken down into its elements. Through a series of questions, the pupil himself must find all or part of the correct answer. He can then go on to the next question. Thus the pupil can in principle learn on his own, assessing himself at each stage and proceeding at his own pace. According to the order and method of presentation, the method of marking, etc., a whole variety [of algorithms, that is to say] of types of programmes can be identified."

Other examples include procedures for extracting square roots, numerical algorithms for determining integrals, etc. Of course, many more exist that are much more complex: for example, there are those used to construct certain geometrical shapes, or those used to determine which numerical category a number belongs to, as defined by a characteristic relationship. There are even those used to calculate the values of a function according to the values of its different arguments, etc.

The statement or specification of an algorithm defines precisely what kind of problem it deals with, and includes all the rules and elementary operations needed to resolve that particular type of problem.

Turing based his mathematical theory on the notion of the algorithm, and created mathematical entities known as *Turing machines*.

WHAT IS A TURING MACHINE?

By definition, a Turing machine is an abstract mathematical device, capable of reading and writing units of information at the most elementary level of its logical analysis on an ideal tape that (ideally) is potentially infinite and in the form of a succession of "squares".

In other words, it is an abstract device called, by convention, a "machine" (in simple terms) which carries out a sequence of elementary operations in a succession of discrete stages.

The device has the following features:

1 a *finite repertoire of symbols;*
2 a *potentially infinite tape* (i.e. extensible as required), divided up into successive squares. Each square may either be blank, or may contain one of the symbols from the machine's repertoire;
3 an *erasing device*, capable of deleting a symbol from a given square;
4 a *reading and writing device*, which can move relative to the tape and is capable of reading from a square, or of writing on a blank square, at which the device is situated;
5 a *displacement device*, capable of staying in a fixed position or of moving the tape one square forward or one square back;
6 a *situation table* which, at each step of the machine's operation, defines the situation (the machine's internal state, the contents of

the tape during the stage in question, and the square being processed as identified by its characteristic symbol), which in turn determines the situation of the machine at the next step;

7 a *control unit* (the actual functional level of the abstract device), capable of interpreting the meaning of symbols in the situation table, in order to put itself and the machine in their respective correct states; thus this unit can command the relevant device to perform an action which depends on the instruction in question.

At each step:

- the behaviour of the machine is determined by the current state of the control unit and by the symbol in the square where the reading/writing device is situated;
- the control unit may either remain in its initial state or change its state;
- the machine may either remain in the same position or move the tape forwards or backwards one square;
- the machine may read, write or erase a symbol in the current square, either replacing it with a different symbol, or leaving the square unchanged.

TURING'S FUNDAMENTAL CONTRIBUTION: THE THEORY OF UNIVERSAL ALGORITHMIC AUTOMATA

These *ideal machines* are capable of modifying their own internal state, and of reading and writing units of information according to what may be called sequential processing. Turing machines simulate information-processing procedures at their most analytical level, according to an item-by-item description of what should be written, read, transferred, and modified in order to carry out a procedure, which at the same time specifies the sequence of corresponding internal states (see J. Bureau).

There are of course an infinite number of possible Turing machines, each one defining the structure of a definite family of artificial devices according to the number of its states and the scope of its repertoire of symbols. These include Pascal's machine, the difference engines, mechano-graphic punched-card calculators, Babbage's Analytical Engine, Jevons's logical piano, the English *Colossus* machines and many others, including computers, which are the most general concrete examples possible of Turing machines.

Turing's concepts served to clarify the workings of algorithmic automata (mathematical models of machines capable of executing algorithms in an automatic sequential manner) and led to numerous developments in the theory of algorithms.

However, there was the immediate problem of the *computability* of mathematical functions: can a function that may be computed by a human also be computed artificially? In other words, if there is an algorithm for the computation of a given function, can the same computation be executed by a Turing machine? Or, in yet other words, can such a machine execute the algorithm in question?

This problem was overcome (in part before the war but only completely resolved a few decades ago) by a logical theory which established that the various mathematical definitions (those of Gödel, Church, Turing, Kleene, Post, Markov, etc.) of the notion of algorithm all lead to the same class of functions, namely *recursive functions* (see p. 4 Chapter 1 above, translator's parenthesis).

The class of recursive functions, therefore, is totally independent of the definition selected of the concept of algorithm.

It gradually became clear that each algorithm itself defines a particular class of recursive functions, and consequently delimits a specific category of problems. These considerations led the logician Alonzo Church to publish a thesis ("Church's First Thesis") in 1936, on heuristic grounds, stating that the notion of a recursive function is the mathematical formalisation of the notion of computable function. Church's Thesis is still generally accepted, no counter-example having been found.

On this Thesis, any recursive function is humanly computable, and vice versa. Turing rounded off the Thesis in the same year by showing that recursive functions are computable artificially, i.e. by Turing machines. Therefore "Turing's Thesis", that all humanly computable functions are also computable by Turing machines, and vice versa, is equivalent to Church's Thesis.

To enable an algorithmic machine to deal with a problem, an algorithm must already exist for solving that particular problem. Likewise, a Turing machine must exist which is capable of executing the algorithm. Thus the "power" of a Turing machine can be measured by its ability to solve complex problems, which can be expressed formally in terms of the number of states it can assume and the richness of its repertoire of symbols.

Two Turing machines possessing the same number of states and the same repertoire of symbols will react in an identical way if both are presented with the same data: it is said that they are equivalent or that they have the same structure, and this equivalence relationship defines the corresponding family of Turing machines. When this structure reduces to a single class of problems, that is to say to a single algorithm, then it is said that the corresponding Turing machines are specialised for the calculation defined by this algorithm.

The logical next step was to combine several simple Turing machines, each one specialised for a particular type of problem, under the direction of a master machine, whose own procedures would incorporate the functions of each of the simple machines.

This stage in Turing's work led to his crowning discovery: the scientific concept of *universal machine*: an ideal device for which Turing demonstrated that it categorised a family of machines capable of executing absolutely any algorithm, completely automatically. A universal Turing machine is a logical automaton which is programmable and capable of executing any symbolic calculation. Such a machine is capable of automatically executing any type of calculation on any data which can be expressed by means of abstract symbols.

Putting it another way, Turing's universal machines are universal algorithmic automata. They are machines for manipulating symbols endowed with a very general structure and are thus capable of resolving a vast range of problems whose solutions are capable of being expressed in the form of an algorithm.

This had one fundamental consequence: at least one universal Turing machine can be conceived which can simulate any other Turing machine when under the control of an appropriate programme. As a corollary, such a universal machine is capable of storing in its memory the material representations of all kinds of information and of transforming them, provided it has been programmed to do so.

THE UNIVERSAL TURING MACHINE: MATHEMATICAL MODEL FOR THE COMPUTERS OF THE FUTURE

As D. Andler observes in his article "La Machine universelle":

Which discipline do Turing machines belong to: engineering, information technology, or logic and mathematics?

How is the notion of the Turing machine – or of a function computable by a Turing machine – less abstract and less mathematical than that of a recursive function?

The answer is twofold: a Turing machine is indeed a mathematical object but it is also, whilst not an actual physical machine, or even a blueprint for a machine, a *functional schema* for an abstract machine.

In other words, a Turing machine is to a whole family of actual physical machines what the principle, or rather notion, of "open the bottle" is to all conceivable types of bottle-openers.

The real genius of Turing's invention is the fact that he invented both the abstract form of a revolutionary device and the mathematical concept which allowed the device to be analysed: the theory and its application were unified from the outset.

It's as if Archimedes had invented the principle of the lever and the lever itself at the same moment. Turing's discovery is without equal in the whole history of science and technology.

From a logical and mathematical point of view, the effect of Turing's work was to address the question of decidability, the *Entscheidungsproblem*, and to put its findings in direct relation with the criteria of computability – a question that the German mathematician David Hilbert had already posed in 1929, namely of determining a method which could decide, without the necessity of finding a proof or a counter-example, whether or not a given mathematical proposition was valid (see pp. 263–4).

One of the results of Turing's theory was that the opposite is the case. He showed that even when there is a fixed prescription for resolving problems of a certain class, there are still "well-posed"[76] problems of the same class which cannot be resolved. In other words, for certain problems, albeit "well-posed", there is no algorithm which will terminate within a finite number of steps. These are known as *non-computable* problems, or *undecidable* problems, in the terminology of the important discovery made by the Austrian logician Kurt Gödel in the early 1930s concerning the undecidability of certain general axiomatic theories.

A logico-mathematical system presupposes at the outset the truth of a certain number of propositions – axioms – relating to mathematical entities, the properties of which are defined by these axioms. Defining new objects, or proposing new properties, the system tries to show, using logical reasoning, whether certain propositions are true or false. All those that have been established as true are then adopted as "theorems", so that a hypothetico-deductive theory is formed, according to a chain of logical deductions linking a given theorem to the system of axioms adopted.

In 1931, Gödel proved that this could not be completely done for a logico-mathematical system containing arithmetic (more precisely number theory; and assuming number theory to be consistent), because in such a system, there is at least one proposition that is *undecidable*; in other words it can neither be proved true *nor* false (see p. 264).

Transposed into the theory of Turing machines, Gödel's findings on the *Entscheidungsproblem* are as follows: a universal *programme U* cannot exist that is capable of answering, within a finite number of steps, the following question, relative to any *programme P*: *will the programme P terminate after a finite number of steps?* If such a programme existed, it would generate a logical contradiction simply by being applied to itself.

In other words, it is impossible to conceive of a general programme

---

76 That is, problems whose statement in the terms defining the class is a valid statement according to the rules of these terms: valid in the sense of being permissible though possibly false (e.g. 2 + 2 = 5 is a well-formed formula in arithmetic). [*Transl.*]

which can predict in advance whether or not the machine will come to an end of its calculations for any given programme P [see J. C. Simon (1984)].

Mathematically, Turing's discoveries confirmed Gödel's findings. They showed that in reducing mathematical reasoning to symbolic calculation (i.e. to a process that could be executed on a Turing machine) it would be impossible to find a logical sequence of elementary operations sufficiently general to determine whether or not a given theorem is demonstrable.

Unfortunately, this discovery had disastrous consequences. Many scientists, including some of the greatest minds of their day, concluded that every operation of the human brain could be described by an algorithm, and that therefore every intellectual process could be performed by a Turing machine (see pp. 313ff. and 362ff.).

It was not until the 1950s that it was shown that only a limited variety of intellectual processes are problems whose solution can be expressed as an algorithm (those that involve calculation). Many types of problems are totally inaccessible to Turing machines (and by extension, to computers which, as von Neumann later proved, are no more than manufactured models of Turing's universal algorithmic automata; see pp. 312–13).

Nevertheless, Turing's discovery, an essential contribution to the development of symbolic logic, was to exercise an extraordinary influence over future progress in automatic calculation. Less than ten years later, following technological advances and with sufficient technical expertise, Turing's work served as the basis for the theoretical justification of the concept of a stored programme, and brought about the revolution in information technology in the second half of the twentieth century.

It was to enable the construction of the first ever machines capable of describing and reconstructing certain complex processes of human intelligence, carrying out elementary logical operations, and manipulating symbols expressed in the form of numerical code.

17. Further Progress Towards the Computer with John von Neumann

WHO INVENTED THE COMPUTER?

After Turing's fundamental contribution, the next great advance took place in the United States at the end of the Second World War, mainly owing to a fundamental synthesis achieved by the American mathematician of Hungarian origin, John von Neumann.

The difficulties and problems with the machine named ENIAC served as a catalyst to research which resulted in this decisive chapter in the history of automatic calculation.

ENIAC, the analytical electronic calculator, had one major fault: its programming system was external, so its sequences of instructions were executed independently of the results of calculation. Nothing which happened during the execution of the instructions was capable of changing the order in which they were executed. In other words, once the process was under way, the machine was incapable of automatically locating a given instruction in the programme so as to "take it out of order". Thus the presence of a human operator was necessary to make any decisions that were required. Moreover, each time a new sequence of events was to be carried out, the programme had to be changed by making radical modifications to the highly complex system of connection panels ("patch boards"). The machine was not usable until this lengthy and tedious manual work was completed (see pp. 220ff.).

Before the building of the ENIAC was completed, the group at the Moore School of the University of Pennsylvania (then led by John P. Eckert and John W. Mauchly, and including, amongst others, the mathematicians Arthur Burks and Hermann H. Goldstine) were well aware of the improvements that were needed for the calculator and were looking to correct the conceptual deficiencies and the imperfections of this machine.

In 1944, a project was begun to construct a new analytical calculator, which was given the name *Electronic Discrete Variable Automatic Computer* (EDVAC).

Numerous working sessions were dedicated to the new machine, and its technical and theoretical problems. A solution which gradually emerged was the *stored programme*.

This solution meant that the calculator no longer had to be modified for each problem it dealt with. The programming system could now modify itself, its programmes, now stored in its internal memory, working on themselves as well as on the data they were given. A system for representing data and instructions using symbols expressed in the form of numerical codes was introduced for the machine to read, interpret, and carry out.

This, amongst other characteristics, is precisely what was described in the *First draft of a report on EDVAC* published on 30 June 1945, over John von Neumann's signature. It was intended to be read by around thirty experts in the field. Identifying the originator of this crucial discovery has caused much ink to flow, and given rise to much ill-feeling and sterile controversy – indeed the distribution of this report led to a fierce argument between von Neumann and the inventors of the ENIAC.

The quarrel was provoked by Eckert and Mauchly who wrongly believed that they should be credited with the discovery of the concept of a stored programme, which undeniably had been brought into being on the product of their own invention.

So they applied for a patent for the ENIAC. Because some concepts of this machine were used in the calculators produced by the Honeywell group, a lawsuit was filed against the latter by the Sperry-Rand group, which had bought the company founded by Eckert and Mauchly to create the UNIVAC Division.

On 19 October 1973, the court of Minneapolis, while fully recognising that Eckert and Mauchly had invented the ENIAC, rightly refused them the title of inventors of the computer, and returned a rather surprising verdict. They attributed the honour of that invention to John Atanasoff, simply for the little binary electronic calculator that he had made in the 1930s at Iowa State College. From a technological point of view this had exercised a certain amount of influence over John Mauchly's ideas. It was, however, a very basic calculator which had never even been operational, nor did it even have the structure of the analytical calculator (see p. 217).

This unfortunate chapter in the history of artificial intelligence is in fact characteristic of the contemporary state of mind which, in the absence of epistemological reflection and in the absence of scientific progress, (yet how could it be otherwise at a time when information technology was not even recognised or named as a discipline?) was unable to assess precisely what was at stake in this major discovery.

The word *computer* is in many ways responsible for much of the disagreement and controversy surrounding information technology in the United States and Great Britain because it is such an imprecise and ambiguous term.

The vast reach of the semantic field that secular usage has conferred upon this word (from the abacus to the expert system, and including along the way ordinary calculators and analogue machines or instruments), has given free rein to dialectic and its wide variety of arguments. That is how the inventor of an artificial calculating device came to be proclaimed as the inventor of the first computer in history. Thus anyone who ever invented a device that had even a few of the characteristics that would eventually come into play when building the first computers (in the "true" sense of the word) could claim that they were the inventor of the first computer. The word caused violent quarrels, some men going as far as to take their claim to court.

In their defence, it is extremely confusing when the same term, *computer* (or *calculator*) is employed to describe every stage in this history. It is never clear exactly what is meant, and worse still, it is difficult to follow the real progress made during the various stages. The history of information technology becomes a history that can be altered at will. When it comes to interpretation the ambiguity of the terms engenders propositions that can only be both true *and* false. True, because every device in question could indeed be the first *computer* or *calculator* of its kind, but false because it would not be the first computer in history.

Ambiguous and imprecise terminology has often masked the truth and the events in this story, obscuring the fact that the history of this discipline has generally been rather linear and reductionist. The history of information technology, followed stage by stage, needs a multidisciplinary approach, a closely reasoned classification of concepts, and an exact terminology.

There have also been several interested parties during the course of this universal scientific journey who have wished to take sole credit for work which is the result of the collective ideas of a great number of scientists from all over the world, believing that all they needed was a US patent to confirm them as sole inventor of the computer.

Von Neumann, however, never made any such claim, neither in writing nor verbally. He had enough understanding to know that in this history there cannot be a single invention, still less an inventor. There was an immense series of discoveries, developments, reflections, and choices carried out by a whole chain of scientists from all horizons, before the final revolutionary concept was reached. All the predecessors of this great mathematician, from Pascal, Leibniz and Thomas of Colmar, to Babbage, Boole, Jevons, Torres and Turing, not forgetting Archimedes, Brahmagupta, Al Khuwārizmī, Al Jazzarī, Fibonacci, Leonardo da Vinci, Galileo, Huygens, Schickard, Descartes, Falcon, Vaucanson, Jacquard, Hollerith, Russell, Hilbert, Gödel, Church, Valtat, Couffignal, Stibitz, Atanasoff, Aiken, and many another, have their place in this story, each having his share in the honour attributed to John von Neumann or whoever else.

One thing is certain, and that is that the concept of the stored programme had been central to the preoccupations of a whole team headed by Eckert and Mauchly, people who brought with them considerable knowledge of and technical experience in electronics. The part they played in the development of the emerging world of information technology (from ENIAC to EDVAC and from BINAC to UNIVAC I) was more than decisive in the technical and commercial history of computers.

"The concept of the stored programme came before von Neumann's participation in the work on the EDVAC." (N. Metropolis, J. Howlett and G. C. Rota)

"Working together, the ENIAC group developed the concept of the stored programme," explains Harry D. Huskey, a member of the ENIAC team. "The conclusions of this work were summarised by von Neumann in a text which, not yet in its final form, made no mention of the other contributors. This report was reproduced and circulated in its initial form. The result was that von Neumann came to be generally recognised as the inventor of this concept."

A. W. Burks, however, claims that the report was circulated without von Neumann's knowledge. "Had a definitive version of the report been written, von Neumann would have certainly given the other members of the team the credit due to them." The fact remains, though, that, as John P. Eckert later stated, "A misunderstanding resulted from the publication of the Draft . . . Dr von Neumann never attempted to correct it."

In retrospect, it is regrettable that von Neumann did not have the scientific instinct to mention the numerous preliminary discussions he had had with Eckert, Mauchly, and the others which led to his highly fertile and powerful idea, precisely because this first edition, even if it was only a rough draft, was liable to get into circulation and appear to establish his priority.

Eckert, however, was gracious enough to acknowledge von Neumann's work: "Dr von Neumann expressed some of the ideas of Dr Mauchly and myself in a form of semi-mathematical logical notation which is of his own devising." Now it is precisely this "semi-mathematical logical notation" of von Neumann's that made all the difference between his own work and the contribution of the other members of the team.

The quarrels that took place over this issue were pointlessly provoked by a misunderstanding caused by confusion between the practitioners' viewpoint and the rigorous justifications contributed by a theoretician.

The following is a quote from an interview with Eckert in 1962 (see pp. 225–6): "The puzzling thing is that all the components that went into the ENIAC were available ten or maybe even fifteen years earlier . . . the ENIAC could have been made ten or fifteen years earlier than it was, and the real question is: why wasn't it?" The answer is yes, the components were all there, but the expertise and most of all the necessary advances in science were not yet in place.

To go back in fact to the start of the story, the process of discovery actually began in 1944 when Mauchly and Eckert perceived the advantages that would accrue, for the calculator of the future, from storing the programme instructions in the internal memory, and therefore from the processing of these instructions in the same way as the data and results of the calculation. This was clearly set out in a report of September 1945:

> In January 1944, a project to make a magnetic calculating machine was begun. The most important feature of the project was the fact that the instructions and the function tables were memorised on the same medium as the data. The invention of acoustic delay-line storage by Eckert and Mauchly in early 1944 provided the means to construct a large and rapid memory from a relatively small amount of material. Around July 1944 it was decided that, as soon as work on the ENIAC permitted, the development and construction of a new machine would begin. This machine was known by the acronym EDVAC . . .

From the second half of 1944 until the present day, Dr von Neumann, advisor to the Ballistics Research Laboratory, was fortunately available for consultation. He participated in numerous work sessions on the logical controls of the EDVAC, developed certain instruction codes and tested the proposed systems by writing the necessary sequences of instructions for the treatment of specific problems.

"However," as R. Ligonnière points out, "this idea was condemned to remain no more than notional so long as a large fast memory remained out of reach." Whilst this is a valid reason, there is another one which is much more profound. Even when the technical difficulties were overcome, the idea in question would without a shadow of a doubt have remained a mere notion if the fundamental problem of a theoretical justification had not been resolved (see pp. 223ff.).

The fundamental problems of the logic of the analytical calculator had to be solved, as well as the questions about the application of the necessary electronic components and circuits needed for the technical realisation of elementary logical functions.

This clearly marked the limitations of these engineers and technicians and brought to the fore the need for a theoretician of high achievement and wide culture who would be capable of grasping these fundamental questions and solving them completely and scientifically.

A unique theoretician thus came on the scene, one not only unrivalled in skill but also at the front of all the latest scientific, logical, and mathematical developments: John von Neumann (1903–1957), one of the greatest mathematicians of his time who, with his extraordinarily quick mind, his faculty for simple reasoning about complex issues, and his astonishing gift for mental calculation, was a real phenomenon. His colleagues were amazed by him and claimed that he did not belong to the human race, rather he only looked like a human being.

He was born Hungarian in Budapest, then part of the Austro-Hungarian Empire. From 1921 until his death, he unceasingly produced great quantities of work in all fields of mathematics: from set theory to quantum physics, from the theory of functions to geometry, from digital computation to hydrodynamics, he knew everything, but he remains celebrated above all others as the inventor of the theory of games and its application to economics. He emigrated to the United States in 1930, and in 1933 joined the Princeton Institute for Advanced Study, where Einstein also worked. Von Neumann was the youngest member of the Institute. During the war he participated in the research into the atomic bomb at Los Alamos and was struck by the length and astronomical quantity of repetitive calculation required to compute the propagation of chain reactions in nuclear fuel ... He made his own contribution to these calculations, being gifted with a prodigious facility for mental arithmetic, which is a rather rare gift for a theoretician. At the end of 1944, as von Neumann had shown an interest in automatic calculators, he was offered a contract with the Moore School and commissioned to work to conceive an entirely new type of calculator (J.Tricot in: SV 741/June 1979).

THE GREAT SYNTHESIS OF JOHN VON NEUMANN

This great mathematician, who made fundamental contributions to numerous branches of pure and applied mathematics and had a perfect grasp of symbolic logic, left a profound mark on the scientific discovery of the concept of the stored programme.

According to H. H. Goldstine, von Neumann's preliminary report was not only "a masterly analysis and synthesis of all the reflections made between autumn 1944 and spring 1945", but also "the most important document written to date on calculation and calculators. Not everything in the report was due to von Neumann, but he had contributed to everything that was essential in it."

Von Neumann was characterised not only by his analytical mind, but also by a mind capable of synthesis. The greatest synthesis in the birth of information technology was derived from his work. He it was who transformed a jumbled mass of disparate ideas into a coherent whole. "Faced with a complex object, the human mind performs the following three acts: 1. takes a confused, overall view; 2. takes a distinct and analytical view of its constituent parts; 3. uses the knowledge of the parts to make a synthetic reconstruction of the whole. These three stages can be called syncretism, analysis, and synthesis, and they correspond to the three stages of knowledge" (E. Renan, *Avenir de l'intelligence*, XVI, p. 301).

This complementarity between analysis and synthesis plays an essential role in the formation and advance of scientific disciplines: As Diderot observed: "I distinguish between two ways of developing the sciences: one is to increase the body of knowledge by discoveries, and that is how one acquires the title *discoverer*; the other is to bring the discoveries together and organise them, so that men may be enlightened, and each man may participate, according to his capacity, in the insights of his time."

While it is important to disassemble, to resolve the whole into its parts, for the advancement of science and knowledge, it is also indispensable to combine, to situate, and to re-assemble these diverse, separated fragments that are collected. This creates a new "whole", now intelligible to the inquisitive mind, which can then proceed by itself with the dissection of this "whole" into its component elements, in order to form in the mind, before even putting them in the mind of others, either the "whole" of the initial dissection, or a different "whole" this time associated with a new reality, according to its functional individuality and its characteristic law of action and development.

This leads to the consideration of the real reasons behind the worldwide repercussions caused by the *First draft of a report on EDVAC*, published by John von Neumann. As R. Ligonnière explains:

> Before von Neumann's masterly intervention, the reports issued by the Moore School group had been limited to the periodical reports required by their sponsors, the American government. The reports concerning the EDVAC were short and very general; Eckert's reports, for example, were usually little more than two pages long. [Later on, even when ENIAC was no longer classified information, and when technical publications on electronics had been authorised, Eckert and Mauchly published a lot less than the mathematicians Goldstine, Burks, and von Neumann.] Due to the excessive brevity of the official documents, von Neumann, as an academic should, undertook a complete description of the theoretical constructs necessary for the logical control of high-speed [analytical] calculation. As far as he was concerned, the logical structure was independent of the machine itself; it had a universal significance.
>
> Rather than a paper on the EDVAC itself, the logician and mathematician was describing the computer of the future, with its five major components that he listed as follows: the arithmetical unit; the control unit; the memory, to record the numerical data and instructions; input devices; output devices (in: *Ordinateurs*, 29 April 1985, pp. 34ff.).

It is not merely by chance, then, that even today, computers are referred to as "von Neumann machines". As Claude Bernard noted: "Sciences are studied *analytically*; they are taught *synthetically* ... When a science cannot be taught synthetically, it is because it is not yet sufficiently developed. A science is not a science until it is *synthesised*. Thus the men who synthesise science are great men indeed: Newton for physics, Lavoisier for chemistry ... (C. Bernard, *Cahier rouge*, pp. 102–44).

Had Claude Bernard lived in the second half of the twentieth century, he would undoubtedly have added the name of von Neumann to his list, for information technology, for his synthesis enabled the development of computer science.

In *Regards sur le monde actuel* Paul Valéry writes that "Science has transformed the life, and increased the power, of those who have possessed it. But by its very nature, science is transmissible; it is made from universal methods. The solutions science gives to one person are solutions that anyone may acquire."

The *First draft of a report on EDVAC* was the first scientific document to give a general overview of the concepts bearing on the new systems of programming and on the theories of numerical and sequential automata with stored programmes. Von Neumann made the concept of the computer scientifically accessible and universally available.

THE ACTIONS AND REACTIONS OF TURING AND VON NEUMANN

Where, then, does this leave Turing? Contrary to popular belief, Turing was not simply a pure theoretician of symbolic logic: he also was a pioneer and practitioner of the emerging science of information technology. In 1937 he was at work on a binary multiplier using electromechanical technology.

It was the war, however, which gave him the opportunity to round off his experience in the field of artificial automatic calculation. In 1939, the Second World War led him into the domain of military applications, above all to put his theories into practice in the service of the Allies. At that time, he was employed by the British intelligence service at Bletchley Park, his mission being to participate in the construction of cryptanalytic machines in order to decipher the enemy codes and gain knowledge of the German army's movements. This is how the *Robinson* machines were first created, incorporating telephonic electromagnetic relays, before being replaced in 1943 by a series of logical machines called *Colossus*, which used electronic tubes and were once again developed from Turing's theories (see pp. 218ff.). Without being directly involved in the construction of these machines, he saw the process at sufficiently close quarters to master the electronic technology and gain competence in engineering.

Once the war was over, Turing devoted himself to putting his theory of 1936 more directly into practice: the construction of a working model of a *universal Turing machine*, which especially enabled him to appreciate the alternating duality between abstract theoretical questions and the physical problems of giving them material form. In October 1945, he presented a project to the government, and then joined the National Physical Laboratory (NPL) at Teddington, becoming an established member of the Scientific Civil Service. Turing's report, *Proposals for Development in the Mathematics Division of an Automatic Computing Engine* (reference number E882), was a detailed project for the construction of a true electronic computer: the *Pilot ACE*, which was not to see the light of day until 1950, after much scaling down of the original concept, Turing having resigned from the NPL in 1948. The project was particularly interesting in that it introduced very new and extremely bold concepts (see B. E. Carpenter and R. W. Dorant).

When von Neumann carried out his enormous work of synthesis, he had been greatly influenced by the ideas of the great British logician and mathematician. Von Neumann first met Turing in person during the latter's time at Princeton between 1937 and 1938, and on subsequent occasions during the war (especially in 1943 when Turing went to the States to work at the Bell Telephone laboratories to develop an encoding system for strategic communications by telephone between the United States and Great Britain), and von Neumann knew Turing's work in depth. Impressed by his intellectual capacities, von Neumann even offered to take him on as an assistant for a year but Turing refused, preferring to return to Cambridge.

The question is, did Turing arrive at the concept of the stored programme before or after von Neumann's publication?

The history of Turing's pre-war ideas has been represented with an exaggerated bias towards a "prospective vision of the development of information technology", and he is credited with having, as early as 1936, "a deep understanding of the implications of the notion of the computer".

It is true that by 1936 Turing had begun to realise, little by little, the essential importance of the concepts he had developed, paying particular attention to the object of his research: the *universal machine*. He began to develop the idea that these mathematical entities could be physically modelled by machines that could be endowed with the ability to undertake numerous intellectual tasks, assuming that their programmes were sufficiently developed (see pp. 272ff.).

But this does not necessarily imply that he had hit upon the concept of the stored programme. The physico-mathematical relationship between the concept of the ideal machine and the all-too-concrete reality of multi-purpose analytical calculators had yet to be established. In Turing's universal machine, the notions of "programme" and of "data" were equivalent, but this only became clear later, since this equivalence existed only in a potential, virtual state before the war. Turing's theory still needed to be translated into operational terms, closer to physical and technical realities than to the abstract truths of pure mathematics. Constructing a real universal machine had been everyone's dream since Babbage's time (see pp. 245ff.).

Hitherto everyone had been heading in confusion down a blind alley, in the absence of a precise logical-mathematical foundation. Turing, on the other hand, had started with a theoretical foundation which rested solidly on mathematical foundations; the idea which he developed would therefore provide the powerful lever which would help to establish the equivalence between the *universal Turing machine* and the *universal analytical calculator with a stored programme*.

Turing's results, whilst most promising from the very beginning, apparently remained locked inside the brain of their inventor for more than a decade, ideas that were not yet operational since the expertise and technical experience of the time still fell short of what was needed to construct a real machine. If, throughout this book, we have stressed the essential role of the theoretical breakthrough, this, however, is the moment to be emphasising the crucial role of technological progress in passing from the abstract universe of mathematical objects to the concrete world of material constructions.

Conceptually, an infinity of Turing machines exist in the mathematical sense, amongst which of course we may perceive the structure of the computer but also, amongst many others, of Babbage's Analytical Engine, of Jevons's and Leibniz's machines, and even of Pascal's calculator. To focus exclusively on Turing's machine is imprecise. Moreover, any discussion of the applications of Turing's theories must carefully examine the structures which can be achieved in reality. If Turing's ideas enabled the *Colossus* machines to be built, this was because of the very specific structure of the code-breaking machines. These analytical calculators were necessarily so highly specialised that it was in fact difficult to use them for any other task; moreover, unlike that of the true computer, their programming system was external.

It is known that in 1945 Turing's project was some months behind von Neumann's Draft of 30 June. Indeed, Turing makes direct reference to the report in the bibliography for his initial project. There are some who have claimed, quite rightly, that Turing's developments were independent of von Neumann, arguing that Turing in his project had developed most audacious concepts of information technology which were far in advance of those of von Neumann. This proves nothing, however, bearing in mind the lapse of time between von Neumann's report and that of Turing.

Certain authors have also claimed that if Turing referred to von Neumann's Draft, this was for tactical reasons, in order to give his own project credibility by associating himself with an international authority. Also with an eye to funding, lest Britain fell behind their allies, but also their rivals, across the Atlantic. This theory cannot be ruled out, but at the same time it has never been proved.

Moreover, why would Turing have waited nine years to the end of 1945 to produce the plans for a computer? And why have similar plans attributable to Turing never been found in the British archives, guarded jealously during the war, but now in the public domain? If Turing thought of the idea of a stored programme during the war, why did he not apply it to the *Colossus* calculators? If such were the case, Turing would surely have convinced his superiors of the advantages of using his discovery and they would have made specialised computers for code-breaking which would have been eminently more efficient than the Babbage-type machines specially constructed for this purpose.

The most plausible hypothesis would appear to be as follows: von Neumann was influenced by the theoretical study of the universal Turing machine, and Turing in turn was influenced by von Neumann's demonstration establishing the mathematical equivalence of the concept of the stored programme with the concept of the universal Turing machine. This hypothesis allows Turing to be viewed as the master of his ideas, gifted with such extraordinary intellectual ability that von Neumann's results, in a short space of time, sparked off in him ideas far more advanced than those of von Neumann himself, leading to results that the latter had not yet foreseen. (It was Turing who had introduced such novel concepts as subroutines, stacks, high-level languages, microprogramming, and so on.)

Turing's influence, however, remained cloistered within the walls of the British secret service code and cipher community, while von Neumann openly published his own findings for the benefit of the world scientific community. His Draft provided the impetus for all subsequent research in automated calculation, with the result that further research along the lines originally proposed by Babbage was soon abandoned. So although Turing's ideas between 1942 and 1944 contributed to the Allied victory both in the battle of the Atlantic and in the code-breaking war, they were all so highly classified that most of his ingenious discoveries were kept under lock and key right up until 1975.

FROM TURING'S UNIVERSAL MACHINE AND CLASSIC ANALYTICAL CALCULATORS TO COMPUTERS

Turing's contribution has been fundamental to this story, his work being the most important source of inspiration for von Neumann. In order to find a mathematical justification for the concept of the stored programme, von Neumann placed the structure of analytical calculators within the general theoretical framework of Turing machines and proved the equivalence of this structure with that of the universal algorithmic automaton.

The proof of this equivalence was von Neumann's great synthesis, the integration of diverse elements into a "whole". It matters little if von Neumann did not arrive at the concept of the auto-reproductive automaton at that time, because he laid the groundwork for others to do so later.

Von Neumann demonstrated mathematically that by storing a

programme in the memory of an analytical calculator that is structurally conceived to carry out a vast category of operations, one obtains a completely new kind of calculator which has the properties of the ideal universal Turing machine, which processes its programme on the same footing as its data.

Von Neumann's essential contribution was to demonstrate that a programmable sequential automatic numerical calculator with a stored programme is a complete concrete model of Turing's universal algorithmic automaton (in terms of abstract mathematical structure), while the universal Turing machine, as a mathematical concept, characterises the family of analytical calculators with a stored programme.

It is this "computer structure" which, by eliminating the main structural defect of Babbage's machines (their external programme) made the leap from the family of analytical calculators to von Neumann machines or computers. Therefore von Neumann's advance can be summarised as the synthesis of the technical parts and functions of the analytical calculators with the mathematical and logical properties of the ideal universal Turing machine.

Thus revolutionary artificial logical automata were born, capable, without human intervention, of executing any kind of algorithm within the physical limitations of such a machine.

The concept of the universal Turing machine already possessed the principal characteristics of the computing machines that were to follow:

1 a large-capacity memory;

2 a limited set of precisely defined instructions;

3 programmes of which each is a translation into terms the machine can understand of an algorithm to solve a given class of problems;

4 automatic sequential operation;

5 symbol-manipulation capability;

6 the ability to solve a vast range of problems;

7 the ability to simulate any other Turing machine when suitably programmed, etc.

In a word, the mathematical notion of Turing's universal algorithmic automaton was destined to provide the theoretical model for all the computers of the future.

Today the concept corresponds to the most basic level of analysis to which a computer's operation can be reduced. In other words, a given operation cannot be performed on a computer unless it can be mathematically demonstrated that a Turing machine exists that is capable of carrying out the same operation (see pp. 244ff.). A machine cannot be a computer unless it can be demonstrated that it is structurally equivalent to a universal Turing machine. Thus all computers, being equivalent to a Turing machine, are equivalent to each other and, mathematically speaking, form the same family.

This same structure has given analytical calculators an almost unlimited flexibility, allowing them, within the limits of available technology, to perform a vast variety of increasingly differentiated tasks, according to the following procedures, which constitutes one of their main properties (see pp. 302ff.):

    a the programme for each calculation contains all the commands needed to perform the calculation in question;

    b by loading a given programme into the internal memory, the machine, once the process is under way, carries out the successive basic commands of the programme according to a procedure that goes from the beginning to the final result of the calculation that has been undertaken;

    c each programme that is executed remains stored in the machine's memory at least until all the operations have been completed;

    d each programme loaded into the machine is completely indepen-dent of the data and, once the work is finished, may be then executed as many times as desired, with either identical or totally different parameters.

These procedures seem so familiar today that we are no longer conscious of the revolutionary importance of their discovery. But it is easy to become convinced of it by thinking of the analogy with writing: compare the old office typewriter with no memory, and no capability of being programmed – again nowadays taken for granted – or the old quasi-manual office calculator. In each case, in the old days, once a task was completed, or if anything needed changing, work would have to start again right from the beginning.

18. The First Generation of True Computers

THE FIRST TRUE COMPUTERS

The age of the computer more or less coincided with the end of the Second World War. The ideas contained in von Neumann's Draft had so much potential that at the end of 1945, on both sides of the Atlantic, a frantic race began to make electronic and binary computers.

Fortified by experience, but disappointed by the publication of von Neumann's Draft, Eckert and Mauchly left the Moore School group and in March 1946 founded their own company, the Electronic Control Company (which, in 1948, became the Eckert and Mauchly Computer Corporation). From 1947 to 1949, they built the BINAC (*Binary Automatic Computer*): the first electronic computer ever built in the United States. Built for an aviation company, this computer was limited in its applications, for its functions were related purely to missile control systems.

Meanwhile, von Neumann returned to Princeton, soon to be followed by Goldstine. From 1946 until 1951, they threw themselves into the making of a machine called the *IAS Computer* (after the Institute of Advanced Study) which was inaugurated in 1952 and was the first scientific American computer, used for complex problems in pure mathematics.

Work still continued on the EDVAC, but at a much slower pace than expected. The construction broke off in 1949 and was not completed until two years later, when it was to serve until 1962 as calculator at the Ballistics Research Laboratory.

At the American National Bureau of Standards in Washington, DC, another von Neumann machine was constructed known as the *Standard Eastern Automatic Computer* (SEAC), which remained in use until April 1950, and was the first non-specialised American computer.

The *Whirlwind I* of the Massachusetts Institute of Technology was another computer based on the plans of von Neumann; it was conceived and completed between 1946 and 1951 under the direction of Jay W. Forrester, and it was the first American computer structurally conceived to solve a vast range of problems by real-time simulation. Installed in 1951 for the US Air Defense, it had a magnetic-core memory, a graphics printer, remote user interaction system, a communication system via a telephone line, on-line documentation (also available via telephone line), and the programming language which was easier to use than machine language, etc.

The first commercial American computer was the UNIVAC I, designed and built by Eckert and Mauchly in 1951, whose company was absorbed by Remington Rand to become the UNIVAC Division. Delivered to the US Census Bureau this machine, the first to be used for non-military purposes, marked the beginning of the industrial era for information technology. UNIVAC I attracted a spectacular amount of interest from the general public as to the potential of computers; it was used on CBS television for opinion polls during the presidential elections of 1952, accurately predicting Eisenhower's victory.

Finally, there was the *Defence Calculator*, commissioned by the American Department of Defence during the Korean War, but not inaugurated until April 1953. It is better known by the name IBM 701, the first IBM electronic computer (that is to say it was built according to von Neumann's plans), and was the first of a long line of computers produced by that company.

However, it was not the United States that witnessed the first operational von Neumann machines. The very first was at the University of Manchester in Great Britain. It was called MADM (*Manchester Automatic Digital*

Machine), designed in 1946 under the direction of Tom Kilburn and Frederic C. Williams, and brought into operation on 21 June 1948. This machine was very limited, however, and had a poor performance, being only an experimental model. Its capabilities were extended in 1949, and the first commercial machines were launched in 1951 by the British firm Ferranti Ltd.

The world's first non-specialised operational computer based on von Neumann's plans, fully functional and in regular use, was the EDSAC (*Electronic Delay Storage Automatic Calculator*). It was designed from the beginning of 1947 under the direction of Maurice V. Wilkes and became operational on 6 May 1949 at the University of Cambridge, where it remained in use until July 1958. This machine already embodied interesting concepts: translation into machine language of commands input in symbolic form; automatic loading of the programme into the core memory; etc. The EDSAC was put on the market in 1950 in an improved form by the J. Lyons Company of London, under the name of *LEO Computer* (Lyons Electronic Office).

As for the *Pilot ACE* of the National Physical Laboratory, this was built in 1950 on Turing's plans but along much less ambitious lines than the ideas presented in the initial project in 1945. Its principal use was in aeronautics.

These early models mark the beginning of a totally new era, characterised by a cascade of technological and conceptual advances. This was the beginning of a profound revolution in the history of civilisation, in which the frontiers of the imaginable would each day recede further towards the most distant horizons.

19. The Pocket Calculator

A few decades ago no-one would have imagined a computer that could be held in the palm of one's hand. This, however, is now a reality, with the arrival of the pocket calculator towards the end of the twentieth century. We can hardly close this chapter, then, without devoting a few pages to the principal stages in the history of this device.

Over the last twenty years, tiny electronic calculators have flooded the international market. Never making an error, working with an extraordinary precision, they can carry out calculations involving hundreds of digits and solve highly complex problems. In order to understand how these tiny machines came to be equipped with such a surprising capacity, efficiency, speed, and power, we must look at the recent progress in electronic technology, which has had considerable influence over their development.

This will also shed some light on the technological evolution of electronic computers themselves.

THE FIRST GENERATION OF ELECTRONIC CALCULATORS

Their story began in the early 1940s, when the first classic electronic calculators were built.

One of the first to be built was the experimental "four-operation" electronic calculator, created in 1942 by J. Desch, H. E. Kniess, and H. Mumma in the NCR laboratories at Dayton, Ohio. It would be another decade, however, before it appeared as a commercial product, notably the famous *National Computronic* and *National Multitronic*.

Based on the use of *vacuum tubes* (the main function of which was to switch electrical signals to enable basic arithmetical functions to be carried out), these calculators were thousands of times faster than the electro-mechanical calculating machines of the 1930s and 1940s.

However, this technology finally turned out to be inadequate and unreliable: the valves were fragile, cumbersome, gluttons for electricity, and gave off a lot of heat which had to be extracted, as well as being very expensive.

THE SECOND GENERATION OF ELECTRONIC CALCULATORS

The next step was taken in the United States at the end of the 1950s, after a discovery made in December 1947 at the Bell Telephone Laboratories in New Jersey: the *transistor*, a semiconductor which considerably shortened the path of the electronic impulses in the circuit, and therefore dissipated much less heat.

Using *printed circuits* (which combine several transistors with diverse capacitors, resistances and other discrete electrical components), the "four-operation" electronic calculators of this second generation were reduced from the size of a wardrobe to the size of a suitcase. They were faster, too, much more reliable and much less costly to produce.

The first calculator based on this technology was *Anita*, built in 1959 by the British firm Bell Punch & Co., and marketed internationally from 1963. Equipped with a complete keyboard and a single display showing first the data, then the results, the *Anita* had twelve neon tubes, set out side by side, to display the luminous digits of the data or the results.

In 1961, the *Friden 130* appeared, with a reduced keyboard, and a cathode ray tube as in televisions, the screen displaying data and results simultaneously.

Many machines of this kind were built and marketed, creating a booming industry which lasted up until the end of the 1960s. These machines were greeted with delight at the time, their silence and speed being such a contrast to the infernal racket created by the old mechanical or electromechanical machines.

However, they were still quite expensive and bulky, being roughly the size of a cash register.

THE THIRD GENERATION OF ELECTRONIC CALCULATORS

Meanwhile, a great advance in semiconductor technology occurred in 1958, with Jack Kilby's invention of *integrated circuits*. Kilby was an American engineer working for Texas Instruments (one of the electronics firms involved in the miniaturisation and standardisation movement – led by the military – which became the first company in the world to manufacture silicon transistors on a large scale).

The elimination of discrete transistors and mechanical connectors in electronic circuits had significant consequences because it meant that thousands of transistors and capacitors could be contained within a very small space. (The Mercury, Gemini, and Apollo space programmes, which sent the first men to the Moon, would probably not have been possible without this fundamental contribution, which won its inventor the National Medal of Sciences.)

The integrated circuit, made of a semiconducting metal, comprises different zones which can perform different functions in an electric device (amplification, logical operations, and switching). This technology brought about a great reduction in the size of these machines, as well as in their cost, giving rise to the first ever *pocket calculators*.

THE FIRST POCKET CALCULATORS

The first electronic pocket calculator was made in the 1960s in the laboratories of Texas Instruments. Research had begun in 1965, under Patrick Haggerty, then chief executive of the firm, and ended in 1966 with the production of the first experimental model (now kept at the National Museum of American History at the Smithsonian Institution in Washington, DC).

This was to revolutionise classic calculators. In 1967, the principle of the little machine and its various components were described in minute detail by Jack Kilby, with the collaboration of Jerry Merryman and James van Tassel, in their application for a patent for the *Miniature Electronic Calculator*, as this prototype was named. Officially registered in 1972, the patent was granted to Texas Instruments by the US Patent Office, with the number 3 819 921 (personal communication of R. Ligonnière).

However, Texas Instruments were not the first company to make commercial calculators of this kind, nor even the first to go into mass production. The first commercial model was a four-operation pocket calculator weighing two and a half ounces and priced $150, launched 14 April 1971 under the name *Pocketronic* by the Universal Data Machine Co. from a warehouse in Chicago. Using MOS integrated circuits (metallic oxide semicondutors) bought from none other than Texas Instruments, in the early 1970s this company produced between five and six thousand pocket calculators a week, sold in American department stores. In reality, it was not until 1972 that Texas Instruments launched its first commercial product to the general public: the *Datamath*.

Since then, the race was on to make smaller and smaller calculators, whose price shrank as fast as their dimensions.

FROM INTEGRATED CIRCUIT TO MICROPROCESSOR

The final step in this technological advance was made when the *microprocessor* was introduced in 1971, a highly advanced integrated circuit, complete with memory, input and output devices, data-processing control and a programming system, and able to specialise in certain types of task.

Widely used since 1976, when high-density components went into mass production (technology known as VLSI, or *Very Large-Scale Integration*, which is still undergoing constant development today), this important innovation has had many applications: washing machines, cars, computers, micro-computers, industrial robots, etc. It became possible to include tens, indeed hundreds, of thousands of memory positions on a circuit occupying less space than a little-finger nail. Its potential revolutionised a multitude of activities, radically changing everyday life for millions of consumers.

THE POCKET CALCULATOR'S NUMERICAL DISPLAY

Another improvement brought to pocket calculators concerns methods of displaying the digits. On the electronic desk calculators of the 1960s, the numbers were displayed by *luminescence* using a certain number of electronic tubes: each tube contained ten electrodes which respectively formed the outlines of the digits 0 to 9; these all shared the same anode, but each one was connected to a different pin. Thus the display was created by an electric charge which brought the relevant electrode to the potential at which a glowing outline of the digit would appear.

With the arrival of the first pocket calculators, the technique changed radically, the digits now being displayed by LEDs (*Light-Emitting Diodes*), using a semiconductor which emits light when subject to a voltage.

Since the middle of the 1970s, however, LCDs (*Liquid Crystal Displays*) have been used which, under the effect of a very weak electric field, change the liquid crystal from transparent to opaque.

BIRTH AND DEVELOPMENT OF POCKET CALCULATORS WITH PRINTERS

Another improvement was to provide these machines with a printer, allowing the intermediate operations and the results of calculations to be permanently conserved and which, for a great number of professions, is absolutely indispensable. An electromechanical system was at first employed, but this was soon abandoned in view of its weight and size.

One of the notable innovations in this domain has been the development of *thermal* printers. A printing head transmits heat onto paper impregnated with a chemical solution which instantly reproduces the characters clearly and distinctly. By doing away with the need for ink ribbons, this technique has made the devices quieter and more reliable.

FIRST STRUCTURAL EVOLUTIONS OF POCKET CALCULATORS

As we have already seen, these calculators, each of whose buttons was connected to a specialised microprocessor, could carry out the four basic arithmetical operations with great speed and total reliability. Some could even square or cube a number, or extract the square or cube root, or calculate the inverse of numbers. But they were still incapable of carrying out the kind of complex calculations that are used in the world of finance, statistics, accounting, physics, and mathematics.

When science and technology had advanced sufficiently, the creators of these calculators had the idea of adding a certain number of buttons that would allow for supplementary functions: algorithms programmed into the internal circuits of the machines would allow the calculation of percentages, simple or compound interest, annuities, trigonometrical functions, logarithms, or exponential functions.

This was to lead to the birth of statistical, financial, and scientific calculators specially adapted to the needs of their users and equipped with more and more memory for fundamental mathematical constants (π or e, for example).

Thus the pocket calculator not only replaced *Curta* calculating machines and desk calculators, but also did away with the need for logarithm tables and slide rules, so long the joy (or bane) of generations of students, mathematicians, physicists, astronomers, engineers, technicians, architects, etc. (see pp. 141ff.).

THE FIRST PROGRAMMABLE POCKET CALCULATORS

Despite all these improvements, none of these machines could be said to belong to the family of electronic computers, because they only solved a very limited kind of problem: the kind for which they were structurally conceived.

As for their technical characteristics, they can be summed up as an *input* device (the keyboard), an *output* device (the display screen, sometimes accompanied by a *printer*) and an *arithmetical unit*, a data-processing device made up of microprocessors integrated into the circuits and set up to carry out predetermined calculations.

But these machines had no memory. Moreover, they lacked flexibility, for any repetitive or sequential process involved the user in a great deal of input.

Whilst they were capable of carrying out quite complicated calculations, they could not cope with mathematical calculations such as finding the roots of a polynomial, calculating a determinant, calculating the coefficients of a Fourier series, and so on. Such calculations had so far remained confined to large analytical calculators or computers, or even to specialised analogue instruments such as the proportion compass, the slide rule, the planimeter, the integraph, the integrator, etc.

Instead of making as many specialised calculators as there are calculations to carry out, why not make a multi-purpose pocket calculator capable of carrying out every type of mathematical calculations? That was the question asked in 1972 at the Hewlett Packard laboratories in California. And that is how the first ever portable programmable electronic calculator was born. The new pocket calculators now came with:

1 an input and output device;
2 a device for storing the instructions of the programmes the machine could carry out;
3 an internal memory in which the machine could store intermediate and final results of calculations;
4 a control unit supervising and ordering the execution of calculations provided for in the programmes;
5 a data-processing device using the data entered into the machine and transforming it in order to obtain the results of the calculation, and performing the required transfer operations.

THE BIRTH OF POCKET COMPUTERS

The above calculators were, even so, not computers technically speaking. True, they could be programmed, but their programming system was external: they did not have the necessary means to locate a given command in the programme being executed, so they could not modify the sequence of operations themselves.

Not until 1975 did Sharp, and 1976 Hewlett Packard and all the others, bring out the first programmable pocket calculators, complete with stored programmes and true programming languages, with which the user could create his own functions.

This progress led to the birth of portable universal machines for mathematical calculations. With input and output devices, a control unit, a memory and a data-processing unit, it was now possible to programme the various stages of a calculation into the calculator, and it could process the programming instructions in the same way as the data. This meant that the mini-computer (in the strict sense of the term) had arrived, specialised for mathematical calculation; hardly bigger than a wallet, but almost as effective in this field as their bigger brothers. These were the first pocket computers.

These pocket computers had the capability of *conditional branching*. In other words, the machine's programme can, at a given point, branch to follow a certain procedure (for example according to the value of a parameter or of an intermediate result) and can even automatically carry out a repetitive calculation without being reprogrammed each time.

The following are some of the technical improvements that have been brought to these little machines over the years: on-screen indicators of pending operations; indicators designating data to be entered (for example a change of units, or when computing statistics on two variables); frequency indicators; indicators of all the information needed to follow the calculation; menus listed by windows; multiple programming levels; the option of file management with alphabetical or alphanumeric codes; branching menus in windows; and so on.

It is thanks to the enormous technological advances of the electronic era that pocket calculators have become such all-purpose instruments of mathematical calculation. Above all, the inspiration has come from the structure of the electronic computer, whose key characteristics will be defined in the next chapter.

CHAPTER 6

WHAT IS A COMPUTER?

A COMPUTER IS NOT NECESSARILY AN ELECTRONIC MACHINE

It is hard to conceive of a computer that does not rely on electronics – such as might be used in underwater equipment, for instance – although the prospect of such computers already lies round the corner. [See G. Verroust (1985)]

It is as well to remember that, contrary to popular belief, a computer does not *have* to be an electronic machine. To construct a physical model, a certain number of elementary functions need to be implemented: the minimal logical functions obeying the laws of Boolean algebra; memory or storage capability; information-processing capability, and communication of encoded information. Such functions may of course be realised through a wide variety of technological means.

Thus one may easily construct *mechanical* computers: we need only clockwork mechanisms and a control mechanism. Equally one may construct *electromagnetic* computers: for this what is needed is the application of electromagnetic relays with their own logic. But the history of analytic calculation proves that, from a purely technical standpoint, the memory of mechanical computers would be far too limited to store all the programming instructions at once, and there would be little point trying to store a programme in the internal memory, for a memory based on such technology would be too slow to access conveniently. As for electromagnetic computers, they would perform their calculations just as slowly, to say nothing of their cost and dimensions (see pp. 189ff.).

Pneumatic computers have, however, been made to meet the specific needs of submarines and space craft, for they are not affected by cosmic, electric, and magnetic interference. They rely on compressed air (or gas), on water or oil, and require the use of air- or gas-jets: the logical decisions are effected by means of taps and fluid relays, and the results of the operations are displayed on manometers.

Optic-electronic computers, or even *opto-optical*, could be made, light being controlled by electronic or optical circuits to realise the required functions. For instance in 1990 the American company AT&T produced a computer working on light, composed of 32 optical switches and 8 laser diodes. The S-SEED (Symmetric-Self-Electro-optic Effect Devices) optical transistors reflect a beam of light emitted by a laser, at the same time modifying its intensity with great rapidity, thus effecting the manipulation and transmission of the elements of a binary code at the rate of billions of units per second. The laser beams can cross each other without interference, so that plane circuits may be realised (cf. D. and M. Frémy, *Quid 1994*, Paris, Laffont).

There is even talk of *biochemical* computers – albeit not for the immediate future – which would rely on the use of enzymes, for instance.

A computer's structure is, in a word, entirely independent of the technology it employs.

However the most efficient computers to date have all relied on electronics. But what made this possible were advances in *microprocessor* technology (in 1971). The microprocessor was a very advanced integrated circuit, constructed within a very small physical volume, possessing a memory, an input and output facility, a data-processing facility and a programming system; it could be applied to the execution of certain kinds of task. It was this advance that in its turn gave rise to the development of what is known as VLSI (Very Large-Scale Integration) technology, with the mass-production of very highly integrated components, much exploited since 1976; this technology radically transformed a number of activities, and hence the daily lives of millions of consumers. It is now possible to include hundreds of thousands of memory locations into a circuit occupying a space smaller than a little-finger nail (see pp. 295ff.).

NOT NECESSARILY A BINARY COMPUTER

Next to the confusion that exists between the concept of the computer and of electronic technology, there exists another, equally widespread, that tends to assume that a computer is necessarily a *binary* machine. In fact any other number base may be suitable. And today there is no technical obstacle to making computers operating in base 3, or whatever other base is preferred. It is however for reasons of convenience and speed, and in particular because of the ease with which physical systems in two states (perforation, passage of current, magnetisation, etc.) may be realised that there has been a general preference, at least up till now, for binary computers (see Fig. 5.2 D, p. 240).

COMPUTATION IS MORE THAN A SIMPLE NUMERICAL CALCULATION

Of course computers are able to carry out all sorts of calculations, from the simplest to the most complicated. But they are a great deal more than mere "calculating machines", in the arithmetical meaning of the word, whose terms of reference are considerably restricted in comparison with the very broad applications of computers. Computers, then, are much more than simple "number crunchers", for the calculations they carry out are symbolic (see pp. 244ff.).

ARTIFICIAL DATA PROCESSING

Another way of looking at the matter is to say that the computer is a *data-processing* machine. It is a machine capable of applying to the most varied data, expressible by means of a numerical code, a certain number of transformations according to clearly defined rules and procedures, with a view to arriving at the solution of problems of a certain class. It is as well to stress, however, that a machine of this type is not necessarily a computer. The old mechanographical appliances also processed data, but this did not make them computers, for their action was subject to purely technical restrictions, rather than being governed by the logic of the problem to be solved; as to their functioning, it was semi-automatic instead of being fully automated (see pp. 185ff.).

Thus the computer is a great deal more than simply a data-processing machine – despite the somewhat misleading terminology applied to them in certain countries.

Similarly, the earlier analogue devices did indeed process data but this did not make them computers, their nature and methods being entirely different (see pp. 154ff.). Indeed, computers are the polar opposite of analogue calculators, in which the data received and processed is represented by a variable physical magnitude which continually changes, at the heart of a system dependent on laws of physics and mathematics that correlate with those of the phenomenon under study. Which means that in these devices the processing supplies a physical model of the input data, the results being extracted by measurement at the end of the process. Mechanisms of this type therefore lack precision. They are, moreover, incapable of committing their results to memory. And since they cannot serve as general-purpose appliances, they can solve only a very limited number of problems (see pp. 160ff.).

On the other hand a computer is a *digital machine* because the data it receives and processes are represented by a variable but discontinuous physical quantity that may be expressed in a one-to-one way, in a discrete form by means of various numbers and symbols following very precise rules that make an entity known as a *code* (see pp. 162ff.).

Now in processing their data in this fashion, computers are able to carry out highly complex operations on data loaded into their memory; to produce results to the required degree of precision; and to execute a vast repertoire of operations.

TO COMPUTE IS TO MANIPULATE
VERY GENERALISED SYMBOLS

Computers, then, handle all manner of data so long as these data can be represented in the form of symbols which in turn are expressible in terms of a numerical code. In other words computers are machines that manipulate symbols, effecting various operations on *character chains* (words, phrases, numbers, etc.), that is, on sequences formed by letters of the alphabet, by numbers, or by other symbols.

So they are the most generalised mechanisms that exist for processing data, because the symbols in question, being those of the algebra of sets and of symbolic logic, are as abstract as could be. Hence the computer's universal scope for calculation, enabling them to address every kind of question; they can sort, merge, control, do mathematical calculations, handle the most varied applications from simple word-processing to analysing words or images, as also solve problems of character- or pattern-recognition (see pp. 106ff. and pp. 244ff.).

COMPUTING IS NOT ABOUT "FABRICATING" DATA

Although in Italy the task of the computer is defined as that of "elaborating data", it has to be said that this is no part of the computer's function, if by that is meant that the machine fabricates data. This designation is manifestly ambiguous for, as P. Morvan points out: "just as machines that work on energy do not create energy but simply transform it as, for instance, when they furnish controlled mechanical energy derived from inchoate calorific energy, so the computer does not create data: it processes data."

THE COMPUTER'S TASK IS TO ORGANISE
AND TRANSFORM DATA

Far from being mechanisms that create or fabricate data, computers are machines that *organise and transform* data, since their data, once processed, are still the same input data which have undergone all kinds of sorting, classifying, grouping, calculation, assembly, compilation, etc.

What, then, is the true meaning of this fundamental concept of *information* or *data*? The need for the question arises from the absence of any general definition of this term, and from the current confusion that exists between the notion of information (in the broadest sense) and its highly specific meanings of *facts* and of *acquiring fresh knowledge*.

A COMPUTER IS NOT JUST A
PROGRAMMABLE MACHINE

The computer is a machine that may be *programmed*, hence its great versatility. And yet it is a great deal more than a machine that may be programmed – there has been, as we know, many a programmable machine that was not a computer (see pp. 174ff.).

A computer, be it remembered, embodies at least the following components: an *input/output* facility; a *memory* for storing computed results, be they intermediate or final; a *control unit* which supervises and controls the execution of the sequences of operations specified by the instructions of a given programme; and a *processing unit* (or *arithmetic-logic unit*), which automatically carries out every necessary transformation on the symbolic representations of the data fed into the machine, and performs all the requisite data-transfer operations.

But it is as well to note that not even these elements are sufficient to define a computer. We know that Babbage's *Analytical Engine* as also the analytical calculators from the Second World War (like the *Harvard Mark I* and the ENIAC) had these facilities without being computers for all that (see pp. 189ff.).

COMPUTERS ARE MORE THAN
SEQUENTIAL AUTOMATA

Moreover, the computer is capable of effecting automatically, and thus without human intervention, linked sequences of operations that follow a procedure pre-ordained by a control unit. In other words, it is able to carry out operations in which each element precedes and follows another, no two of them ever being effected simultaneously. But an automated mechanism that answers to this definition is not necessarily a computer. This is notably the case with the belfry automata in the late Middle Ages, as also the eighteenth-century android automata, and with Babbage's Analytical Engine (see pp. 168ff.).

THE VERY GENERAL NATURE OF THE
SEQUENTIALITY THAT CHARACTERISES
THE WORK OF A COMPUTER

It is in fact in the very nature of the system of programming instructions that the whole difference lies between a computer and an ordinary sequential automaton (see pp. 232ff. and pp. 280ff.).

A *computer instruction* is a logical command that the machine has to obey and that figures as an element in one of the machine's built-in memories. It is a chain of characters that combines, in machine language, all the coded elements of the command defined by it.

When an instruction is called up, the signals composing this character-chain are stored in a register called an *instruction register*. Once this arrives in the command circuits, the instruction itself becomes an *executable command*, each element being interpreted by a logical decoder whose signals activate the component on which the command operates. As J. Bureau explains, "the execution of a command makes a series of calls upon the memory because it is necessary to: derive from the memory the operator in question, derive from it several operands, store the result back in the memory. The co-ordination in time of all this work, making several calls upon the same memory, which may only occur one at a time, is governed by a *pulse* whose period is established by a *phase distributor*. The end of one phase of calculation triggers the start of the next. *The sequencing of commands* consists of this succession, without any overlap."

In analytical calculators endowed with *Babbage's structure*, that is, those whose programming system is external, this sequencing is "such that, the programme being devised on a continuous sequential base, the first phase produces a signal that acts upon the programme reader and moves it on to the next command. This latter is introduced into the command memory, and the sequence of phases of execution begins. Once all the phases have been completed, the programme reader moves on to the next command." (J. Bureau)

Computers, however, are endowed, as we say, with what is called a *Von Neumann structure*; this brings in the concept of a stored programme, and the sequence is

> such that the programme (and thus the sequence of commands) is stored in the main memory on the same footing as the data. This type of command sequencing, therefore, no longer has a strictly sequential structure. The commands are called up in the same way as the data ... In this structure something must assume the function of calling up the next command, and this role may be entrusted to the preceding command in the programme. Thus each command in the programme calls up the one that is to follow it. Such an organisation is known as a *general sequential organisation*. The command register therefore holds the address of the ensuing command. This address is placed on the memory address decoder. The address is located, and the corresponding command, once read, replaces the preceding one in the command register. This new command itself carries the address of the next command to come, and so on. In a

system known as "normal sequential" the address of the ensuing command is generated by a counting circuit known as an *ordinal counter*. The counter advances by one unit as each command executed. The programme is thus initiated manually by feeding in the address of the first command. On the other hand the Von Neumann structure is effective for handling alternatives or choices in the execution of a programme. When faced with an alternative, it is necessary to choose between two addresses. [Thus, where computers are concerned, unlike the analytical calculators endowed with a Babbage structure] it will be a logical circuit that takes this decision. The command that corresponds to the presence of an alternative (and thus where a choice is imposed) is known as a *transfer of control* (J. Bureau)

In a word the computer is not endowed merely with the capability of executing without any human intervention a whole chain of operations leading to a predicted result; it is also capable of taking a certain number of decisions, jumping, for instance, from one command to another. And as it is a logical automaton, it is capable of transfer of control or *conditional branching*. In other words, a given programme may, at certain points, proceed along some other path (according to the value of a parameter or of an intermediate result, for example), and even automatically repeat a command many times without the relevant command having to be re-introduced each time.

Another thing the computer can do is to call up a short sequence of subordinate commands which the machine can carry out at any given point in the programme before returning to the main programme; this brings into play what are called *subroutines*.

Even more, the computer is a system capable of evolving towards a predetermined state by making use of data that allow its procedures to adapt themselves to the end in view.

To get some idea of the absence of such a facility, we need only think of the barrel organ – this is a programmable machine capable of effecting a sequence of operations in accordance with a predetermined procedure, but with no facility for retroactive instructions. Such machines are fed no data in the course of their operation and have no adaptability. This was the case with the ENIAC and with the Harvard Mark I, which had external programming systems: the sequences were carried out independently of the results, so that nothing that took place as the programme was running could affect the programme as such; therefore a human operator had to be present to take all the necessary decisions. The computer is thus endowed with a *general retroactive effect* in the sense that it is capable of modifying its internal action as a function of effects produced on the outside; and in particular it has the capability of homing in on a predetermined command for the programme being run so as to modify its running of its own accord. This is known as *feedback*.

TYPICAL PROGRAMMABILITY OF COMPUTERS

Thus what characterises a computer is that the commands are stored in the memory and are processed on the same basis as the data. And that is what enables it, by virtue of its inbuilt logical structure – simply effecting elementary logical operations – to address and solve, within the limits of its physical capabilities, all kinds of problems whose solution may be expressed in the form of an algorithm (see pp. 274ff.).

But it is necessary to set up these operations correctly if one is to address a given problem in the proper way: this is the role of the *programme*, which is nothing other than the translation, into terms that the machine may understand, of an algorithm for the solution of a given type of problem (see pp. 291ff.). In other words, an algorithm is entered into a computer in the form of a programme, which itself organises what is known as an *algorithmic procedure*. So it is a sequence of clearly determined commands which define for the machine the strict linkage of basic operations it is to carry out automatically. Therefore the elaboration of these programmes (which necessarily requires a considerable knowledge of mathematics and symbolic logic) assumes the existence of a *language*, a means of communication between the human operator and the machine.

The elaboration of a computer programme starts by *analysing* the problem to be solved, and here the successive stages of the solution are expressed in the form of an algorithm, so that they may be treated automatically. The next stage consists in a phase of programming as such, in which each algorithm is broken down into a sequence of basic operations which the computer may carry out, and in which this sequence is encoded in a programming language. The programme is thus written once and for all, independently of the data which it is to address; it may thus be applied to several different sets of data, though it is necessary to predict every possible relevant case which may arise. The programme and the set of data are transcribed onto an input medium, for instance onto diskettes, that the computer may read with the aid of devices for the accessing of data. The written programme is then interpreted by a reader connected to the computer, its contents being translated into the machine's internal language, known as *machine language*, and stored in the memory. The computer may then carry out the programme, which includes instructions for *reading* the data on an appropriate reader, *processing* the data while storing the intermediate results in the memory, and *printing* out the results.

HUMAN COMPETENCE ALLIED TO THE
COMPUTER'S PERFORMANCE

If we wish to assess today's revolution in information technology, all that is needed is to compare the history of this discipline with that of atomic physics in which "the substitution of the term *complementary* for the term *contradictory* [represented] a decisive turning point in physics." [F. Gonseth (1939), p. 165]

"The only way to escape from the problems arising in the physics of matter was to accept that the corpuscular aspect of matter and its undulatory aspect, as they come to light in different kinds of experiment, are two *complementary* aspects (to use M. Bohr's term) of one and the same reality." (L. de Broglie, *Matière et Lumière*, p. 100)

In fact if there is today a unique reality comprising two planes of existence – which, however apparently incompatible, must both be considered as an adequate representation of present-day human activity – it is precisely the one effective union, or more exactly in the *complementarity*, of *man* and *machine*. Therefore it will have taken a great deal of time and thought to realise, after many false starts and errors of judgment, that "Machine and man are *complementary*; the machine is excellent for making a record but is evidently incapable of reason and invention; the human being is a poor record-keeper, but is capable of making his records significant and inventive with an end in view. Thus man, far from being alienated from the machine, finds in it his own functions extended and retrieved to be relaunched, by means of his inventiveness, towards new forms of integration." (H. Van der Leer in *Diogène*, 30 April 1960, p. 19)

An interesting development in this direction, which we have been observing for some years, involves a sort of interaction between man and computer, which supports the notion of combining *performance* with *competence*. Computers can address ever more complex problems at an ever greater speed, their reliability is now close to optimum. So the machine, thus capable of performing operations at speeds far beyond a human's capacities, is endowed with *performance*.

The computer, however, possesses no vision or intuition of the outside world, so it is up to man to furnish it with the knowledge it lacks. Which is as much as to say that man, once he has identified his goal, has to rationalise those operations which he is incapable of effecting on his own, given his physical limitations. He must therefore work out the algorithm enabling the operation to be executed by machine. In other words it is for man to assert his *competence*.

Any old mathematician can calculate the first ten thousand decimals of π, for he knows how to go about it: he has the necessary *competence*.

What prevents him are the inextricable, never-ending calculations. So he is limited not only by his physical capacities and by his lifespan but also by the shortcomings in his mental equipment, which would inevitably get bogged down in a plethora of operations, each one having a bearing on thousands of pieces of data at a time. But if he endows his computer with the necessary competence, he will be able to derive from it all the performance needed to solve the problem under consideration (see pp. 100ff.). This complementarity between man and machine is put into effect when the human identifies the algorithms best suited to solve the problems under consideration, and thence establishes the best procedures to feed into the machine so as to endow it with all the competence needed to bring back the answer.

In a word the human factor remains the dominant one. But in entrusting to the computer all operations of an algorithmic character, man frees himself from activities that involve tedious, repetitive calculation in order to devote himself only to those tasks that simply involve pairing competence with performance.

Now this potential can be supplied in part by the machine learning about a given type of problem. Recent developments in mathematical logic and in programming methods allow for the elaboration of a basic algorithm to enable the computer to build new programmes on its own as it gains experience. So programmes are *developed* from the preceding ones and correspond to a *graded* acquisition of knowledge. It is a little as if, in a game of chess, the computer were to memorise its opponent's moves in a pre-recorded programme, so as to be able to counter them in any subsequent game.

But there is a limit to what can be expected of a computer. If occasionally it makes mistakes this will not be its fault: either erroneous data were fed in or the original algorithm for solving the problem was flawed. The results obtained are as valid as the data which were input.

Now, "if we think that machines are harmful to humans, this may be because we lack the objectivity to judge the effects of such a rapid transformation. What are the hundred years of machine-history against the two hundred thousand years of human history?" (Antoine de Saint-Exupéry, *Terre des hommes*, p. 58). Far from being harmful, the machine (when not employed in murder or in putting people out of jobs) has improved mankind's standard of living to a remarkable degree. And "there are even many cases where the machine has made ordinary work a great deal more sophisticated ... The farm labourer in charge of agricultural machinery has more varied work which requires greater intelligence than was the case for his forbears." (H. de Man, *L'Ère des masses*, p. 73). But this is by the way.

GENERAL DEFINITION OF THE COMPUTER

Let us attempt a precise definition of the machines known as computers, which subsisted in a potential, albeit partial, limited state in all the preceding developments. The properties we have considered hitherto never constituted anything beyond logical consequences arising out of a definition of computers. Here are the prime, typical characteristics: [1]

First definition: A computer is a complete model of universal algorithmic automaton.

Second definition: A computer is a universal analytical calculator with a stored programme.

For a given appliance to be a computer it is thus necessary and sufficient for its structure to be the mathematical equivalent to one or the other of the above propositions.

To recapitulate in a different way what derives from one or the other of the above definitions, a computer is :

1 an automatic machine, one whose functions are autonomous;

2 a sequential machine, thus proceeding in discontinuous manner, stage by stage, that is, by sequences of basic operations;

3 a machine capable of modifying its own internal condition;

4 a machine capable of evolving towards a predetermined state by making use of data that allow it to adapt the means to the end;

5 a machine that handles symbols of a very general nature;

6 a logical automaton capable of carrying out all calculations of a symbolic nature, and thus of effecting operations of all kinds on any data that may be expressed by means of general symbols;

7 a machine capable of executing any algorithm quite automatically, within the limits of its physical capacities;

8 a machine capable of resolving a vast range of problems whose solution may be expressed in the form of an algorithm;

9 a programmable machine capable of storing in its memory all kinds of data that may be expressed in a discrete form, and of organising and transforming such data so long as it has been programmed to this end;

10 a programmable automaton governed by a programme stored in its internal memory, and in which the commands are treated on the same basis as the data;

---

1 To obtain a better idea of the historical significance of the computer, the reader is referred to Chapter 5, which gives a structured, multidisciplinary history of artificial calculation and makes it possible to understand in context the origin and evolution of the relevant concepts. On pp. 272ff. the universal Turing machine is described in detail, this being the theoretical model for computers, and its principal properties in relation to the theory of algorithmic automata.

11 a programmable machine each of whose programmes constitutes the translation, in terms the machine can understand, of an algorithm for the solution of a given type of problem.

Which, in sum, brings us to

Third definition: A computer is an artificial automaton comprising a facility for input and output, a memory, a processor capable of effecting all sorts of transformations on data expressed in the form of character-chains (material representations of encoded data) and which, within the limits of its physical capacities, permits the execution of all types of symbolic calculations (and thus the solving of all problems where the solution may be expressed in the form of an algorithm), governed by a control unit instructed by programmes input into the memory (and thus handling the commands to be enacted in the same way as the data to process).

This then is the most intelligible, the most precise, and the most general definition that may today be given of a computer. It is an algebraic definition, so to speak, independent of all technology and potentially applicable to any computer.

THE MYTH OF THE "ELECTRONIC BRAIN" AND OF THE "THINKING MACHINE"

What the computer is not is an "electronic brain", still less a "thinking machine". These terms, still occasionally current, are unsuitable and very ill-chosen, for they introduce a dangerous notion that equates calculation and logic with all the reasoning faculties of the "thinking being".

Such terms date back in particular to the heady days of the 1930s and 1940s, and were bound up with a highly mechanistic conception of human beings; hence the simplistic idea that the general public still has of computers, as of machines endowed with inductive and creative rational processes, capable of taking all manner of initiatives, and to which mankind may entrust all its problems without exception and obtain the answers almost on the instant. Whence certain persons gifted with a freewheeling imagination passed on to the idea of giant "electronic brains" that may develop their own self-sufficiency, acquire a veritable conscience and even a measure of psychological depth in order to take over the reins of government, abolish democracy and exterminate the human race. (Shades of George Orwell's *1984*!)

In fact the myth of the "electronic brain" and the "thinking machine" started long before the first generation of computers, before the Second World War; it started effectively with the development of the first logical machines that could cope with syllogisms. But it really got off the ground with the publication in 1930 of a play by Lyonel Britton called

Brain which first introduced the notion – a cold, dark, dead thing, creeping over the world. The whole of human knowledge was fed into the brain, where it underwent combinations arising out of logical connections, with the result that the brain could give objective, unemotional answers to every question, not excluding moral and political ones. Over the centuries it grew and grew until it was master of the world.

And yet the myth of the artificial brain was not new even then. It went back to the era of the first android robots, as witness the famous legend of the Golem, a robot evidently built in 1580 and receiving the breath of life thanks to the name of Yahweh, from Rabbi Yehudah ben Loew of Prague. But the fantasy developed especially in 1899 with Ambrose Bierce's *Fantastic Fables*, in which there is a terrifying robot who plays chess and strangles his inventor after losing the game. The movement expanded considerably in 1921 with the staging of the Czech writer Karel Čapek's science-fiction play *Rossum's Universal Robot*. The action takes place in the future on Rossum's island, which features a factory with artificial workers, or robots, made by Rossum's nephew who possesses the secret of living matter.

The vogue only reached its peak after the Second World War, however, with the arrival of the giant analytical calculators, and most of all with the SSEC, the highly impressive machine exhibited by IBM at the start of 1948 in the shop windows of a busy New York avenue. This calculator was taken out of service in 1952 after a short life, but long enough to fascinate the public, who came in their thousands to look at the lights on the calculator blinking away. This spawned a mass of sensational articles in the American and foreign press, as also many a newsreel presentation which made quite an impression. And the strangest thing is that this machine continues to cast its spell over the makers of sci-fi movies, who still draw on the now thoroughly dated picture of space-craft cabins and vast computer rooms full of blinking lights, all harking back to calculators of that type.

But IBM was by no means the only company responsible for this highly charged and dangerous movement; many pioneers of the early information technology revolution took a hand in it as they tried to establish a close anthropomorphic parallel between the circuits of electronic calculators of the day and the nerve cells of the living brain.

Leslie John Comrie is noted for an article he published in the *American Weekly* on 14 October 1944 where he used a term and suggested a reflection that had a disastrous effect on public opinion: "The Harvard robot, a superbrain," he said of the Mark I and its tiny punched-tape memory, "will be able to solve problems in physics, electronics, the structure of the atom that no human being has any hope of tackling and, who knows?, may one day give the answer to the question of mankind's origins." Thus we see how far an ill-chosen terminology has led us down the path of exaggeration when figures of speech are taken at face value.

THE RELATIVE INTELLIGENCE OF MACHINES THAT COMPUTE

Be it noted that a measure of "intelligence", however limited in comparison with the infinite complexity of human mental activity and even with that of other animals, has been accorded to computers, along with a performance capability that lies quite beyond the reach of human beings.

The computer's advantage is obvious enough in a great number of appliances, from the typewriter to the motor car – man-made objects devised to perform given tasks, but only according to instructions. It can, however, on the basis of its programme, perform automatically operations that have exercised human minds since the dawn of time.

In a word, by relieving mankind of burdensome and repetitive calculations, the computer allows him to devote himself more effectively to higher intellectual tasks that involve reasoning, initiative, and decisions.

But the computer's role is limited in that it can address only problems whose solution may be expressed as an algorithm. It carries out its tasks without will, without intuition, unconsciously, unfeelingly, uninductively. Lacking discretion, it obeys commands, never mind how ill-conceived.

BEYOND COMPUTERS: THE AUTOMATA OF A BROADER INFORMATION TECHNOLOGY[2]

Information technology is no longer simply the science of computers. "In overcoming the structural limitations of logical systems, many scholars researching in the most varied fields (psychology, computer studies, cybernetics, neurobiology, etc.) have long been working to conceive and realise new kinds of intelligent systems that work on the model of the living brain ... It is a field of research that starts with no preconceived ideas as to the nature of intelligence, the object being to become better acquainted with it. The point has been reached where we are building complex automata ... whose overall functioning may be studied and predicted, but no method of calculation permits us to recognise within them at any given moment the internal condition of any particular feature ... In philosophical terms we are going beyond [calculation and] mechanistic thinking towards dialectical logic." [G. Verroust (1992)]

2 The author gratefully acknowledges the advice and information received from Gérard Verroust for the section that follows. Several of his publications are listed in the Bibliography.

In fact information technology is discovering new extensions every day thanks to the development of what is known as *general cybernetics* and *artificial intelligence*. Without wishing to enter into the details of this complex field, it is necessary to say a word or two about it in order to give some small idea of this evolution which already gives a glimpse of the extraordinary prospects for the wider field of information technology in the third millennium.

Among the already significant achievements in the field of cybernetics (where research projects are being actively pursued) let us first mention the logical automata of greater or lesser complexity known as *robots*, which are entrusted with specific tasks that would otherwise fall to humans, and which have been endowed with the capability to adapt to their own environment. The science and technology that governs them is known as *robotics*.

> The majority of present-day robots comprise a manipulator (an articulated arm with pincers at the end) and a command system (generally via a microprocessor). Sometimes they are given an element of mobility (as when they are mounted on wheels).
>
> Robots are characterised by their automation (undertaking movements without human intervention), their versatility (effecting a variety of actions), their self-adaptivity (taking account of the external environment), to say nothing of their strength, agility, dexterity, etc.
>
> As a rule we distinguish three generations of robots:
>
> 1 *apprentice* robots, which commit to memory the movements fed into them, and reproduce them faithfully;
> 2 *programmed* robots, which carry out the movements as instructed by their command computer;
> 3 *self-adapting* robots, which take account of their environment thanks to proximity sensors, reactions to force, even television cameras.
>
> At present robots are employed in three main areas: *hostile environments* (nuclear, space, underwater); *medicine* (particularly as an aid to the physically handicapped); and *industry*, with the development of so-called industrial robots working with machine-tools, soldering, painting, assembling, etc. [P. Morvan (1986)]

Another field of expansion is that of *expert systems*, where research aims at broadening the scope of computers, especially in the field of reasoning. Let us remember that an expert system is an *artificial* system of *universal reasoning* which starts from formal initial data and, referring to a body of information previously acquired, automatically builds a programme to address a given question. This is done by a procedure known as *heuristic*, the Greek term denoting search or discovery, as distinct from the notion of algorithm. The theoretical limitation of such systems arises from the fact that they work on formalised data. Their performance is, furthermore, limited by the purely sequential character of the computers which drive them. (And yet important progress is anticipated here through the employment of data-processing appliances running in parallel or networked, under instructions from data specially devised for this purpose.)

Cellular parallel automata have also been envisaged, which would work through the interconnection of a large number of identical parallel electronic circuits, known as "cells", which are photosensitive and have a memory function. Such appliances are in theory meant to study and simulate the phenomena of sight and perception, with a view to solving problems arising out of pattern recognition.

Lastly let us mention, among all these prospects, the *neural automata* or *neural networks*. They simulate the nervous system and are made up of a very large number of elementary organs known as "neurons", endowed with the ability to memorise, transmit and correlate to a highly advanced degree, and are interconnected through a network that serves to promote or inhibit, and allows for the strengthening or attenuation of the data memorised and transmitted.

It is as well to note that *neural automata* are adaptive systems which organise themselves by virtue of impressions previously received. Whence the need to submit them to prior "training". A neural automaton is therefore not programmed, it is *conditioned*. While a computer does not make a mistake if the algorithm fed into it is correct and the data suitable, this type of automaton normally does commit a certain percentage of errors. And when it makes a mistake, it is possible to influence it by means of a series of inhibitory inputs in order to reduce the percentage of errors after a certain interval.

In the behaviour of a neural automaton it is possible thus to establish at a given moment a relationship between: the percentage of errors committed, the amount of prior training, the complexity of the problem posed, and the number of interconnected neurons. But after a certain period of training the system's asymptotic "intellectual" limit is reached. [See G. Verroust (1984 and 1985); also E. Davalo and P. Naïm]

The potential applications for this new domain are in particular: the automatic processing of natural languages, pattern recognition, the reading of manuscript texts, the realisation of man/machine interfaces, the exploration of unknown areas, etc. It is also hoped by these means to achieve a greater speed in the execution of certain operations.

But of course the study and realisation of this kind of automaton involves an enhanced theoretical knowledge and the elaboration of suitably powerful mathematical tools (threshold logic, majority systems, etc.), as also some substantial advances in technology, which, according to G. Verroust, is no longer an insuperable obstacle. Which is to say that it is now possible to envisage data-processing equipment that can solve problems beyond the reach of logical automata and thus of computers.

According to this perfectly suitable generic description, the computer is a particular kind of information-handling machine, which may justifiably be termed a data processor. But it is just as clear that at present a data-processing machine does not have to be a computer: it may be *neural*, *connectionist*, etc. or indeed mixed by virtue of combining two or more data processors of the same type or different, whether computers or not.

Whence we see how the proposed terms are perfectly suited to the designation, with or without any particular specification, of every appliance to be used in this revolutionary technology that lies on the horizon (Fig. 6.1).

To repeat Alain's formula, all that pertains to the mind is embodied in language. But "the meaning of words does not lie within them as by predestination – it applies to them as a function of what we can do with them." [F. Gonseth (1937) p. 42]

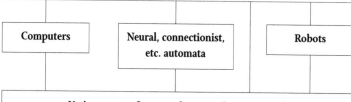

GENERAL CYBERNETICS
Science and technology of the processes of automated command and control, within automated systems for the acquisition and processing of data

ARTIFICIAL INTELLIGENCE
Science and technology of the artificial realisation of various functions of human intelligence

INFORMATION TECHNOLOGY
(in the broadest sense)
Terminology already adopted or conceived for information processing, or informatics

Information Systems
Appliances equipped with automated control and command systems, which realise by artificial means one or more procedures, whether intellectual or not, by virtue of automated systems for the acquisition and processing of information. Informatical machine or system

Computers

Neural, connectionist, etc. automata

Robots

Various types of more or less complex automata [1]

1 These types include mixed automata, machines derived from the combining of one or more automata with one or more other automata, whether or not of the same kind (for example, machines derived by combining a computer with several neural or connectionist automata).

FIG. 6.1. *The machines of an extended information technology*

PART THREE

CHAPTER 7

INFORMATION, THE NEW UNIVERSAL DIMENSION

After so much talk of gesture, number, language, speech, writing, inscription, numeral, sign, symbol, instruction, calculation, data, result, code, computer, cybernetics and information technology, we still cannot come to a conclusion without asking: what is this fundamental thing which we call *information*?

But when we finally ask "what is information?" we shall not find the answer in any of the everyday sources. The notion seems so familiar, yet its true meaning is so profound. Leaf through any publications, general or specialised, official or not, in any language: you will get such vague and varied impressions of "information" that you will not make out any definite idea of it, so much do the various notions of it seem to contradict each other.

What, we may ask, is the relationship between the information input to or output from computers, and the information which circulates amongst the cells of living creatures? What do the information processed inside a computer and the information processed inside an animal have in common?

Is information merely a matter of *facts* – the enrichment of knowledge – such as might be reported in the newspapers or collected by the Police? One might easily be led to think so.

But – supposing we could establish a relationship between them – how can we talk in the same breath of the *enrichment of knowledge* of an animal, or of a machine, or of a biological cell, when each in turn receives a piece of information? And – if we could achieve that – how then can we relate such information to the factual information communicated to the public for social, political or economic purposes?

Then again – considering literature, history, philosophy, epistemology and science, which all teem with information – in what form is information manifest in the memorisation of the facts of past history, or in the discovery and transmission of knowledge and of science?

What, again, is the nature of the information processed by computers, or by the robots and automata of artificial intelligence? What of the information travelling along telegraph wires or in radio waves, the information which emerges from telephones and television sets?

So many vexing questions! Which only goes to show what varied and divergent notions people have of information. So much so that, naively, we really wonder if there is any possible relationship between all these different kinds of "information". What can there possibly be in common between the information communicated by journalists, the information garnered by researchers and put out by information bureaux, the information discovered, transmitted and used in biology, medicine and genetics, and the information processed in computers?

THE CONCEPTUAL CHAMELEON OF THE THEORY OF KNOWLEDGE

For all that – and this is the most astonishing aspect of the question – everybody is correct despite all these apparent contradictions. Everyone has some notion which conforms to the true reality of what we commonly call *information*.

To stimulate your judgment, I give below some very instructive quotations, arranged so as to more effectively lead on to the extreme generality of the definition of information which we shall set down afterwards.[1]

INFORMATION AS FACTS AND KNOWLEDGE

Citation 1*. "*Information*: an item of knowledge, about some more or less well known subject" [A. de Monzie, A. Fèvre et Bercier: article "Cybernetics" in *Encyclopédie française* (Paris: 1937–1959)]

Citation 2*. "*Information*: an item of news, given out by the Press, on the radio or on television" [*Petit Larousse*]

Citation 3*. "*Information*: a message, a communication; something which can bring us knowledge" [*Le Grand Livre des Techniques* (Paris: 1978)]

Citation 4*. "*Information*: more a statement of fact than a mere suggestion" [P. Conso and P. Poulain (1969)]

Citation 5*. "*Information*: the action of imparting knowledge, of bringing someone up to date with events; an item of information which is communicated . . . ; knowledge which one obtains about someone or something . . . ; the aggregate of knowledge which has been acquired about someone or something . . ." [*Grand Larousse encyclopédique*]

Citation 6*. "In general, *information* means the arrangement and

---

1 It is important here for the English-speaking reader to realise that the noun *information* in French has an everyday meaning which is not directly shared in English. In French *une information* (literally "an information") means *a piece of information*, such as a fact, the answer to an enquiry, etc. The noun *renseignement* has much the same sense, but can also refer to making an enquiry as in *prendre un renseignement*. In English, however, the word *information* is already more abstract and generic than this French sense.

In the numbered quotations which follow, the numbers marked with an asterisk denote items where this French sense is the one mainly being considered. [*Transl.*]

communication to an audience of data bearing on any kind of subject (above all political, social and economic, but also scientific and literary)." [*Alpha Encyclopédie* (Paris: Éd. de la Grange-Batelière 1973)]

Citation 7*. "*Information*: a piece of knowledge which can be represented in a conventional way for the purposes of being preserved, processed or communicated" [French Ministerial Order of 22/12/1981 concerning the enhancement of terminology in Informatics]

"INFORMATION": THE REDUCTIONIST VIEW

Citation 8*. "A *piece of information* is a written formula capable of imparting knowledge . . . I distinguish amongst the various sources, and reserve the term *information* solely for those which I can *write down*. Anything which can be measured can be written down; thus I may note down the series of values given by a measuring apparatus. I refuse to give the name *information* to ways of acquiring knowledge which I am unable to describe explicitly. An animal which is taught to perform certain acts by training or by conditioned reflex is acquiring something different from information; likewise the baby in its cradle, learning about the world around it through its eyes and by touch. I do not deny that such other modes of learning exist: I merely call them by some other name than *information*." [J. Arsac (1970)]

This reductionist statement clearly reveals the confusion which continues to reign about the idea of information. As Berger says: "to transcend the narrow vision of specialists . . . , nothing is better than an exchange of views between people of experience and of different backgrounds and responsibilities . . . Out of this encounter between the personal opinions of competent people will emerge a communal vision which, instead of seeing confusion in conflicting views, will instead perceive their complementarity." [*Encyclopédie française*, XX, 54, 13]

This has all the more force in that multidisciplinary encounters would teach these reductionist experts a valuable lesson: human language is not arbitrary. It is not a mere convention which can be fashioned to taste; on the contrary, it is a system of verbal expression which is not particular to a solitary person, nor yet to any special category of individuals, but belongs to an entire people, to a whole country. Within the mind, therefore, "words. . . enter into psychological relationships and are not independent and isolated from each other – except for the grammarian." [H. Delacroix (1922), p. 432] They "embody in their significations, not everything that one could know, but everything that one does know: the community of meaning and the wisdom of a nation as they bear on the facts which the words represent." [Lafaye, *Synonyme*, Introduction, LXX]

INFORMATION AS PRIMITIVE DATA

Citation 9. "*Information* is a series of characters which may be capable of imparting new knowledge, depending on the person who perceives them." [J. Arsac (1987)]

Citation 10. "*Information*, which is at the basis of human knowledge and communication, presents itself as a collection of numerical, alphabetical, or symbolic elements, to which a meaning can be assigned by definition or, more simply, by looking it up in a table." [Boulenger (1968)]

Citation 11. "*Information*: a collection of elementary data supplied by the external environment to a living creature, particularly to a human being, by means of sensations; or to a machine" [H. Piéron (1979)]

INFORMATION IN LIVING CREATURES

Citation 12. "The living organisms of today have three characteristics in common, by which they can be defined: the ability to transmit a given item of information with a certain margin of error; the conversion of energy and matter into complex structures; and physico-chemical processes taking place in an aqueous medium of finite volume. So we may be permitted to think that *life* owes its origin to the earliest systems which proved themselves capable of presenting, preserving and transferring *information*. In order to rediscover these ancestral processes, we face a double problem: first, to conceive of a *primitive form of information*; second, to demonstrate by experiment that this *information can be transmitted* in simple conditions." [A. Brack, *Supports et transferts d'informations prébiotiques* (Representation and transmission of prebiotic information) in: M. V. Locquin (1987), p. 253]

INFORMATION, INFORMATICS, AND CYBERNETICS

Citation 13. "*Information*: data which have been processed by a computer and produced as output in a form meaningful to a user" [*Webster's New World Dictionary of Computer Terms*)]

Citation 14. "*Information*: the meaning assigned to an item of data, using known conventions. Example: '2' is a data item; when '2' is defined as representing 'February', the second month of the year, the '2' becomes information." [P. E. Burton (1982)]

Citation 15. "*Information*: (in data processing and office machines) the meaning that a human being assigns to data by means of the conventions applied to that data. *Data*: a representation of facts, concepts, or instructions, in a formalised manner suitable for communication, interpretation or processing by human beings or by automatic means" (ISO, 2382/1-1984, E/F, 01.01.02)

Citation 16. "*Information*: intelligible data of any kind. It is in particular a concept which embraces any abstract formulation which is capable of reflecting or describing some element – situation, property or event – in some structured system." [J. Bureau (1972)]

INFORMATION: QUANTITATIVE AND MATHEMATICAL

Citation 17. "*Information* (in cybernetics): a quantitative factor which defines the state of a system and which may possibly be transmitted from this system to another" [*Grand Larousse encyclopédique*]

Citation 18. "*Information* is defined as the degree of uncertainty which is resolved by the arrival of a signal. It is proportional to reciprocal of the probability of this particular signal." [R. Escarpit (1978)][2]

Citation 19*. "A piece of *information* is, by definition, one or several possible events amongst a collection of possible events." [D. Durand (1987)]

Citation 20. "The *information* belonging to a system is measured by the ratio of the number of possible responses of the system prior to acquiring the information, to the number possible after acquiring it." [G. Verroust (1984)]

Citation 21. "Information, in the sense employed in Cybernetics [and, more generally, in Informatics] is not directly related to the content or meaning of a message. This has been stressed by all the specialists in the subject, beginning with C. E. Shannon [founder of the quantitative theory of information]. *There is no relationship between the semantic aspect of information and the engineer's view of information.* In fact, information in Cybernetics is not concerned with 'what we say' in a message, but with all the things we could say: our quantitative theory of information is about *choice*, that is, the variety of possible messages . . . All we can study is how the information of the recipient *will* vary *once he has received* the message." [G. T. Guilbaud (1954)]

Citation 22. "*Information* in Cybernetics has a purely statistical definition which excludes any human content such as moral meaning, scientific importance, artistic merit, or even the speculative values of stocks and shares or political, social, or financial information. Not one of these aspects, albeit essential to our common notion of 'information', plays any

2 This definition differs from Shannon's accepted definition in the Theory of Communication, which is the logarithm (to base 2) of the reciprocal of the probability. However, the two are equivalent in the sense that the value of either can be deduced from the value of the other. A similar remark applies to some other definitions cited in this section. [*Transl.*]

role in our definition. We arbitrarily draw a line between 'information' and any notion of science, learning or knowledge." [L. Brillouin, cited by G. T. Guilbaud (1954)]

INFORMATION: PHYSICAL ACTION AND PSYCHOLOGICAL EVENT

Citation 23. "In Cybernetics, *information* means any physical action with which a psychological event is associated. The definition includes the usual meaning of the word 'information', namely the communication or acquisition of some piece of knowledge about the external world. However, it goes beyond this meaning. In order to have a terminology which distinguishes the psychological event inherent in information, we shall use *information* to mean the combination of the medium of communication and the semantics of the message." [L. Couffignal (1963)]

Citation 24*. "An entity which is *information* is composed of two parts. One of these is a physical phenomenon which we shall call the *medium* of the information; the other is a psychological phenomenon which we shall call the *semantics* of the information." [L. Couffignal and M. P. Schutzenberger, article "La Cybernétique" in *Encyclopédie française*]

INFORMATION: SYNTAX AND SEMANTICS

Citation 25. "Any message can be studied from two points of view. One is its *meaning* – the ideas which it transmits, and which we shall call its *semantic content*. The other is its *structure* – the grammatical rules used to construct it and which constitute its *syntactic aspect*.

"These two aspects are fundamentally different: the same message may be written in French or in Chinese, using two quite different syntaxes but conveying the same meaning.

"In Informatics, *information* is the medium for the communication of knowledge; in Informatics, therefore, we are only interested in its syntactic aspects and we exclude its semantic content. Consequently, two messages with the same meaning, one in French and the other in Chinese, would be considered as two different messages.

"To simplify, we may say that in any given language a dictionary can supply the semantic content of the language, while the grammar of the language dictates its syntax (the rules of conjugation, agreement, formation of plurals, etc.).

"More generally, if we are given a collection of elements, in Informatics we are only interested in the rules which govern how these elements should be combined amongst themselves: that is, in the syntactic structures defined on this collection and in the operational rules which govern the transitions from one structure to another, in abstraction from all semantic considerations." [J. Hebenstreit, article "Informatique": *Encyclopaedia Universalis*]

INFORMATION: REPRESENTATIONS AND INTERPRETATIONS

Citation 26*. "A piece of *information* consists of a pair of components: one is a *concrete representation*; the other is a set of one or more possible *interpretations*." [J. C. Simon (1980)]

RECONCILING THE IRRECONCILABLE

In these quotations we can clearly see how the notion of information has been progressively stripped of any factual or cognitive connotation, in order to become a quantitative and measurable concept applicable in Cybernetics and Information Sciences (as illustrated in Citations 17 to 20 above).

But of course the general concept of information goes well beyond the purely quantitative aspect considered in Informatics (where it was long believed that it could be reduced to a purely mathematical concept).

How therefore are we to reconcile a notion which concerns a psychological act (the acquisition or transmission of knowledge) with an apparently contradictory notion which concerns a measurable physical process with no intrinsic cognitive content?

We have to face "one of the most vicious conceptual chameleons" in the words of the American Von Foerster, who devoted many years of study to this daunting quest.

In order to arrive at a solution, we must consider a notion of information which lies well beyond the specific realities of any particular domain, and extend our vision to a much wider conceptual field.

First, we have to understand that "*Shannon Information*, the purely quantitative concept used in Cybernetics and Informatics, is but the tip of a very large iceberg indeed" and that "a true theory of information must be *meta-informational*, i.e. can be developed only as an integrated and articulate whole, embraced within a complex theory of *organisation*." [E. Morin (1977), p. 317]

In Cybernetics "we have gone down the wrong road," admits Brillouin "by considering information on its own. It is always essential to consider together both *information* and *negentropy*." [L. Brillouin (1959)] Regarding negentropy, see later in this chapter.

In fact, the only escape from these difficulties is to admit that these two aspects are not contradictory but complementary. To grasp the essence of this deep and apparently elusive concept, we must grasp that the physical process and the psychological act are two inseparable aspects of the single reality that is information, like the two sides of a coin which we could not separate without destroying its essence.

Thus it is that we progressively arrive at definitions such as Citations 21 to 25 above, and especially J. C. Simon's (No. 26), which, while imperfect, nevertheless clearly rises to a higher level of generality.[3]

ETYMOLOGICAL ASPECTS OF "INFORMATION"

These two complementary aspects of information can already be discerned in the etymology of the word itself.

The word comes from the Latin *informatio*, corresponding to the action *informare* "to inform", of which one meaning is indeed "to conceive", "to explain", "to suggest", "to give [someone] to understand", "to impart knowledge", "to cause [someone] to take [something] into account".

From this source we derive our common notions of information such as:

1 gathering information about someone or something (as for an archive, or for a dossier of confidential information);

2 the action of informing someone, or oneself, about someone or something (an inquiry, an investigation, etc.);

3 the action of imparting knowledge, of instructing, of learning, of clarification, etc.;

4 bringing a fact or an event, or a collection of facts or events, to the notice of someone or of a group (announcements, notices, bulletins, newspapers, radio or television, for example);

5 announcing something to the public (government announcements, information bureaux, etc.).

Despite this, the most absolute meaning of the word *information*, its true etymological meaning, can be perceived for example in religious phrases such as "informed by Divine Grace" which gives form to that which has not form; or "the soul informs the body".

"The immaterial principle is the eternal being, which *informs*; matter is the eternal entity which *is informed*." [Diderot, *Opinion des anciens philosophes*, "Egyptians"]

"One does not lose ones faith; it simply ceases to *inform* one's life. That's all." [G. Bernanos, *Diary of a Country Priest*, p. 151]

---

3 We take the opportunity to express our deepest thanks to Jean-Claude Simon, who provided us with his most important publications and the notes for his lectures, which have been a great inspiration to us; and also to Erik Lambert and Gérard Verroust who by their observations and criticisms proved invaluable collaborators.

"The scientific spirit, we said; better perhaps to talk of the spirit *informed* by science." [J. Ullmo, in F. Le Lionnais (1962)]

This seems to be the original meaning of the word, and etymologically it derives from the Latin *informatio*, the action of "forming", "fashioning", "giving form to", etc.

Informatio itself comes from *in-formatio*, combining the Latin preposition *in-* (in its sense of "inwards", "towards", and not its negational meaning) with the word *formatio* which means "formation", "shaping", "composition", as well as the actions of "organising", "structuring" etc.

INFORMATION SCIENCE AND TRANSFORMATION

The meaning of *In-formatics* therefore corresponds closely with that of the Latin *formaticum*, "forming", especially denoting "arranging", "regulating" and "structuring" according to a particular method of organisation. This, by the way, is the sense of the French word *formage*, which is used particularly to denote the shaping of metal pen-nibs and which, like its cousin *fromage* (cheese) derives from the Latin *formaticum*:

"The final operations required to make a metal pen are performed in the following order: 1 cutting out; 2 piercing; 3 marking; 4 reheating; 5 *formage* (shaping); 6 quenching . . ." [*Journal officiel*, 18 August 1875, p. 6925, col. 1]

Thus the Latin compound *in-formatica*, in the sense of "shaping" and "structuring" along with a notion of "movement towards" (denoted by the preposition *in*) expresses the idea of forming a new structure from an assemblage of elements. It therefore implies the notion of *transformation* of the group of elements with the aim of forming a new object. In short, we have the idea of transforming one piece of information into another piece of information. This interpretation, which while fictitious is in perfect accord with the principles of Latin syntax, corresponds to the meaning of the neologism *Informatics* (an importation of the French word *informatique* invented by Philippe Dreyfus in 1962 from the French words *informa(-tion)* and *(automa-)tique*).[4]

Informatics, in fact, is the discipline concerned with defining and studying the operations which can be applied to combinations of symbols of a very general kind, and of the transformations by which the operations convert them into new combinations of symbols, the symbols themselves being vehicles of information.

4 The word *Informatics* has become current in English usage only recently. On the other hand, *Informatique* has been in common use in French for some decades. The standard English equivalents for domains denoted by *Informatique* have been *Computer Science*, *Information Technology*, *Data Processing*, etc. A French *informaticien* is what the English speaker would call a *computer scientist*. Though English may now use *Informatics* as it uses *Mathematics*, for instance, it does not yet have *informaticians* even though it has had *mathematicians* for a very long time. [*Transl.*]

THE TWO INSEPARABLE COMPONENTS
OF INFORMATION

The Latin verb *informare* – to fashion, to shape, also to form an idea of – in turn comes from the noun *forma* which can mean "shape" – the sum of the external features which characterise a being or a thing; or again, a picture, an image, an idea, a symbol or a formal representation of a concrete or abstract object. In other words, it is *information* which gives a thing its appearance, its sensible character, from the structure of its parts, and which makes it identifiable. That is to say that *information* corresponds to the manner in which we structure a *form* in order to achieve an effect which depends on organisation according to a given norm.

To *inform*, therefore, is also to mould a form, to fashion a tangible expression, rather as one builds up a thought or an idea according to some meaningful structure: according, that is, to an intelligible – or at least perceivable – arrangement with a view to making it graspable, interpretable, comprehensible and knowable.

So we see that the idea of information implies two things at the same time: the idea of *form*, of *concrete representation*, organised in such a way as to reflect the structure of what it represents; and the idea of the acquisition, the comprehension, which takes place when this concrete form is perceived, and of the interpretation and acquisition of knowledge on the part of the recipient, which are provoked by this perception according to rules laid down in a predetermined code.

INFORMATION, AND THE "SIGN" IN LINGUISTICS

Understood as explained above, the concept of information is the structural analogue of the concept of *sign* in the linguistic sense of the word, which underlies all science of language.

The logicians and grammarians who contributed to the *Logique de Port-Royal* in the seventeenth century had already pointed out that the *sign* embraces two ideas: the thing that represents, and the thing that is represented; and it is the essential nature of the sign that the first invokes the second. [See Arnauld and Nicole (1662), Introduction, I, 4]

In 1910, the linguist Ferdinand de Saussure gave a definition of *sign* as a compound of *signifier* and *signified* (*Cours de linguistique générale*), of a sensible fact (phonetic, optical etc. – the *signifier*) and of what that fact evokes (the *signified*). This (though we cannot go further into it here) implies that the character of a sign may be arbitrary.

The relation which constitutes the association between signifier and signified is the *signification* of the sign.

We must nevertheless distinguish between *sign* and *signal*. "The signal elicits a certain *reaction* but bears no signification. Communication between animals is often a matter of signals. In human speech, an imperative may indeed function as a signal; however, it is still possible to hear and understand *Shut the door!* and yet not do it: the sign has worked, but not the signal." [O. Ducrot and T. Todorov (1972)]

We must also note that the signification of a sign is only meaningful when it is governed by a *predetermined code* whose rules allow the combination of signifiers and make messages intelligible. The signification of a sign can, therefore, only exist within the context of a domain of possibilities already foreseen and prepared [R. Jakobson (1963)]. On this domain we carry out "transformations which are conventionally agreed, usually term-by-term, and reversible" [E. C. Cherry (1957)], whose rules are exactly the code in question.

Thus it is, that when verbal messages are exchanged between members of the same linguistic community "the detached spectator, the outsider, finds himself in the position of a cryptanalyst who receives messages not intended for him, and to which he does not know the code ... for as long as he is ignorant of the things signified in the language, and perceives only the signifiers, he must for better or worse act as a detective, drawing as much as possible on external data for clues as to the structure of the language ... But once the linguist knows the code, that is to say once he has mastered the system of transformations by which a collection of things signified is converted into a collection of signifiers, then there is no point any longer in playing at Sherlock Holmes (unless he is particularly interested in studying to what extent this artifical procedure is capable of yielding valid results)." [R. Jakobson (1963), pp. 92–93]

SO WHAT *IS* INFORMATION?

Returning to our task of eliciting the concept of information, this structural analogy with the notion of *sign* in linguistics leads us to propose the following very general definition:

An item of information is a composite of

- a *formant* element
- a *formed* element

whose nature may possibly consist of a change of state such that when the formant occurs, the formed element is activated according to the rules of a pre-established code.

The first component of an item of information therefore refers to that which *fashions* and *structures* the information, while the other refers to that which is *fashioned* and *structured*.

By definition, a *formant* is a concrete representation, a physical thing: a sound, a smell, a word, a drawing, a photograph, a series of characters written on paper, a succession of signals picked up by a microphone or a radio or television receiver, etc.

Furthermore, an item of information can only exist if it has a *formant element*, since this is what is required in order for it to be physically *perceived*, through perception of the concrete representation which constitutes the formant.

As to the *formed element*, this is what is called the *interpretative domain* of the item of information – the collection of all possible corresponding *interpretations*.

Similarly an item of information can only exist if it has a *formed element*, for it needs to be *received* by a receiver which has to interpret it and grasp the concrete representation within the context determined by the rules of a pre-established *code*.

In this way we are led to the following alternative view of the matter: An item of information is a composite of

- a *concrete representation* (the formant element)
- a collection of *possible interpretations* (the formed element)

whose nature may possibly consist of a change of state such that occurrence of the concrete representation provokes the activation of the corresponding interpretative domain, according to the rules of a pre-established code.

INFORMATION, AND THE RULES OF THE CORRESPONDING CODE

We may note that "information, even when lost and forgotten, may be regenerated, provided it has remained stored, and provided a generative apparatus is available. When we rediscover a lost inscription from the distant past, if we can then reconstitute the *code* as Champollion did for hieroglyphics thanks to the Rosetta Stone, then the message which has slept for millennia may speak again. The Dead Sea Scrolls have been revived; the Maya inscriptions now speak to us. These texts snatched from death will have a whole new life, a new negentropy, by finding a place in libraries, by being reproduced, printed, translated, photocopied and commented.[5] In the new modality of cultural history (and not mere belief in myths) the same process continues: *transformation of information into negentropy and of negentropy into information.*" [E. Morin (1977)]

It is thanks to the *formant* – the concrete representation as characters

---

5 And in the modern day, we may add, by appearing on Internet Web sites. [*Transl.*]

in the corresponding script – that such an item of information may be copied and recopied at will onto the most diverse media and can be transmitted across space and time. The preservation of this same formant element, lost and forgotten though it was, enables the regeneration of the information provided the appropriate "generative apparatus" can be found. So the condition of regeneration is that we can find a decoder which, by giving the key to the code of the corresponding writing, will thenceforth enable any future reader to activate, in his own psyche, the formed element as a result of perceiving the formant of this item of information.

Here we see the structural analogy between the linguistic concept of *sign* and the general concept of *information* which we have just defined. Information can link language to communication according to the various corresponding kinds of message – gesture, speech or writing.

It is precisely this particular kind of information that we have in mind when speaking of a notion like "a collection of signs received and interpreted within the complex system of a language". [J. C. Simon]

TRANSMISSION AND DURABILITY OF INFORMATION

Clearly, an item of information is not necessarily of the kind we have just been considering. But in any case the formant, by its very nature, is that one of the two components of an item of information which gives it its entire *structure* and *organisation*, which are the only concepts whose transmissibility and duration can be guaranteed, and without which we could not conceive of information.

FORMS AND ORGANISATION OF INFORMATION

"Organisation is the fundamental concept which makes information intelligible, implants it in the natural world, breaks down its isolation and respects its autonomy. The most strange and remarkable characteristics of information can only be understood through the idea of organisation." [E. Morin (1977)]

The general theory of information therefore includes *Gestalttheorie*, the "theory of form", which studies organisations where the properties of the parts (or of sub-processes) depend on the formal representation of the whole and therefore on the entire structure considered as given [see P. Guillaume (1939)]. This discipline is a kind of theory of information formants oriented towards interpreting the physical, biological or mental world, and it enables the observation and measurement of the physical or psychological effects of concrete representations of the most varied kinds – geometrical figures, passages of music, intelligent acts, forms of reasoning, etc.

STRUCTURE AND SYSTEM OF INFORMATION

The formant, because it determines the appearance, the sensible aspect, the conformation of the information it represents, gives the item of information its *structure* which enables it to be identified uniquely. The formant is therefore *organised* as a *coherent whole*, as a function of its formal constituents alone and quite independently of the elements which are formed. From this point of view, the general concept of information unites structuralism and systematics, contemporary ethnology and the contemporary theory of knowledge.

INFORMATION AND NEGENTROPY

As it turns out, there is a very close relationship between the general concept of information and the concept of *entropy* (from the Greek *en*, "in" and *tropê* "turning", i.e. intended to represent "transformation content"). Entropy is a mathematical quantity which expresses the degree of degradation of energy into an inaccessible form.[6]

Entropy is a fundamental notion in physics. The Second Law of Thermodynamics postulates that the entropy of an isolated system can only increase.[7] Entropy, because it is essentially a statistical or probabilistic concept, is also encountered in a broader sense in such varied domains as statistics, biology, genetics, anthropology and anthropo-sociology, and population genetics, where it corresponds to the degradation of structure, i.e. of established order, implying the idea of disorder and dispersion.

From entropy, by a kind of "change of sign",[8] we arrive at *negentropy* which, for a given system, gives a "measure of its degree of organisation"

6 As a simple intuitive example – a system consisting of a container of hot gas communicating with a container of cold gas through a tiny aperture is in a relatively *organised* state: all the hot gas on one side and all the cold gas on the other. With the passage of time, however, molecules percolate from the "hot" container to the "cold" container and vice versa, until they are mixed up and the system is relatively *disorganised* with molecules, originally "hot" and "cold", equally mixed up on either side, and the two containers at the same temperature. Originally, the difference in temperature between the two parts could have been utilised to drive an engine and so the energy in the system was accessible to perform work. Finally, this is no longer possible; and while the total energy has not changed it is no longer available for this purpose, as a result of its redistribution or "degradation". [*Transl.*]

7 According to the Second Law of Thermodynamics, such systems always evolve in this direction. The increase in "disorganisation" corresponds to the increase in *entropy*, a numerical quantity defined as a mathematical expression involving the distributions of molecular velocities in the system. The statistical aspect of the Law relates to the mathematical fact (considering the previous example) that there are enormously many more ways of distributing the "hot" and "cold" molecules more or less equally between the two containers, than there are ways of having mostly "hot" ones on one side and mostly "cold" ones on the other. In the context of the random movements of the molecules, the "organised" states of the system constitute a very small target, and the "disorganised" states constitute a large one: as time passes, the system will tend to migrate towards ever larger targets. [*Transl.*]

8 In mathematical treatments, negentropy is literally entropy (in the mathematical sense) with the sign changed. This is usually defined as the meaning of *information* (in Shannon's sense). [*Transl.*]

[O. Costa de Beauregard (1963)]. By extension, in philosophical language, *entropy* denotes "degradation" itself while *negentropy* is synonymous with "organisation".

In other words, *information is negentropy*: an "ordered structure" is, in our terminology, the organisation of a system as fashioned by the corresponding formant.

MATTER, ENERGY, AND INFORMATION: UNIVERSAL DIMENSIONS

We are looking upon undoubtedly the greatest discovery of history, whether in the theory of knowledge or the theory of the physical universe. This equivalence of information and negentropy *exhibits information as the third fundamental dimension of the Universe, after matter and energy*.

INFORMATION IN LIVING SYSTEMS; SELF-ORGANISATION

Information is at the basis of what we call *life* (see Citation 12 above). "Life is characterised by [self-]organisation. The presence of organisation is sufficient to affirm the existence of life" [J. Piveteau, in *Revue de synthèse*, (July 1951), p. 170].

"The fundamental and specific characteristic of living organisms is ... that they construct themselves *from within*, through their own structure and function and not by means of an external being acting as constructor ... Living organisms function from the very beginning, while machines only function once their construction is *finished*, i.e. after their organisation [imposed by humans] is complete." [G. Matisse, *Le rameau vivant du monde*, III, pp. 6–7).

"If we can describe life as a phenomenon of self-organisation of matter evolving towards ever more complex states, in well defined circumstances which cannot be considered rare, then life itself is a predictable feature of the Universe. It is a natural phenomenon as inevitable as the fact that heavy objects fall under gravity." [I. Prigogine and I. Stengers (1979)]

BIOLOGICAL INFORMATION AND THE ORIGIN OF LIFE

As it is now believed, history began about 15,000,000,000 years ago when the "Big Bang" occurred. This cosmic explosion produced a large number of very hot clouds of gas out of which in due course emerged Hydrogen – a simple atom destined to become the most abundant element of the Universe. Some hundreds of millions of years later, oxygen was in turn constructed in the interiors of the stars as a result of thermonuclear reactions taking place at temperatures of millions of degrees. Finally, the chemical combination of these two elements under certain conditions gave birth to water in the Cosmos perhaps 12,000,000,000 years ago, in the form of a vapour which the cold of outer space turned into ice.

Into this primaeval chaos were born the planets of our galaxy and among these we find the Earth, whose magma, as it cooled and formed rocks, released trapped water in the form of vapour. This vapour then condensed as rainfall and thus it came about that water first appeared on the surface of our planet – the only one of our Solar System known to be endowed with liquid water. Its accumulation formed the earliest oceans and, by virtue of the properties of its three states – solid ice, liquid and vapour – water has ever since played a dominant role in shaping the structure of our physical and biological world.

By its mechanical energy, its chemical activity and its capacity to erode and to transport materials torn up or sedimentary, water first fashioned the landscapes: water was the architect of the Earth. For a long time, the oceans held neither seaweeds nor fish, and the lands were arid deserts.

But, one day, a *completely new structure* organised itself within the oceans as the result of a series of natural events. Drawing on the surrounding elements for the energy required, a particle of inanimate matter became animate, in the form of unicellular animals akin to the bacteria of our day. This occurred about 3,500,000,000 years ago: *life was born on the Earth* as a consequence of complex chemical reactions and physical processes.

In these favourable conditions matter had thus given rise to completely new pockets of information whose effect, after preservation and transmission, was to form negentropies of a completely new kind destined to become the negentropies of *living systems*. These were, therefore, particular items of biological information; the structure of the formants would later confer on the resulting systems the ability to self-organise, to regenerate lost parts, to assimilate non-living matter, to reproduce and, by the same token, to resist increase of entropy and regressive development.[9]

INFORMATION AND EVENTS

When information is transmitted, the phenomenon has the nature of an *event*, that is to say it is a form whose appearance is an *outcome* – the realisation of one particular possibility out of several possibilities which

9 In case the reader may wonder how the increasing level of organisation of living matter is compatible with the Second Law of Thermodynamics, it must be remembered that the "isolated system" to which the Law applies includes not only the organism but also its environment. A living creature achieves organisation by disorganising its environment (and on a larger scale . . .); also, when a creature dies its own form becomes disorganised, often by other forms of life. [*Transl.*]

might have occurred. When the outcome is realised, it is manifest in its action of establishing a relationship between the formant and the formed elements, according to the rules of the pre-established code, as a result of the appearance of the corresponding formant.

Its realisation is an emergence from the domain of the *possible*, and it is unpredictable and indeterminate from the prior viewpoint of the receiver. So the establishment of the relationship is but one particular event amongst the many which were possible. Therefore it is, in the mathematical sense, an event whose uncertainty may be measured by the probability of the occurrence of the formant. It is in this sense that the notion of information has given rise to the abstract theory which considers information in quantitative terms, as a branch of probability theory applied to mathematical entropy.[10]

INFORMATION AND COMMUNICATION

As we have already emphasised, however, the concept of information extends far beyond this purely quantitative point of view. It is a very general concept, which not only corresponds to a very abstract mathematical theory but also conforms to all possible meanings of transmission and communication.

Recall, if need be, that *it is only the material component of an item of information that can be communicated* (words, a book, images, or any other medium of transmission). It is not the *meaning* of the information which

---

10 Arrange 8 objects in line and ask someone to mentally choose one without telling you, at random. Ask "is it in the left-hand four?" (i.e. a question with a Yes/No answer). Then you know which group of four it is in. For this group, ask "is it in the left-hand two?". Then you know which group of two it is in. Ask of this group "is it the left-hand one?". Then you know which one it is.

So three Yes/No questions can discover which out of eight equally likely cases has occurred. Also, this is the best you can do: there is no other way which can consistently, or on average, discover the truth with less than three. Likewise, four questions can discover which one of 16, five questions which one of 32, and so on. "Twenty Questions" – if well chosen – can discover which one of 1,048,576 objects is being considered!

Define the answer to one Yes/No question, with Yes and No equally likely, as one *bit* of information (i.e. the unit of measurement of information). Reducing your uncertainty from "any one out of 8 equally likely possibilities but I don't know which" to "this one" is equivalent to acquiring 3 *bits* of information; reducing it from 1 amongst 1,048,576 possibilities requires 20 *bits*.

When the possibilities are not equally likely, some probability is required. If your friend chooses, not with equal likelihood but with respective probabilities $p_1, p_2, p_3 \ldots, p_8$, then learning that he chose, say, number 3 is equivalent to acquiring $- \log_2 p_3$ *bits* of information (logarithms to base 2). The average amount of information you would acquire per go is $- p_1 \log_2 p_1 - p_2 \log_2 p_2 - p_3 \log_2 p_3 - \ldots - p_8 \log_2 p_8$, and this is the *entropy* of his strategy of choice. In a long run of plays of this game, you could arrange your questions so that the average number of questions you need to get the right answer for each play is numerically equal to the entropy (but not less).

This entropy is greatest when he chooses with equal likelihood: his game is at its most "disorganised". It decreases as he introduces more inequality between his choices, and the entropy is zero if he always chooses the same one (i.e. his game is at its most "organised" – note that in this case you don't need to ask any questions, so you can get it with zero questions in the zero-entropy case). This illustrates the basis of the mathematical theory of information and entropy, as developed by Shannon. [*Transl.*]

is transmitted but its concrete representation, its *carrier*; this is a strictly physical entity. In other words, what we transmit is the *formant*: in a telephone line this consists of electrical signals, in a letter or a publication it takes the form of a series of written or printed characters.

Only when this arrives at the point of reception do such material representations take on "meaning", provided of course that the receiver – human or artificial – is present in order to perceive and acquire them and, if need be, to process them and to communicate them.

INFORMATION: INTERPRETATIVE STRUCTURES IN THE RECEIVER

When an item of information is transmitted and received, its material representation is *grasped*, and then *identified* by appropriate organs (in humans) or mechanisms (in a machine); this leads to an action which associates the formed element with the formant element, by *activation* of a potential which is inherent in the structure of the material representation.

The formant element, therefore, over and above the purely formal and sensible aspect of the information which it provides by virtue of being its carrier, also activates the formed element within the receiver as a result of occurring as an event within the receiver. This activation is mediated by the receiver's *interpretative structures* or *functions* which, so to speak, translate the *structure* of the received formant into *meaning* according to the rules of a *pre-assigned code*.

INFORMATIONAL PROCESSES IN HUMAN BEINGS

In human beings, what we have just described corresponds to the triggering of what may be called an *informational process*, capable of enriching knowledge with or without an accompanying emotional event. Its result may be to impart *factual knowledge* which, when it impinges on the *cognitive faculties*, becomes stored in memory along with the innumerable other facts already stored there. These facts are processed and transformed by the same cognitive faculties into new formants which in turn may be transmitted to others through one or more communication channels.

This is the point at which the general concept of information subsumes the various special concepts: the concepts of factual data, of the acquisition, the formation and the transmission of factual data, of science and of knowledge.

Obviously, however, representations of information only have "meaning" when a living creature or a machine is capable of interpreting them through having appropriate aptitudes or as a result of mechanisms specifically designed for the purpose.

In human beings, representations are perceived by the *sense organs* and are at once distributed within the brain, where the intellectual faculties generate an interpretation which is not unique, but depends on the environment, on the existing knowledge, and on the psychological and even the emotional state of the subject. They are received and stored by the human being from its very first moments, the earliest stages of foetal existence, even though the newly emergent structures of the foetus do not yet enable it to grasp their true meaning.

We must also emphasise that one and the same representation of information could give rise to many different interpretations in someone whose structures have matured. We need only consider how, at different levels of interpretation, our acts, our words and our writings may give rise to different consequences.

For instance, we can see in any dictionary that the same word may have many different meanings according to context. Think, for example, of the wide semantic scope of the word *computer*. If this is true for one word, how much more variable can be the effect of a phrase, or of a discourse. Roland Barthes's study *S/Z* of Balzac's *Sarrazine* shows that there are five levels of interpretation – which he calls *lexies* – according to five corresponding *codes*. It is also obvious that if many different people receive identical concrete representations of a message, they will probably understand as many different things.

So far, we are supposing that the human being acting as receiver has properly functioning sense organs, and is endowed with psychological capacities and interpretative and cognitive faculties which have followed a normal development, in a normal social environment, since birth.

In *Eye and Brain* (Oxford University Press, 1998) Richard Gregory describes the sad case of a man, blind from birth, whose vision was restored at the age of fifty-two years:

> This case, a man of fifty-two, whom we may call S. B., was when blind an active and intelligent man. He would go for cycle rides, with a friend holding his shoulder to guide him; he would often dispense with the usual white stick, sometimes walking into parked cars or vans and hurting himself as a result. He liked making things, with simple tools, in a shed in his garden. All his life he tried to picture the world of sight: he would wash his brother-in-law's car, imagining its shape as vividly as he could. He longed for the day when he might see, though his eyes had been given up as hopeless, so that no surgeon would risk wasting a donated cornea. Finally the attempt was made, and it was successful. But though the operation was a success, the story ends in tragedy.

When the bandages were first removed from his eyes, so that he was no longer blind, he heard the voice of the surgeon. He turned to the voice, and saw nothing but a blur. He realised that this must be a face, because of the voice, but he could not see it. He did not suddenly see the world of objects as we do when we open our eyes.

But within a few days he could use his eyes to good effect. He could walk along the hospital corridors without recourse to touch; he could even tell the time from a large wall clock, having all his life carried a pocket watch having no glass, so that he could feel the time from its hands. He would get up at dawn, and watch from his window the cars and lorries pass by. He was delighted with his progress, which was extremely rapid.

When he left the hospital, we took him to London and showed him many things which he never knew from touch, but he became curiously dispirited. At the zoo he was able to name most of the animals correctly, having stroked pet animals, and enquired as to how other animals differed from the cats and dogs he knew by touch. He was also of course familiar with toys and models. He certainly used his previous knowledge from touch, and reports from sighted people to help him name objects by sight, which he did largely by seeking their characteristic features. But he found the world drab, and was upset by flaking paint and blemishes on things. He liked bright colours, but became depressed when the light faded. His depressions became marked, and general. He gradually gave up active living, and three years later he died.

Another typical example is the case of *enfants sauvages* – children reared since infancy by wild animals, remote from other humans – who exhibit an almost total incapacity to acquire human behaviour patterns or to learn how to use human language. The two "wolf-children" of Midnapore, in India, are famous. Captured in 1920, they were two little girls who were given the names Amala and Kamala, four years old and eight years old, and they were found living in a cave with wolves. They ran about on all fours, had no knowledge of human language, and expressed their distress by howling like wolves. They were given into the care of missionaries, who made totally vain efforts to teach them words and human behaviour. They failed to adapt to a human way of life and soon died: the younger in 1920, the elder in 1928. [N. Sillamy (1967)]

For human beings, this shows how important it is for their development to grow up in a human social environment from birth, for it is this factor which determines the differentiation of their individual characteristics, allows their fundamental organic structures to mature correctly, and provides the necessary appropriate stimuli for their mature functions and aptitudes. If they do not experience this essential factor, then the psychological faculties and interpretative structures needed to acquire and interpret information will fail to adapt and even remain atrophied, despite having a potential or virtual existence. Of this, there is no doubt.

INFORMATION, LIVING CELLS, THE GENETIC CODE

At first sight, genetics and the biology of cells are domains with which the notion of information, as we have been discussing it, may seem to have nothing to do. Nevertheless, we can observe processes of exactly the same kind on the genetic level, mediating the reception and transmission of information by living cells, regulating and maintaining life, and participating in the transmission of hereditary characteristics.

The process of inheritance depends on *genes*, which are segments of *chromosomes* made up of a macromolecule of *deoxyribonucleic acid* (DNA) and structured as a pair of strands formed into a double helix. Arranged in sequence along these two strands are complementary pairs of nucleotides whose bases are of four kinds: the purines *Adenine* (A) and *Guanine* (G); the pyrimidines *Thymine* (T) and *Cytosine* (C). Adenine on one strand must always pair with Thymine on the other, and Guanine always with Cytosine. Therefore the sequence of nucleotides along either strand determines the complementary sequence along the other.[11]

These nucleotides are like the letters of an alphabet (which we may call the *genetic alphabet*) which has only four characters (A, T, G, C), and their arrangement into sequences is analogous to words, and sequences of words. The organisation of this sequencing for the purpose of genetic information is called the *genetic code*. We know that each hereditary character is *encoded* according to a specific structure, whose formal (concrete) representation is situated on a particular segment of DNA as a sequence of letters from the above genetic alphabet. The translation, or decoding, of this sequence gives rise to a protein whose nature depends on the particular hereditary character.

---

11 This is the basis of the duplication of DNA when a cell divides into two, creating two "daughter" cells. In the "mother" cell, the two complementary strands of DNA are joined into the double helix which is folded up in the nucleus of the cell: say these strands are a and b where, as it might be,

> a: ... A C C T G G T A A G C ...
> b: ... T G G A C C A T T C G ...

When the mother cell divides, the two strands in the double helix unwind and separate. The a strand goes to one daughter cell, the b strand to the other. Within each daughter cell, its respective strand is used as a template to rebuild the complementary strand, which then binds once again to it. Thus in one daughter cell the old a is used to make a new b which recombines with the a to form the ab double helix; while in the other daughter cell the old b is used to make a new a, which again recombines to form the ab double helix. [*Transl.*]

There is DNA in both the nucleus and the cytoplasm of a cell, as well as another nucleic acid RNA (*ribonucleic acid*) of similar structure.

In order to maintain themselves and to reproduce, cells require proteins, and the synthesis of proteins is an essential component of the process of life. This protein synthesis takes place in the cytoplasm through the mediation of a particular RNA called *messenger RNA* which functions as a carrier of genetic information. The following explanation, albeit technical, clearly shows how important a role the nucleic acids play in protein synthesis, and gives us an idea of the role of *information* in genetics.

"The DNA acts as a mould or template for the formation of three kinds of RNA. Several segments of *ribosomal RNA* (r-RNA) combine with proteins to form the two parts of the ribosome. The code is transcribed from the DNA onto *messenger RNA* (m-RNA), which acts as a carrier of the code; m-RNA passes to a ribosome where translation of the code leads to formation of a polypeptide, by addition of specific amino acids brought to the ribosome by *transfer RNA* (t-RNA) whose base sequences are complementary to those on the m-RNA chain." [M. Sleigh, *Origin and Primitive Evolution of Cells*, in: M. V. Locquin (1987), p. 133]

In plainer words: protein synthesis requires the intervention of a messenger RNA, which bears a copy (from the DNA) of the structure of the concrete representation of the information. The messenger RNA then transfers this representation to another organite[12] where it gives rise to production of the particular protein corresponding to the code.[13]

Therefore, through the intermediary of the messenger RNA, there does indeed occur the transmission of an item of *genetic information*, whose acquisition and subsequent interpretation trigger the appropriate consequence. This information consists of a *formant element* (the encoded representation of the hereditary character in question) and a *formed element* (the domain of interpretation of the sequence of "base-letters" which make up this encoded representation) whose nature is that of an *event* which comprises a *change of state which, upon the realisation of the formant element, triggers the formed element according to the rules of the genetic code.*

The formant element, once it has been received by the appropriate components of the cell, is identified and then interpreted, its structure being matched to a particular one of the possible interpretations; and this then gives rise to the process which produces the requisite protein. Thus we

---

12 An *organite* is any of the distinct biologically active components of a cell.
13 An excellent short summary of the elements and processes involved in molecular biology can be found on the World Wide Web, thanks to IACR, Rothamsted Experimental Station. See:
http://www.res.bbsrc.ac.uk/molbio/guide/
This is also listed in the BBC Education Web Guide, for which see:
http://db.bbc.co.uk/education-webguide/
[*Transl.*]

can see how the molecular genetic processes in the cell fall within our general definition of information.

THE UNIVERSALITY OF OUR DEFINITION

We can now see how our unifying definition of information gives the greatest generality to the very meaning of the concept of information.

This definition can be applied to any domain. Whatever topic we are concerned with can be brought under it, whether genetics, cybernetics, informatics, artificial intelligence, media or telecommunication, the judicial system, philosophy, psychology or the theory of knowledge. In any of these one may study merely the structures of the formant elements of the information, or one may study the possible relations between formant and formed elements, or from these relations one may proceed to all possible informational or cognitive acts.

This definition is therefore, in its way, an *algebraic proposition*, since it is an abstraction independent of any particular aspect of the completely general notion of information. These particular aspects, such as those we cited at the beginning of the chapter in order to build up a view of the concept, have usually served in the past only to obscure a clear view of it so long as the general concept was lacking. This general concept at which we have now arrived had nevertheless long been suspected as the true "universal dimension".

INFORMATION AND INFORMATICS

We shall close by considering the kind of information which computers receive in the form of data, on which they carry out various transformations. These data are, in fact, sequences of symbols (numbers, characters, various graphical signs) from some finite alphabet decided upon in advance. Within the machine, these symbols are in turn represented by sequences of signals, known as *code*, which obey precise rules.

Provided it has been programmed with appropriate instructions, a computer can carry out any transformation on an encoded *formant*, and store it as an appropriate series of symbols. The computer has been endowed with certain structures, and programmed with certain instructions; it reads the input data supplied to it, encodes their representations, and identifies them and interprets them according to its own language; then it manipulates and transforms them, giving them a new form which is the *encoded representation of the result*. This is then stored in memory prior to being delivered to the outside world in one of many possible formats according to the possible interpretations which a human being may wish to bring to it.

In short, a computer, to quote J. C. Simon, "interprets sequences of characters, either as *data* to be transformed or as *instructions* which will give rise to precise and definite operations within the machine, and then yields *results* which are themselves represented as sequences of characters which can be interpreted by a human observer."

In this context, according to Simon, the *quantity of information* is defined as *the amount of memory occupied by a representation, relative to the amount, occupied by a standard representation, adopted as a unit of measure*. This is a quite precise notion, completely independent of all possible interpretations, of the measure of the concrete representation of a piece of information – of its formant, in other words, which is its only directly observable, quantifiable and measurable element.

We may, therefore, adapt the words of G. Verroust (which we have quoted as Citation 20 above) to the vocabulary of the present context: *the quantity of information carried by a system* is equal to *the ratio of the number of possible states of the system prior to acquiring the information, to the number of possible states after acquiring the information.*[14]

It is in this context that our general concept of information meets Shannon's quantitative theory of information (formulated in 1948), a theory in which information is treated as an observable measurable quantity with a view to maximising the performance of a physical information-transmission system consisting of an emitter, a transmission channel, and a receiver, in order to minimise the cost of transmission. And in this context, the content of the formed element of the information – that which corresponds to its possible interpretation – has little part to play in the principal concerns of the information scientist, the cybernetician or the telecommunications engineer.

Our manufactured machines have interpretative structures. They are provided with physical receptors which can interpret the external world into the internal phenomena of the machine: a microphone responds to the pressure waves in the air, and transforms these into electrical signals representing sounds; a television camera captures intensities of light and transforms these into electrical signals representing images; the behaviour

---

14 Strictly speaking, for an accurate concordance of these passages on the measurement of information it is appropriate to consider the quantity of information, in the "Verroust" sense, as measured by the *logarithm* of this ratio. As an example to illustrate this, consider a machine whose memory consists of N binary digits ("bits"). Such a block of memory can have a total of 2^N states, corresponding to all the possible combinations of 0–1 states for each of the N bits. Now suppose that the machine has stored a representation of information which occupies a total of k bits of memory (this being Simon's definition); then these k bits may be considered as fixed, and only the remaining $(N - k)$ are free to vary, so the number of possible states is now, after storing the information, only equal to $2^{(N-k)}$.

According to Simon, there are k bits of information stored. According to Verroust, the information stored is the ratio of the numbers of possible states prior to storing the information (2^N) to the number possible after storage ($2^{(N-k)}$), namely $2^N / 2^{(N-k)} = 2^k$. These agree if we use the logarithm to base 2 of Verroust's measure, since $\log_2 2^k = k$. [*Transl.*]

of the electrical circuit by which one or other of these electrical phenomena is processed and transmitted is usually, by design, such that the resulting signal can be made to regenerate a sound by vibrations of a loudspeaker, or an image by scintillations of light on the surface of a screen, as the case may be (though it is perfectly possible to generate an image from the electrical signals produced by a microphone, or sound from those produced by a television camera).

These receptors, and their associated circuitry, create representations of information which can be identified and interpreted by human beings; but also, more and more, computing machines themselves can be endowed with internal structures which can automatically perform such functions of identification and interpretation by making use of operations which belong to the domain of *pattern recognition*.[15]

Information-processing systems, therefore, will in future be increasingly endowed with cognitive faculties, and with interpretative structures, which are adapted to the recognition and interpretation of *signs* in the linguistic sense of the word. In the words of J. C. Simon: "They are revolutionising the manipulation, transmission and interpretation of representations. They are becoming partners of humankind."

---

15 An example of which, now commonplace in the computerised office, is the use of software which can take as data the digital representation of a printed page (as produced by an *optical scanner*) and generate in the computer's memory the digital equivalent, i.e. the character encodings, of the printed characters which the human reader would recognise when looking at the printed page – in other words Optical Character Recognition (OCR). [*Transl.*]

INTELLIGENCE, SCIENCE, AND THE FUTURE OF MANKIND

INFORMATION, KNOWLEDGE, AND UNDERSTANDING

From information to the formation and transmittal of knowledge, of science and learning there is evidently but a single step which mankind has been taking readily enough ever since it has been able to acquire, store, and transmit knowledge.

To unravel this story of human intelligence our best option in this concluding chapter has been to have a number of significant authors speak in their own words, which we have collated into a sort of mosaic. Thus we arrive at some idea of what constitutes human intelligence in terms of psychology, philosophy, and history, with an eye in particular on future developments. (The source references are grouped at the end of the chapter.)

Knowledge is the ability to penetrate the forms and structures of nature, the physical world, and of humankind with all that affects it. Consequently knowledge touches upon every kind of learning. Thus "to know is in a certain sense to be other than what one is; it is to become other than oneself." (Jacques Maritain[1])

On the other hand, to know is to give ourselves a proper idea of what we are, as moral beings, of what constitutes our temperament and character, and particularly to have an awareness of our intellectual capabilities, of our strengths and limitations. "Man's greatness lies in his recognition of his wretchedness. A tree does not know that it is wretched. Thus to be wretched we have to recognise the fact. But to recognise our wretchedness is what makes us great." (Pascal[2]) And yet, "if we are too concerned to know ourselves we risk preventing ourselves from existing." (Jean Rostand[3])

It is true that "*to know* is to penetrate our being to the point of effecting an inward coincidence with it by dint of thought, and to discover in ourselves the full variety of its manifestations." (J. Vialatoux[4]) But "strictly speaking, knowledge defies definition ... Trying to 'know knowledge' is like trying to see one's eyesight. To know what vision is one has to look at things, then think; no point closing one's eyes." (G. Berger[5]) In fact "knowledge is at once distance and fusion." (J. Wahl[6]) Which is why knowledge and learning are the result of *information-structure* and its *transmission*.

LANGUAGE AND THOUGHT

Therefore men living in society first developed the faculty of expressing their thoughts, starting of course with this "social institution of a particular type founded on the use of *speech* to communicate thought", that is, *language*, which "rests on the association of *thought-content* with *sounds* produced by speech." (J. Perrot[7])

Now, as Plato[8] said, "thought is inner speech." Or, as Egger[9] suggests, it is "the life of consciousness", allowing for the fact that "to know is not simply to have something lodged in the consciousness, it is to be *aware*." (H. Duméry[10])

LANGUAGE AND HUMANKIND

What is at any rate certain is that speech is inseparable from *thought* and *reasoning* in humans. For "speech has been given to man to explain his thinking; and just as thoughts are the pictures of things, so are words the pictures of our thoughts." (Molière[11])

It is true that "thought and language are interdependent . . . sometimes thought searches language for a form that suits it, and sometimes language finds itself enacted in thought." (L. Lavelle[12]) But language in particular is a means of communication, allowing contact between humans to take place according to a *predetermined code*. "It is, however, a great illusion to believe that, when two people communicate, they convey to each other that which they already possess. All they convey in fact is the ability to acquire through each other that which up to that point neither of them possesses." (L. Lavelle[13])

Nonetheless, "the communication of ideas, as a science . . . must learn . . . to express each idea as clearly as possible." (D'Alembert[14])

WRITING AND CIVILISATION

Now, "if we say that *language* is what makes *man*, we have to add that *writing* is what makes for *civilisation*." (H. Piéron[15]). Which explains why history does not go back to prehistoric times, but rather to the era of the first *written* documents, which constitute the true beginnings of what we call civilisation.

WRITING AND THE SPACE/TIME FACTOR

Writing is a conventional system associating *graphic signs* with the actuality of the spoken word and is realised, as is articulate speech, in a wide variety of forms. Born of the need to fix thought, which is of its nature fleeting, the invention of writing has permitted mankind not only to conserve and transmit the spoken word but indeed to enable it to travel across space and time.

This is what makes writing the great invention it is, for it defies these inevitable factors, especially time which "advances irreversibly, in accordance with the evolution of all that the universe contains." (J. Matricon and J. Roumette[16]) And assuredly "there is nothing more precious than time, for it is the price of eternity." (L. Bourdaloue[17]) "Those who rely upon the action of time show that they have no inkling of the price of time." (P. Peeters[18])

WRITING AND THE WORLD OF IDEAS

But writing is essentially a means of direct access to the world of ideas, it apprehends thought in order to organise it by setting it down. It is "a mental discipline that affords it greater firmness and constancy; it prevents thought remaining in an inchoate, dreamlike condition." (L. Lavelle[12]) In other words what writing fosters is not speech "but thought, which it keeps forcing us to bring out, to test, prolong and extend indefinitely". (L. Lavelle[12])

WRITING AND CULTURE

So, by dint of study and reflection which language and writing permit, man has gradually achieved a social and intellectual formation that goes by the name of *culture*. "Culture is what remains when all else has been forgotten." (E. Herriot[19]) True, but "culture is something quite different from learning. It needs poets as much as it needs geometers." (Alain[20])

CULTURE AND UNIVERSALITY

And in fact "there really is no culture until the spirit has expanded out to the dimension of the universal." (J. Leclercq[21]) Besides "what is a cultivated person? It is the person who has completed a great number of apprenticeships in reflection, who can see from a wide number of viewpoints. Culture is a function of the number of categories available to a person's intelligence." (H. F. Amiel[22])

CULTURE AND TOLERANCE

Thus, as Alain[23] observes, "the essence of thought is its universality." And that is why culture, in the highest meaning of the word, has no truck with any restrictive nationalism nor with any patriotic dogma promoted as a political ideology. Rather it respects the legitimate right of each culture to differ from the next, by virtue of its very nature, that is, it respects *tolerance*, which consists in allowing others the freedom to express opinions with which we ourselves may not hold, and to live in conformity with such opinions. Tolerance cannot, however, be exercised except in reciprocity. And "it is important to distinguish the spirit of *tolerance*, which means not persecuting anybody, from the spirit of *indifference*," which tends to reduce all ideas to the same level. (cf. D'Alembert [24])

Herein lies the basic characteristic of modern-day humanism.

WHAT IS INTELLIGENCE?

There is no need to stress that if mankind has been able to experience learning and, better still, science and culture, with all their potential to push back the frontiers of knowledge, this stems, first and foremost, from the development of human *intelligence*, and especially from the progress of its native wit, its reasoning and capabilities.

To count, number, measure, calculate, weigh, estimate, meditate, consider, reflect, believe, opine, hazard a thought, reason, deduce, infer, invent, discover, imagine, memorise, abstract, elaborate, create, experience intuition or feeling, generalise, particularise, guess, choose, discern, determine ... these are just facets of those various faculties connected with human intellectual, rational activity which is applied to general as well as to specific ideas, and goes under the name of *human thought*.

And it is precisely here that the possibilities of language and thought combine perfectly with those of the *intellect*, which is nothing other than "the ability to engender and employ abstract ideas expressed in words". (C. Spearman[25]) But "to be intelligent is not only to be a reasoning being; it is to add to reason this facility, this flexibility that leaves us free to penetrate other minds and often understand them better than they understand themselves." (Jean Guitton[26])

"Watch out for those teachers who always complain that their pupils are stupid. Let them blame only themselves! Those who have faith in other people's intelligence stimulate it and bring it to birth. Those who doubt it and mistrust it make it timid to the point of destroying it altogether." (Jean Guéhenno[27]) This is because intelligence is "the ability to adapt oneself rapidly and rationally to a change of circumstances". (G. Morf[28]) Those with little faith in their pupils' or colleagues' intelligence risk destroying their intellectual potential until they are incapable of placing themselves in harmony with their fellow men as well as with nature.

But intelligence is also "the ability to think through novel problems". (E. Claparède[29])

Which is why "intelligence is *discernment* and *choice*. Not a predetermined, automatic adaptability, but the application of thought to a given situation, to resolve a difficulty that has already been seen, with its solution to follow, unless the solution has been identified in the problem itself." (H. Delacroix[30])

Now, "intelligence *feels* its way forward, too. It proceeds simply by trial and error. But instead of trying things out that will not achieve anything ... intelligence makes mental experiments to identify those leading to a dead end. It reduces the possibilities to a small number; [by virtue of *inductive reasoning*] it chooses those paths that have some chance of reaching the goal ... ; [and by *deductive reasoning*] it sometimes discovers the only one that can attain it ... The workings of the intelligence are thus those of *selection*, but a selection that kills off ideas and hypotheses instead of killing off what has life in it." (E. Goblot[31]) Be it remembered in this connection that "*induction* proceeds from the particular to the universal." (Aristotle[32]) And it "has to be *intuition* that teaches us principles, for it is in this way that the feeling itself attains the universal in us." (Aristotle[33])

As for *deduction*, it is "the passage from one intuition to the next so as to connect the two extreme terms of a series by an evident and necessary linkage." (C. Serrus[34]) In other words, "deduction supplies *intelligibility*, not certainty; induction, conducted methodically, produces *certainty*, not intelligibility. True science stems from their collaboration." (G. Heymans[35])

Be it noted, too, that "inductive reasoning is far more difficult to analyse [than deductive reasoning]. It relies on analogy and intuition, on flair rather than on any spirit of geometry, it seeks to *guess* at what is not yet known so as to establish new principles which may serve as the basis for further deductions." (Louis de Broglie[36]) Which is as much as to say that in the intelligence "there is an active, personal, inventive tendency which explains that which is most essential to it. It is [as it were] "*constitutive* reasoning in contrast to those publicly constituted rules and accepted truths which, however valuable, express only the imperfect acquisitions of civilisation." (A. Lalande[37]) In fact, while *passive* reason "consists in those principles and norms on which our reasoning is based, norms that will incline to vary with times and peoples (while tending at the best towards unity)," *constitutive* reason is nothing other than "the permanent and universal faculty to form necessary, universal principles out of identifiable relationships." (P. Foulquié[38])

Moreover "intelligence is the art, when confronted with a situation or a universal, of constructing *systems of abstraction* and inserting them into this situation or into this universal. The definition applies equally to practical as to theoretical intelligence." (H. Delacroix[39])

It is true that "*theoretical intelligence* derives general laws from particular instances; it goes on to invoke these laws to resolve new cases as they arise; it is conceptual, discursive. *Practical intelligence*, though, is intuitive: it is in the concrete instance itself that it finds the solution it is seeking." (P. Foulquié[38])

ABSTRACTIONS, INVENTIONS, AND DISCOVERIES

It should be noted that "practical life, which at first glance is the realm of the concrete and singular, relating to particular circumstances, is in fact the realm of the abstract and the universal, since it comprises distinct representations and typical methods." (J. Segond[40])

Nevertheless "the *abstract* properly speaking starts with the realisation of similarity and difference. Abstraction is to distinguish the common characteristics shared by several objects or the characteristics exclusive to a single object." (A. Burloud[41]) Which means that "in strictly scientific terms abstraction has to discover the characteristics which are the substitutes for a group ... which represent a group and allow us to think of it as such." (T. Ribot[42])

Now it is precisely owing to the fertility of the spirit and the imagination, whose work is undertaken by every faculty and through these different abstractions, that mankind has realised all of its *discoveries* and *inventions*; in the former it finds things already in existence but hidden or unknown, or existing only in a potential form, or in the process of elaboration, and for the latter, things completely new, which had no previous existence, or existed in only a virtual state.

SCIENTIFIC INITIATIVE AND KNOWLEDGE

But "in science no initiative is pursued from a starting point of total ignorance, a condition in which the scholar would be in a position to obtain data in a perfectly purified state, and would rely upon methods of absolute certainty. A *scientific initiative* must start from a point where the scholar disposes of a measure of *pre-existing knowledge* and a certain *predetermined language*." (F. Gonseth[43])

It is, however, to be noted that "we do not arrive at a *new idea* by virtue of a conscious labour of logic. It crops up one day, fully formed, after a long period of gestation which has taken place in the *unconscious*." (A. Vandel[44]). "The essence of scientific discovery lies in its not happening to order ... The spirit lights upon the discovery only at a moment when it has nothing else on its mind." (J. Duclaux[45])

TECHNIQUE, KNOWLEDGE, AND CAPABILITY

But from the start intelligence has "the ability to construct artificial objects, in particular tools, and to alter their construction indefinitely". (H. Bergson[46]) Moreover, "every *technique*, from splitting stone to all those that have succeeded it, implies something that is to evolve into *learning*, and thence into *science*." (A. Rey[47]) Indeed "either technique becomes automatic and goes into decline, or it is vested with intelligence, and goes on to adapt or improve what it has reconstructed. The greater its degree of invention, the more it will serve to educate." (R. Le Senne[48])

And this is where the *acquired* comes in, which, unlike *innate* reason, relates to that which builds up over the whole length of a person's life and corresponds to that which lies within mankind's collective memory.

As for *capability*, that is what we acquire by dint of *knowledge, practice* and *experience*.

But as everything in man is linked, "experience moulds reason and reason moulds experience. Between the real and the rational there is therefore a mutual interdependence along with a relative independence, which makes it singularly difficult to establish what part in the advancement of learning is due to outside pressures and what to the demands of the intelligence." (J. Piaget[49])

In any event, "technique . . . consists in a collection of rules, founded on reason, but [determined by the intuition, then] proved in practice and becoming the collective property of a civilisation, rules whereby a tool is brought into use for a particular end." (A. Siegfried[50]) And it is in this sense that technique relates to a craft when these rules are manual, and to an industry when tools are made with the aid of machines.

TECHNIQUE AND TECHNOLOGY

"Historically speaking, *technique* came before *science*, it was known to primitive man . . . [But] technique would not start to develop until science took a hand. So technique had to wait on the progress of science." (J. Ellul[51]) Or as Siegfried[52] says, technique eventually became a "rationalised art", to the point of constituting what we now call *technology*. And "what most essentially characterises *scientific technique* is the fact that it proceeds from experiment, not from tradition." (Bertrand Russell[53])

SCIENCE AND LEARNING

This is all the truer in that "*science* is learning that is authenticated by *criticism*." (S. Bachelard[54]) Now what gives science its precise character is that "in science invention becomes discovery, thanks to an intellectual process of *verification*" (L. Brunschvicg[55]) "In a certain sense, invention within *scientific theory* has the character of a discovery, but a discovery in the realm of the spirit." (L. de Broglie[56])

SCIENCE AND MATHEMATICAL KNOWLEDGE

Now if there is one science in which invention possesses precisely this characteristic, in the most absolute sense of the term, then it is *mathematical science*. It is, moreover, "by the study of mathematics and only by this study that we may obtain a true and deep understanding of what a science is." (A. Comte[57]) "Mathematics is the name given to a system of closely interrelated scientific acquisitions, based on notions common to every mind, and conveying rigorous truths which reason is capable of discovering without need of experiment and yet truths which may always be confirmed by experiment within such broad limits as experimentation requires." (A. Cournot[58])

This is because "mathematics of its nature is conditioned by intelligibility, and its fundamental feature is that it is perfectly intelligible and completely certain; which is not to say that mathematics affords knowledge of any part of nature." (E. Goblot[59]) "Mathematics is the science that applies to exact reasoning . . . Therefore in measure as human knowledge becomes more exact and precise so does mathematics have an increasing role to play in it." (J. Hadamard[60])

Of course "the *experimental sciences* are unable to offer such a complete satisfaction – or so it would seem – but on the other hand, they alone reveal to us the world in which we live." (E. Goblot[59])

Physical science is not of itself an application of mathematics to physical phenomena; and yet the essential basis for the physical sciences, for any science whose object is to penetrate the secrets of nature, lies in calculation, measurement, and mathematical operations and logic. Thus mathematics constitutes "the most powerful instrument at the disposal of the human mind in its research into the laws of natural phenomena." (A. Comte[57])

IMAGINATION AND THE PRETENCE OF LEARNING

"It has often been said that the perfection of a science is measured by the quantity of mathematics it comprises; it might be argued, conversely, that the imperfection of science is a function of the quantity of *imagination* it comprises. Where the human mind cannot establish or explain a thing it necessarily invents, preferring a pretence of learning to a complete lacuna." (T. Ribot[61]) Now few pretences of learning are more typical in this regard than *numerology*. Here numbers are exploited for mercenary ends, on the strength of practices and interpretations based on nothing and leading nowhere. It is the domain of preconceived ideas ever since "numbers cultivated by mythologies and endowed by them with occult meanings have varied the world over." (H. Poincaré[62]) Hence numerologists, not content with spreading their obscurantism and their ignorance, exploit simple souls quite unscrupulously for their own ends. While ready to take advantage of what contemporary science has to offer, they are equally ready to attack what they call "official science" which, they allege, is incapable of grasping the subtlety of the power they claim to find in numbers. Numerologists, they claim, are heirs and initiates of a tradition going back to Adam and Abraham.

It is true that "the pretentiousness of the stupid conspires with the meekness of thoughtful souls to procure the triumph of obscurantism." (J. Rostand[63]) Numerology follows in the steps of its cousins astrology and divination, whose predictions, ostensibly supernatural in origin, are nothing but the products of the imagination, the conjectures of certain people admittedly well versed in the psychological subtleties of neat coincidences. And like them, numerology is a pseudo-science which has long held up the progress of scientific thought.

In fact, just as "the concrete interest attaching to particular stars hindered the birth of astronomy", so equally "the special fascination exercised on the human mind from time immemorial by numbers taken individually has proved a major obstacle to constructing a collective theory of numbers, in other words, arithmetic." (T. Dantzig[64])

Thus "it will have taken a considerable intellectual effort to strip objects and words [as the heavenly bodies and the numbers] of the magical virtues and properties with which they were spontaneously associated, to arrive at the idea we have of phenomena seen objectively." (J. Pelseneer[65])

THE NATURE OF SCIENCE

"In fact, our knowledge is directed *against* a previous knowledge, by destroying [erroneous doctrines or] ill-conceived opinions, by overcoming that which, in the mind itself, is an obstacle to a spiritual dimension." (G. Bachelard[66])

Furthermore, "all science is nothing more than a *clarification* of day-to-day thinking." (A. Einstein[67]) "The object of *science* and science itself differs from the object of *opinion* and opinion itself. For science relates to the general and proceeds by necessity, and it is impossible for necessity to be other than what it is." (Aristotle[68]) In fact "thought does not assume a scientific character until it possesses a universal application . . . A known fact is not scientific except to the extent that it applies to the *whole* of the mind [and is *accepted* by every mind]. Opinion, and even collective belief, is at odds with science if it has no means to be applied universally." (E. Goblot[69])

"But . . . we must not forget that science has a more immediate, a higher destiny, which is to satisfy our need to understand the *laws of phenomena*." (A. Comte[57]) Whence this essential distinction that has to be drawn between *empirical initiative* and *scientific method*. "Empiricism records evident facts, science publishes this evidence to arrive at hidden laws. There is no science apart from that of what lies *hidden*." (G. Bachelard[70]) "In a word, *science gives birth to prediction, prediction to action*; this is the very simple formula that expresses the exact relationship between science and art, taking these two terms in their broadest meaning." (A. Comte[57])

SCIENCE, TECHNIQUE, AND CIVILISATION

And it is precisely by virtue of the establishment of such relationships with the other areas of human life that "scientific knowledge has transformed life and increased the power of those in possession of it." (P. Valéry[71])

Thus the development of electronics, in pursuit of a broad range of needs, and offering a wide number of potential applications, has led to all manner of technical breakthroughs, giving us a glimpse of extremely vast horizons in which the solution to unimaginable problems, or problems long considered insoluble, was soon to become a daily reality: the invention of radar, advances in methods of communication, the development of the radio, television and telecommunications, the construction of supersonic aircraft, rockets, etc.

Furthermore, acquisitions made over several centuries in automation, in programming, in procedures for automatic commands and data processing, have brought this technology, thanks to mathematics and symbolic logic, within reach of ever bolder achievements. Among these achievements feature the present-day computer, of course, whose rapid development will without any doubt have been facilitated by this technology. But it will also have generated impetus inasmuch as it is the needs arising out of information technology (in the broadest sense of the term) which today provide the main impulse in the development of electronics.

There is, of course, an essential difference between *science* and *technology*, but the one may not progress without the help of the other, *scientific theories and models* being elaborated by the one, and *technical instruments* being furnished by the other, in accordance with an interaction that characterises fundamental science today.

"But by its very nature, [fundamental science] is *transferable*; it is resolved necessarily into methods and recipes of universal application. An advance that benefits one may serve to benefit all." (P. Valéry[71]) And yet, "from regarding science as some sort of exercise in miracle-working, a way to solve every problem with a wave of the magic wand, we have shown our disappointment by falling into the opposite extreme and seeing in technological progress the root of all evil. This is no different to the behaviour of the priests of Athens who, after sacrificing an ox, visited judgment on the knife that slew it and condemned it to be thrown into the sea as being guilty of its death." (P. M. Schuhl[72])

Indeed "all that comes to perfection through progress also perishes through progress." (Pascal[2]) Let us look again at the knife-blade: it is the very thing for cutting into living sinew when it is a matter of an ox, but to apply it to that of a human being is criminal. True, many a modern invention possesses the same cutting edge and the same ambivalence, conferring on the one hand power and on the other hand utility. Clearly all depends, when deciding on the use to which to put the blade, on the sense we have of our own existence and actions, a sense which, from a moral standpoint, we identify with duty.

And in this sense it is right to call a halt at the point where the conscience comes into play. "To the individual as to the nation science appears a threat only if the individual and the nation place themselves beyond the reach of science and civilisation, like those savages mentioned by Montesquieu who satisfy their hunger by chopping down the tree from which they want to gather the fruit." (L. Brunschvicg[73]) "Scientific thought", moreover, "is neither a concomitant nor a condition of human progress, it *is* that progress." (W. K. Clifford[74])

MANKIND AND TECHNICAL PROGRESS

It must also be stated that "technical progress is not an *end* . . . A machine . . . is a *means* . . . , a means for promoting in mankind that which is essentially human . . . In other words . . . technical progress obliges mankind to specialise in what is *human*." (J. Fourastié[75]) Thus technological progress offers us the possibility of bringing into play, discovering and exploiting whatever there is that offers mankind the greatest value – human ingenuity, which is a spirit of invention, "the assembled faculties of the mind raised to their highest level of expression". (V. d'Indy[76])

MANKIND AND HUMAN GENIUS

But, as Balzac puts it, "the good thing about genius is that it looks like everybody and nobody looks like *it*."[77] The reason for this is that "genius seeks obstacles, and it is obstacles that make for genius" (R. Rolland[78]). Which is why "all men possess genius if they are capable of discovering it in themselves. But that is the difficulty: for we scarcely ever do anything but envy others for what they possess, imitating and striving to better what others have instead of exploiting our own resources." (L. Lavelle[13]) Now not even the world's greatest geniuses have exploited their resources except in the tiniest measure, considering their potential.

MANKIND AND HIS MEMORY

But if genius is precisely the ability to invent, *memory* is the faculty allowing us to preserve what is invented. It is a collection of functions – fixing, preserving, recalling, recognising, and identifying – thanks to which we can call to mind elements earlier in time and fix them, memorise them, re-evoke them and recognise them, be it as experience recollected or as facts experienced by others in the course of history.

MEMORY AND HISTORY

Besides, "when the genius of invention withers the genius of history is stirred to life." (J. Barbey d'Aurevilly[79]) "We live above all by imagination and by memory . . . Whatever element of richness and joy is to be found in what is real and present depends chiefly upon what these retain from the past and points towards the future, upon all that our imagination and our memory insinuate therein." (C. Blondel[80]) It is as well to stress the special importance played by memory and history in our present responses and in our understanding of contemporary civilisation.

HISTORY AND THE SEARCH FOR THE PAST IN THE FUTURE

Of course "history is the knowledge of mankind's past." (H. Marrou[81]). And it "never is anything but a historian." (J. Barbey d'Aurevilly[79]) But the study of history which, from every point of view, offers the greatest measure of advantage and of interest is not a nostalgic exploration of the past nor the mere chronicle of lost times but incontestably the study vested by the historian with an identifiable structure by means of a pertinent co-ordination of the facts that will have given rise to it. This is achieved by a historian who "to the extent that he lives historically, focuses on action and looks for the past in the future". (R. Aron[82]) Herein lies the whole interest of history, "the concrete representations of the past are bound up with a thrust towards the *future*." (R. Aron[82])

Seen from this angle, history's ultimate goal is to "furnish us with an explanation, to demonstrate the mechanisms of cause and effect whereby human society has developed at every stage." (L. Halphen[83]) "That is what history is: a means of understanding and, precisely thereby, of influencing the course of events. It does not mean appropriating the 'lessons of the past', which even military men, after studying them so diligently, have not always turned to their advantage. If what we look for are recipes for success, the past has no lessons to offer." (L. Fèbvre[84])

In fact "the true character of history is itself to take a role in history. The idea of the past is meaningless and of no value save for the man who nurtures passion for what the future holds. The future, by definition, presents no picture. History confers on the future the means of being pictured. It provides the imagination with a table of situations and catastrophes, a gallery of ancestors, a tabulation of acts, of expressions, of attitudes, of decisions, all made available to us in our vacillations and doubts, to help us in the business of becoming." (P. Valéry[85])

History, then, is a means permitting us to measure the road travelled by mankind, it enables us the better to grasp the extent of mankind's actions and initiatives, failures as well as successes, in accordance with the potential available to connect the elements of the past to those of the present in view of a conscious realisation of our own future. For "up to a point and within the limits of certain generalities history may be defined, interpreted or deciphered – to the extent that we [man and society in possession of its own history] succeed in eliciting from it meanings and intelligible directions and laws that throw light on events without inducing those events." (J. Maritain[86]) So "history will assuredly not tell us what is to be done, but it will perhaps help us to find out." (Fustel de Coulanges[87])

THE HISTORY OF SCIENCE: A MEANS OF INFLUENCING THE EVOLUTION OF SCIENCE

And that is why the history of science, as a bountiful gallery of actions, ideas, and discoveries, is susceptible, by linking up with our present state of knowledge, to showing us just how the science of the future is to develop. Here is a typical example out of history. Around 1621 a new edition of Diophantus's *Arithmetica* was published; it contained a development of the properties of Pythagorean numbers, according to which there exists an infinity of whole numbers x, y and z which satisfy the relationship: $x^2 + y^2 = z^2$. A copy of the book came into the hands of the French mathematician Pierre de Fermat who, considering the possibility of generalising the theorem, wrote in the margin of one page: "It is impossible to divide a power of 3 into two powers of 3, a power of 4 into two powers of 4 and, generally, to divide any power greater than 2 into two powers of the same degree. I have found a truly admirable proof for this proposition, but this margin is too small for it." (cf. T. Dantzig[64])

Thus was published Fermat's famous "last theorem", which was to intrigue scholars, and whose demonstration was to exercise the wits of reputable mathematicians to the point of turning it into "one of the greatest challenges posed to human intelligence". (In modern-day terms the question was in fact to demonstrate that there exists no triple of whole numbers x, y, and z that satisfies the relationship $x^n + y^n = z^n$, where n represents a whole number greater than 2.) In the course of generations a certain number of demonstrations were invented, but only in the context of particular propositions, and up till 1993[1] nobody discovered a solution to the problem that was of absolutely general application. Indeed there were munificent sponsors, like Dr Wolfskeol in 1908, who even went so far as to offer sums in the region of 100,000 marks to whoever discovered the complete solution to the puzzle. Euler, Lagrange, Riemann and many another eminent mathematician tackled the problem without fully getting to the bottom of it.

Nevertheless, the German mathematician Ernst Eduard Kummer (1810–1893) in trying to penetrate the enigma posed by Fermat, created new mathematical entities, the *ideals*, which were to constitute one of the most remarkable, fundamental, and productive creations in nineteenth-century mathematics, without however bringing a complete solution to the thorny problem any closer.

Thus in trying to solve a problem which in fact possessed no real interest for the science of mathematics, and without even achieving the goal that was being pursued, Kummer, by his knowledge of a fact of history and by dint of all the reflection resulting from it, gave birth to an important branch, indeed one of incalculable consequences, that was destined to give a tremendous fillip to contemporary mathematics. To paraphrase J. Guehénno's[88] words, this tells us to what point the exploration of the ideas of the past may enable each one of us to go to the altar of history to light the lamp of the future.

---

1 In June 1993 Andrew Wiles announced the first full proof in a lecture at the Isaac Newton Institute in Cambridge. In August 1993 a flaw was found, finally made good in October 1994 following an essential step established in collaboration with Richard Taylor. (See S. Singh, *Fermat's Last Theorem*. London: Fourth Estate, 1997.) [*Transl.*]

FROM HISTORY TO THE PHILOSOPHY OF HISTORY

Thus a philosophy of history presupposes that "human history is not a simple sum of juxtaposed facts . . . but that it is at each and every moment a single entity moving towards a privileged state that gives a meaning to the whole." (M. Merleau-Ponty[89])

Note, however, that "like history itself, like all that is permanent, the philosophy of history is nothing but so many ideologies, projections. Past, future have no existence; only the present exists and, basically, it is by getting out of this present that we establish history and project ourselves as it were into the future." (G. Gurvitch[90])

FROM PHILOSOPHICAL REFLECTION TO THE CONSTITUTION OF HISTORY

This clearly illustrates the importance, the necessity of thinking deeply on all that mankind has accomplished, on all that mankind has experienced. This is no less than the objective of *philosophy* (in the highest meaning of the term), whose key role in the development of ideas as in the evolution of science and technology, by virtue precisely of its universal and multi-disciplinary nature, is well recognised. (Let me take this occasion to acknowledge my particular debt to P. Foulquié and his remarkable *Dictionnaire du langage philosophique* [Paris: PUF, 1982]. It has offered me authoritative guidance on the language of philosophical concepts; its definitions are always concise yet limpid and include many pertinent quotations from other authors. Not that others whom I have consulted in this domain – Balmes, Boutroux, Cuvillier, Julia, Russell, etc. – are to be undervalued, for I have derived enormous profit from their works too.)

REFLECTIONS ON COMPUTERS AND ARTIFICIAL INTELLIGENCE

Now if at this dawn of the third millennium there is one domain in which philosophical reflection is absolutely essential to effective research and a positive outlook on all that we see and do, that is assuredly the domain of *computers* and *artificial intelligence*. What exactly are the prospects for today's computers? Do they or do they not constitute "all-purpose machines"? Are they machines conceived in the image of the human brain, machines capable of fulfilling all manner of tasks deriving from human intelligence? If not, will it eventually be possible to elaborate programmes enabling them to find solutions to problems in the same way that humans do? These are questions we may justifiably ask ourselves at the end of a book such as this, at the end of this long intellectual, scientific, and technical evolution that is the story of human calculation and its various artificial realisations, the principal stages of which we have been following.

WHAT COMPUTERS CAN DO

Although these questions have been addressed in detail in Chapters 5 and 6 there is still something to be gained by recording here that what in particular characterises the computer is its *universal application* in the field of calculation, thanks to its distinctive programmability which results in the programming instructions being registered in the internal memory and treated in the same way as the data (see pp. 312ff.).

Once suitable programmes have been elaborated, these are in fact capable of addressing a considerable variety of problems, that go far beyond simple problems for standard arithmetical calculations: mathematical computations, sorting, combining, organising, managing texts, managing graphics and pictures, synthesising words, recognising shapes, aids to instruction, automatic translation, musical composition, help in graphic and artistic creation, animation, synthesising pictures, giving commands to machine tools, and so on.

In other words, by dint of appropriate programmes – each one of which symbolically expresses an "algorithm" that solves one class of problems – they can execute a wide variety of tasks, sometimes extremely complex ones, by a series of elementary operations which today they carry out at great speed and quite automatically (see pp. 274ff.). So the computer, which is vastly more powerful, more competent and effective than any numerical calculator, is the most general-purpose of machines that exist for automatic data-processing. The "calculations" which they carry out are applied, be it said, not to numbers but to symbols of any description, symbols thus susceptible of translating quite different realities beyond those very specialised ones in the field of numbers. This means that the computer is a machine that manipulates very generalised symbols, on the basis of a prior organisation of "treatments" in successive stages, according to a technique of "calculus of thought", in a very advanced degree of generalisation deriving from classic logic.

In fact, neither the functions nor the structure of computers are based on the rules of formal classic logic; they are governed, rather, by the rules of a far more universal logic: the *symbolic logic* of today. This type of logic, instead of starting from the everyday language that tends to be imprecise and applying it restrictively to specific inclusive or inherent relationships, rests upon a system of symbols whose meaning is meticulously precise, but is unrestricted in its nature, and applies it to every conceivable logical relationship (see pp. 268ff). Thus they are based upon a sort of ultimate abstraction of Aristotelian logic, on an intellectual process in which symbols are combined according to a pre-established code and in which a concrete or specific meaning of the symbols in question is never at issue.

Thus the functioning of computers corresponds to a superior form of what we commonly call "reasoning" (see pp. 244ff.). *Reasoning*, in fact, is an operation of the mind that consists in a logical sequence of connections tending towards a given end. Therefore it is the application of procedures aimed to effect transformations on data with a view to obtaining a *result*. It is an operation whose end depends, consequently, on the nature of the data under consideration.

In the particular case of classical reasoning, these logical connections are nothing other than judgments, while the result is nothing other than the "conclusion". Therefore the quality of a chain of reasoning is a function of its premises. But with a computer such reasoning is neither analogical nor inductive, in the sense that it does not start from a particular case in order to arrive at another particular case, and never extrapolates from the particular to the general. This reasoning, on the contrary, takes place in a sort of extension of classic deduction, by moving from the general to the particular.

And it is precisely in this that the reasoning of computers structurally returns to the idea of calculation and computational procedures, because the algorithms that they bring into play give rise to procedures of a mechanical character, procedures that may be artificially modelled – by applying rules of transformation that serve data of any kind (expressible symbolically by means of a code) – and necessarily arrive, at least in theory, at a result. In a word, *symbolic calculation* thus effected artificially by computers is the most abstract and general kind possible.

Now these general characteristics effectively possess traits also possessed by human intelligence, which "in its abstract forms, works well on symbols (and even on certain of their concrete forms if we recognise the symbolic character of pictures or pictorial models). It proceeds, also, to logical sequences that are particularly manifested in reasoning." (P. Oléron[91]) So computers are a good deal less stupid than is often alleged, since they are good at carrying out intelligent tasks.

It must be added that these computers are endowed with *general retroactive effect*, in the sense that they are able to modify their internal activities as a function of the effect they produce outside. In other words, computers are capable of evolving towards a predetermined condition by making use of data that allow them to adapt their methods to the end in view. Thus they may find a predetermined instruction in the programme being executed in order themselves to modify the process as it develops (see pp. 302ff.).

On the other hand, recent developments in programming allow for the elaboration of basic algorithms that enable the computer to construct new programmes on its own little by little as it gains *experience*. It is thus a matter of the ability to construct *evolving programmes*, according to a graded acquisition of knowledge benefiting from the experience of previous programmes. Thus the computer is endowed with the potential to serve an *apprenticeship* relating to a given type of problem (see pp. 310ff.).

To conclude: among so many other aptitudes, the computer – with its enormous scope for examining millions of cases in a matter of seconds – has been endowed with the possibility of carrying out certain procedures akin to those carried out by humans: such procedures rely upon programmes that take, of course, a global view of possible moves, although they do not automatically ensure success in the pursuit of the calculations.

WHAT IS ARTIFICIAL INTELLIGENCE?

But if the activities of computers are of their nature intelligent, do they not thus carry out the full range of the human brain's operations? By no means! Let us consider what exactly is a computer's "intelligence" and what effectively distinguishes it from human intelligence.

Without confining this complex field within any narrow, constricting definition that is liable to be overtaken all too quickly, we should specify that what we call "artificial intelligence" is used to study and to understand the procedures of human intelligence and to reproduce them artificially by means of systems of control and automated command. Such a prospect has been the dream of mankind for many millennia, but only in the last fifty years has it progressed in concrete terms; even now we are only at the outset, but some spectacular successes have already been achieved.

On the other hand, the venture has also been marked by some spectacular failures. Now it is precisely the absence of philosophical reflection, or at any rate the development of a mindset that reduces the scope of artificial intelligence to within the limits of the computer sciences, that goes some way to explain the reason for these failures.

WHEN METAPHOR WAS IDENTIFIED WITH REALITY

The discoveries underlying computer science and the theory of logical automata tended from the start to lure a good number of scientists (indeed some of the best) into an attitude of sheer enchantment, with baneful consequences. In their enthusiasm these scholars were quick to believe that every operation of the human mind was exclusively concerned with *calculation* (that is, it was *algorithmic*), and thus that every mental process could be computerised. They used the term *neuron* and *synapsis* to describe modules which in reality had only a very distant kinship with the far more complex components of a living brain; and thus they inevitably invoked a simplistic and necessarily reductive anthropomorphism to establish the tightest parallel between the circuitry of an electronic calculator and the neurological components of the living brain. Once the central organ of a simple electronic computer was considered to be the equivalent to the human brain, the public thenceforth assumed that the activities of a computer and those of the human brain were interchangeable.

Whence, by a mistaken identification of metaphor with reality, by a sort of convergence with the myth of the "electronic brain", the idea developed of a machine theoretically endowed with inductive and creative thought, endowed with the ability to take all manner of initiatives, a machine to which we could entrust all our problems without exception in order to obtain from it then and there every solution required (see pp. 313ff.).

FUNDAMENTAL DIFFERENCES BETWEEN THE HUMAN BRAIN AND THE COMPUTER

Of course the computer's "brain" does possess certain factors in common with the structure of the nervous system of living creatures, but the analogy can only be taken so far. In the words of P. Demarne and M. Rouqerol, "while computers are built with materials relating as a rule to the relatively simple chemistry of metals and metalloids, the physiology of the nervous system can be studied only in the light of the complex chemistry of large molecules, that is to say, organic chemistry. While the functioning of computers may be understood by reference to mechanical laws and physical phenomena, the nervous system can be approached only in the light of *biological chemistry*. Moreover, while a real scientific continuity exists between the physical and the chemical worlds, the fact is that in the one case we are looking at phenomena born out of *inert matter* while in the other, we are looking at phenomena born out of *living matter* – a fundamental difference. The mechanisms of inert matter are far less complex than those of living matter, especially where interaction is concerned. [Now, given the present state of science] living matter remains difficult to analyse, and to scrutinise the nervous system is to study living matter in the noblest state in which an attempt has ever been made to penetrate its secrets. At this stage, each cell seems to play a distinct role; each element of substance, however small, in this cell or surrounding it, can alter the balance of the structure of which this cell forms a part. Thus, knowing as little as we do about the precise molecular reactions going on within each nerve cell, we are bound to undertake a description of the nervous system, however precarious and provisional, only with the most extreme caution. So our understanding of computers and our understanding of the human mind are still a long way apart.

"To get some idea of the progress of the human mind it is vital to look closely at the work of physiologists, psychologists and physicians. Several crucial points transpire from this documentary, and equally descriptive, evidence. First, in order to think, we have to receive a certain number of messages from outside, we have to record data. In one way or another such messages result from the material stimulus (shock, wave, etc.) exerted on one or several sense-organs which act as mechanical, optical, or chemical transformers. The stimuli are transformed into nervous input which is carried to the brain. At this level the data 'enters' the conscious field. A number of complex operations are set in train; decisions are taken 'in' the brain; orders transmitted through the channels of the nervous system will affect the relevant muscles, which will come into play – the externalisation phase . . ." (P. Demarne and M. Rouquerol[92])

Although these indications are summary and incomplete, it is easy to see how complex are the phenomena brought into play in the typical functioning of the human brain. And it goes to show how limited is the structural analogy between man and computer, despite all the functional analogies that it has been possible to establish as to their respective activities.

The computer is in fact well able to carry out a certain number of tedious or repetitive tasks, where the human invariably gets stalled. Moreover these tasks are sometimes so complex that a single person, even one who devoted his whole life to it, could never achieve them in the computer's place. Which shows that the computer is endowed with capabilities that often go far beyond that which any human could achieve.

WHAT COMPUTERS CANNOT DO

These machines, however, do only what is prescribed to them by the programmes introduced into them; the data input into them might be eminently sensible or positively absurd, but they will work in strict obedience and exercise no discretion. The computer is incapable of thinking or reasoning, it is incapable of forming an intention or of feeling – variables that often interfere in human thinking. Procedures for instance that involve an understanding of one's own existence, of one's acts, of the external world, or which bring into play the ability to realise one's own intentions, according to need, the outcome of prior reflection or the fulfilment of an undertaking. Which is as much as to say that the computer is a machine incapable of free will or creativity. It is a work-horse from which not the smallest spark of genius is to be expected. To quote P. Demarne and M. Rouquerol[92] again, it is a first-rate intellectual drudge, nothing more. [Cf. observations on this topic by and concerning Blaise Pascal, on p. 122. *Transl.*]

COMPUTERS AND THE HUMAN FACTOR

This should reassure all those already worried at the prospect of the computer dispossessing humankind, hijacking our initiative, our creative genius and our capacity to keep inventing. "Those concerned about the intrusion of these machines into the highest spheres of human activity, Art for instance, should bear in mind that the mechanisation of draughts-manship, the chemical means for reproducing reality, be it in photography or the cinema, have not yet killed off painting, which has simply developed new forms. Unlike imitators, the most creative talents have in fact benefited from this situation." (P. Demarne and M. Rouquerol[92])

No need to add that in the field of computers the human factor remains in charge and remains the motive force for every step, be it conceptual or technical, that is accomplished.

It certainly seems therefore, to revert to a theme once greatly cherished by G. Berger,[93] that "these machines will help us more and more in our research, but they will never dispense us from being humans. They are like habits: they enslave the weak, but they liberate those who really have something to say or do." Besides, today man and computer truly play complementary roles, the human providing all the *competence* lacking in the machine in order to derive from the machine all possible advantage, and the computer contributing all the *performance* and physical capacity, such as speed and reliability, which are so lacking in us in our methods of addressing problems. These complementary roles translate in practice into human research aimed at determining the algorithms best suited to the solution of given problems, and at perfecting the best procedures which may be fed into the machine, enabling it to furnish us with the answer to our questions in the most effective manner (see pp. 310ff.).

FROM CLASSIC INFORMATION TECHNOLOGY
TO ARTIFICIAL INTELLIGENCE

But the limitations of computers are just as clear: as general-purpose algorithmic automata they are by definition incapable of carrying out any intellectual processes other than those relating to computation (see pp. 302ff.).

Be it also noted that calculation and deductive reasoning form only one very particular class of mental processes, for the solution to a problem is not necessarily of that nature. In other words, there is a vast range of problems whose solution may not be expressed in the form of an algorithm. There is therefore a considerable variety of tasks beyond the scope of computers. Whence of course the structural limitation of these machines and of their type of logic.

And it is in order to get round these inadequacies that researchers from a broad range of disciplines (psychologists, information technologists, cyberneticians, neurobiologists, etc.) have been at work for some decades trying to invent new kinds of artificial systems that function on the model of the living brain. Which means that after developing the different forms of our intelligence as substance of the spirit, and after seeking to "mechanise" computational processes along with all the forms of reasoning that characterise them, we have weighed up the limitations of symbolic calculation and of the corresponding formal logic, and are moving today towards the construction of artificial models of at least a part, hopefully an appreciable part, of our own cerebral activities, with a view to elaborating a new technical organisation and new theoretical systems.

Hence the origin of the new science known as *artificial intelligence*, whose development is due not only to progress in technology but also and particularly to the great strides lately taken in mathematical thought (work by Thom, Prigogine, Mandelbrot, etc.), which has supplied us with several new epistemological models such as the theory of catastrophes, fractal theory, and so on. "The point has been reached", says G. Verroust,[94] "where mankind is conceiving and building complex automata that escape the structural limits of classic logic [symbolic, henceforth], whose overall functioning may be studied and predicted, but which does not admit of any method of calculation that will allow us to assess the internal condition of any given constituent at any given point."

TOMORROW'S ARTIFICIAL INTELLIGENCE

Which means that in the not too distant future, extraordinary *artificial automata* will be constructed, whose "intelligence" will be greater and more real each day. They will be "brains" with ever greater capacities, which can imitate an ever greater number of cerebral activities not in the nature of calculation, and which will be capable of executing for mankind ever more complicated and delicate tasks. But in all probability those brains will be endowed with an "intelligence" very different from our own, For, if we are to judge on the basis of our own state of knowledge, it seems likely that "these 'brains' that we shall quite soon be building will be slave brains, entirely conditioned by their builders and their users. And it will be possible to make them ultra-specialised, as are the brains of those prodigious monsters (calculators and others) which are able to do only one single thing but are able to do it better than the normal run of mortals, being in all other respects truly mentally deficient." (G. Verroust)[94]

There would therefore seem no need to be afraid of artificial brains that will be built in the coming decades. For they will be "brains" that will dispense us from those complex one-dimensional tasks which cannot be achieved without their help, and which, when all is said and done, will procure us greater freedom than we have today, more leisure, and a better grasp of our world by enabling us to blossom out fully in all directions. "Employing our new mastery of the most complex laws of nature, we shall continue to evolve individually and socially towards superior forms of life." (G. Verroust[94])

In any event, looking to the immediate future, these "brains" will not be anything but machines: disembodied, lifeless machines with no past, no social institutions, no collective memory, no history. Built out of inert elements, they will be only lifeless artificial implements, engaged in none of the activities – such as assimilation, growth, reproduction – that fundamentally characterise sentient beings.

QUESTIONS RELATING TO A MORE REMOTE FUTURE

But will it ever be thus? Will mankind succeed one day in making a conceptual conquest of its highest mental functions and in melding them, as it were, into the mould of artificial models? And will we, further down the road, succeed in combining these functions in one and the same brain with the same capacities as the human one? Will we get there by dint of miracles of a technology (say biochemistry) of which we are still a long way from unlocking the secrets, by dint of considerable advances in mathematics, and by bringing together into a common effort the science of artificial intelligence with the disciplines of psychology, neurophysiology, biology, and genetics? Will we get to the point of constructing living beings endowed both with a body and a mind that is comparable, even superior, to our own? In a word, does the future hold for us the appearance and development on Earth of a new breed of artificial implements made into beings that can organise themselves, and that are endowed with intelligence in the sense of the noblest grey matter, thus dispensing us from being humans? It is not absurd to imagine it, but to all these questions it is of course impossible to furnish an answer in the present state of our knowledge.

Should these problems, however, one day be resolved, the great difficulty to overcome will undoubtedly be that of assembling the nexus of conditions which have permitted the development of human intelligence in its infinite complexity and which, let us not forget, is part and parcel of the physical body, its matter, its genetics, its social structures, bound up as it also is with its culture, its individual memory, with the collective memory and with the history of the whole human race.

Assuming that the challenge is not insurmountable and leaving aside the ethical and metaphysical problems it comports, what is certain, in any case, is that to come to fruition these developments will require such intellectual efforts and the convergence of so many technologies that they will perforce have to correspond to a genuine human need.

THE MOULD FOR ARTIFICIAL MODELS OF HUMAN INTELLIGENCE

Coming back to what is presently at stake, the problem posed by artificial intelligence is to know whether the full tally of the activities of the human mind may or may not be simulated by artificial means. This is a fundamental question for the present generation and those to come, and touches on that of the origin of our intelligence as also on the intrinsic nature of the human race. Only a healthily objective philosophical and epistemological approach, based upon plausible arguments and on a pragmatic mindset, will be capable of dissecting the problem, affording us a better appreciation of what is at stake in core science today, and allowing us some glimpse of what lies ahead. We shall have to take everything into account if we are to give an objective assessment of what is at stake: we shall have to distinguish between a realistic and achievable goal on the one hand, and pure Utopia on the other. If not, millions are liable to be spent on what turn out to be dead-end projects.

Of course "a hypothesis is . . . the necessary point of departure for any experimental reasoning . . . If we *experimented* without a preconceived idea we would be groping in the dark. But on the other hand . . . if we *observed* with preconceived ideas, our observations would be worth little and we would risk confusing imagination with reality." (C. Bernard[95])

Conversely, we must not reject wholesale every element of a past theory simply because the organisation resulting from it led to a dead end. Let it not be forgotten that "structure is simply a unity possessing its intrinsic law of action and development, having its own causality and achieving a functional individuality." (R. Mucchielli[96]) For "in the first place, a structure is basically a system. It comprises elements, so that any modification of one element will entail a modification of them all." (C. Lévi-Strauss[97]) Indeed, as has frequently been the case, there are ideas which are entirely negligible when applied to certain given dispositions, but when they are applied to quite different structures, they bear fruit to remarkably good effect.

ETHICS AND PHILOSOPHY OF SCIENCE

As for the ethical problems that these questions are bound to raise, a moral conscience and absolute vigilance will be needed if we are to avoid, notably, the development of certain societies on a basis of eugenics even more terrifying than those for which the twentieth century furnished the ghastly models we all know about. As Rabelais[98] puts it, "science without conscience is simply the ruin of the soul."

To paraphrase Einstein,[67] we shall have to agree on certain fundamental ethical propositions. Such ethical premises, and others that may derive from them, will play a role in the human conscience that is comparable to that played by axioms in mathematics: they will be essential propositions to be placed beyond discussion and *a fortiori* beyond all initiative and all action.

"To pursue philosophy is, in a nutshell, to treat the universe as though nothing were self-evident." (V. Jankélévitch[99]) For, to pursue philosophy is to establish as goal that of knowing the universe, both the physical and the mental universe.

"Philosophy is at once *knowledge* and *wisdom*, *understanding* and a *testing out*: it is a 'comprehensive' knowledge that requires the effective participation of the one who gives his mind to it. In a word, to reflect is to meditate, and meditation is as much the work of conscience as it is the elaboration of knowledge, it has as much to do with ethics . . . as with logic." (D. Julia[100]) And those with no time for philosophy and history mistrust the apprehension of facts and the intellectual faculty itself; worse, they show their disdain for mankind and all that relates to it.

References

1 J. Maritain, *Les degrés du savoir: distinguer pour les unir*, p. 218. Paris: Desclée de Brouwer, 1982.

2 B. Pascal, *Pensées et Opuscules*, 371 and 509. Brunschvicg ed. Paris: Hachette.

3 J. Rostand, *Journal d'un caractère*, p. 19. Paris, 1931.

4 J. Vialatoux, *Le Discours et l'intuition*, p. 14. Paris: Bloud et Gay, 1956.

5 G. Berger, *Recherche sur les fondements de la connaissance*, pp. 41–2. Paris, 1941.

6 J. Wahl, in *Encyclopédie française*, XIX, 12, 13.

7 J. Perrot, *La Linguistique*. Paris: PUF, 1986.

8 Plato, quoted by D. Julia, see Ref. 100 below.

9 V. Egger, *La Parole intérieure*. Paris: Baillière, 1881.

10 H. Duméry, *Critique et religion*, p. 205.

11 J-B. Molière, *Le Mariage forcé*. Paris, 1654, VI.

12 L. Lavelle, *La Parole et l'écriture*, pp. 30, 162–3, 172. Paris: Artisan du Livre, 1942.

13 L. Lavelle, *L'Erreur de Narcisse*, pp. 42, 123. Paris: Grasset, 1939.

14 D'Alembert, *Discours préliminaire à l'Encyclopédie*, p. 42. Paris, 1751.

15 H. Piéron, *L'Homme, rien que l'Homme*, p. 72.

16 J. Matricon and J. Roumette, *L'Invention du temps*. Paris: Presses Pocket, 1991.

17 L. Bourdaloue, *De la perte du temps*. Quoted in P. Foulquié, see Ref. 38 below.

18 P. Peeters, *Sentences intemporelles*, No. 116.

19 E. Herriot, quoted in P. Foulquié, see Ref. 38 below.

20 Alain, *Préliminaires à la mythologie*, p. 142.

21 J. Leclercq, *Culture et personne*, p. 26.

22 H. F. Amiel, *Fragments d'un journal intime* (14 August 1877). Paris: Stock, 1949.

23 Alain, *Propos*, 923. Paris: Gallimard.

24 D'Alembert, *Maximes diverses*. Tolérance.

25 C. Spearman, *Aptitudes de l'homme*, p. 182.

26 J. Guitton, *Invitation à la pensée*, p. 19. Quoted in P. Foulquié, see Ref. 38 below.

27 J. Guéhenno, *Sur le chemin des hommes*, p. 28. Paris: Grasset, 1959.

28 G. Morf, *Éléments de psychologie*, p. 162.

29 E. Claparède, *Genèse de l'hypothèse*, p. 3.

30 H. Delacroix, *Le Langage et la pensée*, p. 118. Paris: Alcan, 1930.

31 E. Goblot, *Le Système des sciences*, p. 142. Paris: A. Colin, 1922.

32 Aristotle, *Topiques*, I, 12.

33 Aristotle, *Seconds analytiques*, II. 19.

34 C. Serrus, *Méthodes de Descartes*, p. 17.

35 G. Heymans, *Psychologie des femmes*, p. 22.

36 L. de Broglie, *Sur les sentiers de la science*, p. 210. Paris: Albin Michel, 1960.

37 A. Lalande, *La Raison et les normes*, p. 187. Paris: Hachette, 1949.

38 P. Foulquié, *Dictionnaire de la langue philosophique*. Paris: PUF, 1982.

39 H. Delacroix, *Les Grandes Formes de la vie mentale*, p. 153. Paris: PUF, 1934.

40 J. Segond, *Traité de psychologie*, p. 179.

41 A. Burloud, *La Pensée conceptuelle*, p. 163. Paris: Alcan, 1927.

42 T. Ribot, *L'Évolution des idées générales*, pp. 232–3. Paris: Alcan, 1897.

43 F. Gonseth, *La Géométrie et le problème de l'espace*, p. 583. Paris: Griffon, 1978.

44 A. Vandel, *L'Homme et l'évolution*, p. 123.

45 J. Duclaux, *L'Homme devant l'univers*, p. 289.

46 H. Bergson, *L'Évolution créatrice*, p. 44. Paris: Alcan, 1907.

47 A. Rey, *L'Apogée de la science technique grecque: l'essor de la mathématique*, p. 6. Paris: Albin Michel, 1963.

48 R. Le Senne, *Le Devoir*, p. 566. Paris: PUF, 1949.

49 J. Piaget, quoted in P. Foulquié, see Ref. 38 above.

50 A. Siegfried, *Aspects du XXe siècle*, pp. 210–11.

51 J. Ellul, *La Technique*, p. 5.

52 A. Siegfried, in *Progrès techniques et progrès moraux*, p. 12. Geneva: Rencontres, 1948.

53 B. Russell, *The Scientific Outlook*. London: Allen & Unwin, 1931.

54 S. Bachelard, *La Logique de Husserl*, p. 135. Paris: PUF, 1957.

55 L. Brunschvicg, *L'Idéalisme contemporain*, p. 163. Paris, 1921.

56 L. de Broglie, *Continu et discontinu en physique moderne*, p. 81. Paris: Albin Michel, 1941.

57 A. Comte, *Cours de philosophie positive*, I, pp. 55, 56, 86 and 99. Paris, 1830.

58 A. Cournot, *De l'origine ...* p. 355. Quoted in P. Foulquié, see Ref. 38 above.

59 E. Goblot, *Essai sur la classification des sciences*, p. 21.

60 J. Hadamard, in *Encyclopédie française*, I, 52, 1.

61 T. Ribot, *Essai sur l'imagination créatrice*, p. 201. Paris: Alcan, 1900.

62 H. Poincaré, *La Science et l'hypothèse*. Paris: Flammarion, 1902.

63 J. Rostand, *De la Vanité*, p. 14.

64 T. Dantzig, *Number, the Language of Science*. New York: Macmillan, 1967.

65 J. Pelseneer, *Esquisse du progrès de la pensée mathématique*. Paris, 1965.

66 G. Bachelard, *Formation de l'esprit scientifique*, p. 14. Paris: Vrin, 1930.

67 A. Einstein, *Conceptions scientifiques*, pp. 35 and 66. Paris: Champs-Flammarion, Gallimard, 1990

68 Aristotle, see Ref. 33 above. I, 33, 88b.

69 E. Goblot, see Ref. 31 above, p. 15.

70 G. Bachelard, quoted in P. Foulquié, see Ref. 38 above.

71 P. Valéry, *Regards sur le monde actuel*. Paris: Stock, 1931.

72 P. M. Schuhl, *Machinisme et philosophie*. Paris: PUF, 1938.

73 L. Brunschvicg, *La Physique du vingtième siècle et la philosophie*. Paris: Hermann, 1936.

74 W. K. Clifford, quoted in M. Boll, *Histoire des mathématiques*. Paris: PUF, 1963.

75 J. Fourastié, *Le progrès technique et l'évolution économique*, II, pp. 247–8.

76 V. d'Indy, *Cours de composition musicale*, p. 15.

77 H. de Balzac, quoted in Alain, *Propos*, 924. Paris: Gallimard.

78 R. Rolland, *L'Humble Vie héroïque*, p. 61.

79 J. Barbey d'Aurevilly, *Les Idées et les histoires*. Historiens, 4. I, 3. Paris: Amyot, 1860.

80 C. Blondel, *Psychographie de Marcel Proust*, p. 1.

81 H. Marrou, *De la connaissance historique*, p. 32. Paris: Le Seuil, 1954.

82 R. Aron, *Introduction à la philosophie de l'histoire, essai sur les limites de l'objectivité historique*, p. 315. Paris: Gallimard, 1938.

83 L. Halphen, *Introduction à l'histoire*. Paris: PUF, 1946.

84 L. Fèbvre, Foreword to C. Morazé, *Trois Essais sur histoire et culture*, p. vii.

85 P. Valéry, see Ref. 71 above, p. 19.

86 J. Maritain, *Pour une philosophie de l'histoire*, p. 45. Paris: Le Seuil, 1959.

87 Fustel de Coulanges, in *Revue des Deux Mondes*, 15 February 1871, p. 671.

88 J. Guéhenno, see Ref. 27 above, p. 217.

89 M. Merleau-Ponty, *Phénoménologie de la perception*, p. 165. Paris: Gallimard, 1945.

90 G. Gurvitch, in *L'Angoisse du temps présent*, pp. 164–5. Geneva: Rencontres, 1953.

91 P. Oléron, *L'Intélligence*. Paris: PUF, 1986.

92 P. Demarne and M. Rouquerol, *Les Ordinateurs électroniques*. Paris: PUF, 1961.

93 G. Berger, in *La Nef*, XI, no 6, p. 22.

94 G. Verroust, "Vers une civilisation de robots qui prendront la place de l'homme sur terre?" in *Sciences et Avenir*, Paris, Autumn 1992.

95 C. Bernard, *Introduction à l'étude de la médecine expérimentale*, I, II, 2. Paris, 1865.

96 R. Mucchielli, in *La Notion de structure*, p. 344. XXe Semaine internationale de synthèse.

97 C. Lévi-Strauss, *Anthropologie structurale*, p. 306. Paris: Plon, 1958.

98 F. Rabelais, *Gargantua et Pantagruel*. Lettre de Gargantua à Pantagruel.

99 V. Jankélévitch, *La Mauvaise conscience*, p. 3. Paris: Aubier-Montaigne, 1982.

100 D. Julia, *Dictionnaire de la philosophie*. Paris: Larousse, 1964.

List of Abbreviations

ACLHU   *Annals of the Computation Laboratory, Harvard University* Cambridge, MA
AD   *American Documentation*
ADO   *The Annals of the Dudley Observatory* Albany, N.Y.
AGM   *Abhandlungen zur Geschichte der Mathematik* Leipzig
AHC   *The Annals of the History of Computing* IEEE, New York
AIM   *Artificial Intelligence Magazine*
AJ   *Accountants Journal*
AJS   *American Journal of Science*
AM   *American Machinist*
AMM   *The American Mathematical Monthly*
AMP   *Archiv der Mathematik und Physik*
AS   *Automata Studies* Princeton University Press, 1956
AT   *Annales des Télécommunications* Paris
BAMS   *Bulletin of the American Mathematical Society*
BARSB   *Bulletins de l'Académie royale des Sciences et Belles-Lettres* Brussels
BDSM   *Bulletin des Sciences mathématiques* Paris
BFT   *Blätter für Technikgeschichte* Springer, Vienna
BIMA   *Bulletin of the Institute of Mathematics and its Applications*
BLR   *Bell Laboratories Records* Murray Hill, N.J.
BMB   *Bulletin for Mathematics and Biophysics*
BMFRS   *Biographical Memoirs of Fellows of the Royal Society*
BSA   *Bulletin de la Société d'Anthropologie* Paris
BSEIN   *Bulletin de la Société d'Encouragement pour l'Industrie nationale* Paris
BSMM   *Bulletin de la Société de Médecine mentale* Paris
BST   *Bell System Technical Journal* Murray Hill, N.J.
CA   *Computers and Automation*
CENT   *Centaurus* Copenhagen
CJ   *The Computer Journal*
CRAS   *Comptes rendus des séances de l'Académie des sciences* Paris
CS   *Computing Surveys*
CTF   *Computers and their Future* Llandudno: Richard Williams, 1970
EBR   *Encyclopaedia Britannica, 14th edition* London 1929
EEng   *Electrical Engineering*
EMDDR   *Entwicklung der Mathematik in der DDR* Berlin: VEB Deutscher Verlag
der Wissenschaften, 1974
ER   *Electrical Review*
GIES   *Glasgow Institute of Engineers and Shipbuilders*
GLE   *Grand Larousse encyclopédique* Paris: Larousse, 1962–1970
HIF.2   *Actes du 2e Colloque sur l'Histoire de l'Informatique en France* Paris: 1990
IAF   *Interface AFCET* Paris
IEEE   *The Institute of Electrical and Electronic Engineers* New York
IJES   *International Journal of Environmental Studies*
ILN   *Illustrated London News* London
InPr   *L'Informatique professionelle* Paris
IS   *Isis, Revue d'histoire des sciences*

JAOA	*Journal of the American Ordnance Association*
JASA	*Journal of the American Statistical Association*
JFI	*Journal of the Franklin Institute*
JIA	*Journal of the Institute of Actuaries*
JIEE	*Journal of the Institute of Electrical Engineering*
JOSA	*Journal of the Optical Society of America*
JRSA	*Journal of the Royal Society of Arts* London
JRSS	*Journal of the Royal Statistical Society*
JSI	*Journal of Scientific Instruments* London
MAF	*Mémorial de l'Artillerie française* Paris
MAS	*Memoirs of the Astronomical Society*
MGA	*The Mathematical Gazette*
MNRAS	*Monthly Notes of the Royal Astronomical Society*
MTAC	*Mathematical Tables and Other Aids to Computation*
NA	*Nature* London
NAM	*Nouvelles Annales de Mathématiques* Paris
NAT	*La Nature* Paris
NS	*New Scientist*
OE	*L'Onde électrique* Paris
PA	*Popular Astronomy*
PACM	*Proceedings of the Association for Computing Machinery* Pittsburgh, MA
PEJCC	*Proceedings Eastern Joint Computer Conference, IRE* New York, 1951
PLMS	*Proceedings of the London Mathematical Society* London
PLS	*Pour la Science* Paris
PR	*Physical Review*
PRS	*Proceedings of the Royal Society* London
PSSC	*Proceedings of the Seminar on Scientific Computation, November 1949* New York: IBM Corp., 1950
PTRSL	*Philosophical Transactions of the Royal Society* London
RACE	*Real Academia de Ciencias Exactas, Fisicas y Naturales, Revista* Madrid
RBAAS	*Report of the British Association for the Advancement of Science* London
RCACM	*Report of a Conference on High Speed Automatic Calculating Machinery, June 1949* Cambridge: University Mathematical Laboratory, 1950
REDC	*Review of Electronic Digital Computers, Joint AIEE-IRE Computer Conference, December 1951* New York: American Institute of Electrical Engineers, 1952
RHSA	*Revue d'Histoire des Sciences et de leurs Applications* Paris
RR	*The Radio Review*
SA	*Sciences et Avenir* Paris
SCAM	*Scientific American* New York
SJ	*Science Journal*
SMQ	*School of Mines Quarterly*
SPRDS	*Scientific Proceedings of the Royal Dublin Society* Dublin
SV	*Science et Vie* Paris
TAIEE	*Transactions of the American Institute of Electrical Engineers*
TAPS	*Transactions of the American Philosophical Society*
TSM	*Taylor's Scientific Memoirs* London

Bibliography

ABBREVIATIONS USED ARE LISTED ON PP. 376–377 ABOVE

(*Dates in brackets are the first edition when a more
recent edition has been published*)

1. WORKS AVAILABLE IN ENGLISH

H. H. AIKEN, "Proposed Automatic Calculating Machine". IEEE Spectrum: pp. 62–9. New York, August 1964. See also B. RANDELL, 1973, pp. 191–7.

H. H. AIKEN and G. H. HOPPER, "The Automatic Sequence Controlled Calculator [ASCC]". EEng 65: pp. 384–91, 449–54, 522–8. 1946. See also B. RANDELL, 1973, pp. 199–218.

S. N. ALEXANDER, "The National Bureau of Standards Eastern Automatic Computer (SEAC)". PEJCC: pp. 84–9. New York, December 1951.

F. L. ALT, "A Bell Telephone Laboratories' Computing Machine". MTAC 3/1948: pp. 1–13, 69–84.

E. G. ANDREWS, "A review of the Bell Laboratories Digital Computer Developments". REDC: pp. 101–5.

E. G. ANDREWS, "Telephone Switching and the Early Bell Laboratories Computers". BST 42/63: pp. 341–53.

R. C. ARCHIBALD, "Georg Scheutz, Publicist, Author, Scientific Mechanician, and Edvard Scheutz, Engineer – Biography and Bibliography". MTAC 2/1947: pp. 238–45.

R. C. ARCHIBALD, "Martin Wiberg, his Tables and his Difference Engine". MTAC 2/1947: pp. 371–3.

W. R. ASHBY, *An Introduction to Cybernetics*. London: Chapman/Hall, 1956.

W. R. ASHBY, *Design for a Brain*. London: Chapman/Hall, 1960.

G. ASHURST, *Pioneers of Computing*. London: Frederic Muller Ltd., 1983.

J. V. ATANASOFF, "Computing Machine for the Solution of Large Systems of Linear Algebraic Equations". See B. RANDELL, 1973, pp. 305–25.

J. V. ATANASOFF and A. E. BRANDT, "Application of Punched Card Equipment to the Analysis of Complex Spectra". JOSA, 26–2/ February 1936: pp. 83–8.

S. AUGARTEN, *Bit by Bit: an Illustrated History of Computers*. London: George Allen and Unwin Ltd, 1984.

G. D. AUSTRIAN, *Hermann Hollerith, Forgotten Giant of Information Processing*. New York: Columbia University Press, 1982.

CHARLES BABBAGE, "A Note Respecting the Application of Machinery to the Calculation of Astronomical Tables". MAS, 1/ June 1822: p. 309.

CHARLES BABBAGE, "Letter to M. Quetelet". BARSB, 2–5/ May 1835: pp. 124–5.

CHARLES BABBAGE, "On the Mathematical Powers of the Calculating Engine" (ms. dated December 1837. Museum of the History of Science, Oxford: Buxton MS7). See B. RANDELL, 1973, pp. 15–52.

CHARLES BABBAGE, *The Exposition of 1851; or, Views of the Industry, the Science and the Government of England*. London: John Murray, 1851.

CHARLES BABBAGE, *Passages from the Life of a Philosopher*. London: Longman

Green, 1864; new edition by Augustus M. Kelly. New York, 1969.

H. P. BABBAGE, "On the Mechanical Notation, as Exemplified in the Swedish Calculating Machine of Messrs. Scheutz". RBAAS, 1855: pp. 203–5.

H. P. BABBAGE, "On the Mechanical Arrangements of the Analytical Engine of the Late Charles Babbage F.R.S.". RBAAS, 1888: pp. 616–17.

H. P. BABBAGE (editor), *Babbage's Calculating Engines: Being a Collection of Papers Related to them, their History, and Construction*. London: Spon, 1889.

H. P. BABBAGE, "Babbage's Analytical Engine" (1910). See B. RANDELL, 1973, pp. 65–9.

G. H. BAILLIE, *Clocks and Watches. An Historical Biography, I*. London: The Holland Press, 1978

R. BAKER, *The Science Reference Library: New and Improved. Inventors and Inventions that have Changed the Modern World*. London:The British Library Board, 1976.

J. BARDEEN and W. H. BRATTAIN, "The Transistor, a Semiconductor Triode". PR, 74/ July 1948: pp. 230–1.

A. BARR, P. COHEN and E. FEIGENBAUM (editors), *The Handbook of Artificial Intelligence*. Los Altos, CA: William Kauffmann, 1981–1982.

C. J. BASHE, "The IBM SSEC in Historical Perspective". AHC, 4–4: pp. 299ff. New York, October 1982 .

C. J. BASHE, L. R. JOHNSON, J. H. PALMER and E. W. PUGH, *IBM's Early Computers*, Series in the History of Computing edited by I. Bernard Cohen. Cambridge, MA / London: MIT Press, 1986.

D. BAXANDALL, *Calculating Machines and Instruments*, Revised by J. Pugh. London: Science Museum, 1975.

D. BAXANDALL, "Calculating Machines. EBR: pp. 545–53.

T. G. BELDEN and M. R. BELDEN, *The Lengthening Shadow: The Life of Thomas J. Watson*. Boston: Little, Brown and Co, 1962.

E. T. BELL, *The Development of Mathematics*. New York: McGraw-Hill, 1945.

E. T. BELL, (1937), *Men of Mathematics*. London: Gollancz , 1937; New York: Simon and Schuster, 1965.

E. C. BERKELEY, *Giant Brains or Machines that Think*. New York: John Wiley, 1949.

J. BERNSTEIN, *The Analytical Engine: Computers Past, Present and Future*. New York: Random House, 1981.

G. BIRKHOFF and S. MACLANE, *A Survey of Modern Algebra*. New York: Macmillan, 1977.

J. H. BLODGETT and C. K. SCHULTZ, "Herman Hollerith: Data Processing Pioneer". AD 20–3/July 1969: pp. 221–6.

E. DE BONS, *Eureka! An Illustrated History of Inventions from the Wheel to the Computer*. New York: Holt, Rinehart and Winston, 1974.

G. BOOLE (1847), *The Mathematical Analysis*. Oxford University Press, 1948.

G. BOOLE, *The Laws of Thought*. London, 1854.

D. J. BOORSTIN, *The Discoverers*. London: Dent, 1984.

A. D. BOOTH and K. H. V. BOOTH, *Automatic Digital Calculators*. London: Butterworth, 1956.

K. E. BOULDING, *The Organizational Revolution*. New York: Harper and Row, 1953.

B. V. BOWDEN (editor), *Faster than Thought*. London: Isaac Pitman, 1953.

B. V. BOWDEN, "He Invented the Computer Before its Time". *Think*, July 1960: pp. 28–32.

D. M. BOWERS, "The Rough Guide to Today's Technology". *Datamation*, September 1977.

C. B. BOYER, "Fundamental Steps in the Development of Numeration". IS 35/1944: pp. 153–69.

C. B. BOYER, *The History of Calculus and its Conceptual Development*. New York: Dover Publications Inc., 1949.

C. B. BOYER, *A History of Mathematics*. New York: John Wiley and Sons, 1968.

C. V. BOYS, "A New Analytical Engine". NA, 81/2070, July 1909: pp. 14–15.

D. B. BRICK, J. S. DRAPER and H. J. CAULFIELD. "Computers in the Military and Space Sciences". *Transactions on Computers*. New York: IEEE, October 1984.

A. G. BROMLEY, "Charles Babbage's Analytical Engine, 1833". AHC, 4–3/October 1982.

B. BUCHANAN and E. SHORTLIFFE, *Rule-Based Expert Systems* . . . Reading, MA: Addison-Wesley, 1984.

A.W. BURKS and A. R. BURKS, "The ENIAC: First General-Purpose Electronic Computer". AHC, 3–4/October 1981.

A. W. BURKS, H. H. GOLDSTINE and J. VON NEUMANN, *Preliminary Discussions of the Logical Design of an Electronic Computing Instrument*. Princeton, N.J: Institute for Advanced Study, 1947.

P. E. BURTON, A *Dictionary of Minicomputing and Microcomputing*. Chichester & New York: John Wiley, 1982.

V. BUSH, "The Differential Analyzer. A New Machine for Solving Differential Equations". JFI 212/193: pp. 447–88.

V. BUSH, "Instrumental Analysis". BAMS 42/1936: pp. 649–69.

V. BUSH, *Pieces of the Action*. New York: Morrow, 1970.

R. CALDER, *The Evolution of the Machine*. Washington, D.C.: American Heritage/Smithsonian Institution, 1968.

M. CAMPBELL-KELLY, "Programming the EDSAC: Early Programming Activity at the University of Cambridge". AHC 2–1/July 1980.

M. CAMPBELL-KELLY, "The Development of Computer Programming in Britain, 1945–1955". AHC 4–2/April 1982.

A. E. CARPENTER and R. W. DORANT, "The Other Turing Machine". CJ 20–3/August 1977: pp. 269–79.

A. CAVE-BROWN, *The Bodyguard of Lies*. New York: Harper & Row, 1975.

P. E. CERUZZI, "The Early Computers of Konrad Zuse, 1935–1945". AHC 3–3/July 1981.

P. E. CERUZZI, *Reckoners, The Prehistory of the Digital Computer, 1935–1945*. Greenwood Press, 1983.

P. E. CERUZZI, *A History of Modern Computing*. Cambridge, MA/London: MIT Press, 1998.

P. E. CERUZZI and P. A. KIDWELL, *Landmarks in Digital Computing*. Washington, D.C.: Smithsonian Institution Press, 1994.

O. CESAREO, "The Relay Interpolator". BLR 23/1946: pp. 457–60.

N. CHAPIN, *An Introduction to Automatic Computers*. Princeton, N.J: Van Nostrand, 1955.

S. CHAPMAN, "Blaise Pascal (1623–1662): Tercentenary of the Calculating Machine". NA 150/ 31 October 1942: pp. 508–9.

A. CHAPUIS and E. DROZ, *Automata. A Historical and Technological Study*. London: B.T. Batsford, 1958.

G. C. CHASE, "History of Mechanical Computing Machinery" (1952). AHC 2–3/July 1980.

E. C. CHERRY, *On Human Communication*. Cambridge, MA: MIT Press, 1957.

L. J. COMRIE, "Computing by Calculating Machines". AJ 45/1927: pp. 42–51.

L. J. COMRIE, "On the Application of the Brunsviga-Dupla Calculating Machine to Double Summation with Finite Differences". MNRAS, 88–5/March 1928: pp. 447–59.

L. J. COMRIE, "The Application of Calculating Machines to Astronomical Computing". PA 33/April 1931: pp. 2–8.

L. J. COMRIE, "The Application of the Hollerith Tabulating Machine to Brown's Tables of the Moon". MNRAS 92–7/1932: pp. 694–707.

L. J. COMRIE, "The Application of Commercial Calculating Machines to Scientific Computing". MTAC 2/1946: pp. 149–59.

L. J. COMRIE, "Babbage's Dream Comes True". NA 158/1946: pp. 567–8.

J. CONNOLLY, *Chronology of Computing in Europe, Africa, Asia and Latin America*. Armonk, N. Y.: IBM World Trade Corporation, no date.

J. W. CORTADA, *Historical Dictionary of Data Processing*. London/New York: Greenwood Press, 1987.

L. COUFFIGNAL, "Report on the Machine of the Institut Blaise Pascal". MTAC 4–32/1950: pp. 225–9.

J. CRANK, *The Differential Analyzer*. London: Longman, 1947.

T. DANTZIG (1930), *Number, the Language of Science*. London: George Allen & Unwin: 1967.

M. DAUMAS, *A History of Technology and Invention. Progress Through the Ages. The Expansion of Mechanization 1725–1860*, vol. 3. New York: Crown Publishers, 1979.

W. K. DAVID, *David's New Lightning Calculator*. Buffalo, N.Y.: W. K. David, 1881.

H. DAVIES, *A History of Recorded Sound*. New York, 1978.

R. DAVIS, "Expert Systems: Where are We ? And Where Do We Go from Here ?". AIM, July–August 1982.

H. L. DREYFUS, *What Computers Still Can't Do: A Critique of Artificial Reason*. Cambridge, MA: MIT Press, 1992.

H. L. DREYFUS, S. E. DREYFUS and T. ATHANASIOU, *Mind over Machine: The Power of Human Intuition and Expertise in the Era of the Computer*. Oxford: Blackwell, 1986.

E. DROZ, "From Jointed Doll to Talking Robot". NS 14/1962: pp. 37–40.

J. M. DUBBEY, "Charles Babbage and his Computers". BIMA 9–3/1973: pp. 622–69.

C. EAMES, R. EAMES and G. FLECK, *A Computer Perspective*. Cambridge, MA: Harvard University Press, 1973.

W. H. ECCLES and F. W. JORDAN, "A Trigger Relay Utilising Three-electrode Thermionic Vacuum Tubes". RR 1/1919: pp. 143–6.

J. P. ECKERT, "In the Beginning and to What End". CTF 4–3/24: pp.3–4.

J. P. ECKERT, J. W. MAUCHLY, H. H. GOLDSTINE and J. G. BRAINERD, *Description of the ENIAC and Comments on Electronic Digital Computing Machines*, Contract no. W-670-ORD-4926, Philadelphia, PA: Moore School of Electrical Engineering, University of Pennsylvania, 30 November 1945.

W. J. ECKERT, *Punched Card Methods in Scientific Computation*, Thomas J. Watson Astronomical Computing Bureau. New York: Columbia University, 1940.

W. J. ECKERT, "Electrons and Computation" (November 1948), in B. RANDELL, 1973, pp. 219–28.

D. B. G. Edwards, "Computer Developments in Great Britain". CTF 1–1/21: p. 1.

G. Estrin, "The Electronic Computer at the Institute for Advanced Study". MTAC 7/1953: p.108.

A. Feldman and P. Ford, *Scientists and Inventors, The People who Made Technology, from Earliest Times to Present Day*. London/New York, 1979.

D. E. Felt, *Mechanical Arithmetic or the History of the Counting Machine*. Chicago, IL: Washington Institute, 1916.

J. W. Forrester, "The Digital Computation Program at the Massachusetts Institute of Technology". *Proceedings of the 2nd Symposium on Large-Scale Digital Calculating Machinery*. Cambridge, MA: Harvard University Press, 1951.

N. Foy, *The IBM World*. London: Methuen, 1974.

M. Gardner, "Logic machines". SCAM 183/March 1952: pp. 68–73.

M. Gardner, *Logic Machines and Diagrams*. New York: McGraw-Hill, 1958.

J. C. Giarratano, *Foundations of Computer Technology*. Indianapolis, IND: Howard W. Sams, 1982.

C. Gillespie (editor), *Dictionary of Scientific Biography*, 16 vol. New York: Charles Scribners, 1970–1980.

J. Ginsburg, BAMS 23/1917: pp. 368.

J. Ginsburg, "Napier on the Napier Rods". See D. E. Smith, 1929, pp. 182–5.

A. Glaser, *History of Binary Numeration*. Philadelphia, PA: Tomash Publishers, 1971.

S. E. Gluck, "The Electronic Discrete Variable Automatic Computer (EDVAC)". EEng, 72/1953: pp. 159–62.

H. H. Goldstine, *The Computer from Pascal to von Neumann*. Princeton, N.J.: Princeton University Press, 1972.

H. H. Goldstine and A. Goldstine, "The Electronic Numerical Integrator and Computer (ENIAC)". MTAC, 2/July 1946: pp. 97–110.

H. H. Goldstine and J. von Neumann, "Planning and Coding of Problems for an Electronic Computing Instrument" (Part 2, vol. 1, 1 April 1947; Part 2, vol. 2, 15 April 1948; Part 2, vol. 3, 16 August 1948), in A. H. Taub, vol. 5, pp. 80–235.

I. J. Good, "Some Future Social Repercussions of Computers". IJES, 1/1970: pp. 67–79.

I. J. Good, "Early Work on Computers at Bletchley". AHC, 1–1/July 1979.

G. B. Grant, "On a New Difference Engine". AJS, 3rd series, 1-8/August 1871: pp. 113–18.

G. B. Grant, "A New Calculating Machine". AJS, 3rd series, 4-8/1874: pp. 277–84.

R. L. Gregory (1966), *Eye and Brain: The Psychology of Seeing . . .* Oxford: Oxford University Press, 1998.

E. J. Haga, *Understanding Automation*. Elmhurst, IL: The Business Press, 1965.

D. Halacy, *Charles Babbage, Father of the Computer*. New York: Crowell-Collier, 1970.

R. A. Harris, "The Coast and Geodetic Survey Tide Predicting Machine". SCAM 110/June 1914: pp. 110.

D. R. Hartree, "The ENIAC, an Electronic Calculating Engine". NA 157/20 April 1946: p. 527.

D. R. Hartree, "The ENIAC, an Electronic Computing Machine". NA 158/12 October 1946: pp. 500–6.

D. R. HARTREE, *Calculating Machines Recent and Prospective Developments*. Cambridge: Cambridge University Press, 1947.

D. R. HARTREE, *Calculating Instruments and Machines*. University of Illinois Press, 1949.

D. R. HARTREE, "Automatic Calculating Machines". MGA 34–310/December 1950: pp. 241–52.

F. G. HEATH, "Pioneers of Binary Coding". JIEE 7–81/1961: pp. 539–41.

S. HEIMS, *John von Neumann and Norbert Wiener*. Cambridge, MA: MIT Press, 1980.

T. HIDEOMI, *Historical Development of Science and Technology in Japan*. Tokyo, 1968.

A. M. HILTON, *Logic, Computing Machines, and Automation*. Cleveland, OH: Meridian, 1963.

A. HODGES, *Alan Turing: The Enigma*. London: Burnett Books, 1983.

H. HOLLERITH, "An Electrical Tabulating Machine". SMQ 10–3/April 1889: pp. 238–55.

H. HOLLERITH, "The Electrical Tabulating Machine". JRSS 57–4/1894: pp. 678–82.

S. H. HOLLINGDALE, "Charles Babbage and Lady Lovelace – Two 19th-century Mathematicians". BIMA 2–1/1966: pp. 2–15.

G. M. HOPPER, "The Education of a Computer". PACM, 1952.

G. M. HOPPER, "Computer Software". CTF 3–7/26: p.7.

E. M. HORSBURGH, "Calculating machines". GIES 63/1920: pp. 117–62.

E. M. HORSBURGH, *Modern Instruments and Methods of Calculation. A Handbook of the Napier Tercentenary Celebration Exhibition*. London: G. Bell and Sons, 1914; Los Angeles, CA: Tomash, 1983.

C. C. HURD, "The IBM Card Programmed Calculator". PSSC: pp. 37–41.

C. C. HURD, "Special issue: IBM 701". AHC, 5–2/April 1983.

H. D. HUSKEY, "Electronic Digital Computing in the United States". RCACM: pp. 109–11.

H. D. HUSKEY and V. R. HUSKEY, "Chronology of Computing Devices". IEEE *Transactions on Computers*, C 25, no. 12, December 1982.

A. HYMAN, *Charles Babbage, Pioneer of the Computer*. Princeton, N. J.: Princeton University Press, 1982.

W. S. JEVONS, "On the Mechanical Performance of Logical Inference". PTRSL 160–2/1870, pp. 497–518, and plates 32–4.

B. JOHNSON, *The Secret War*. London: BBC Publications, 1978.

J. JULEY, "The Ballistic Computer". BLR 24/1947: pp. 5–9; repr. B. RANDELL, 1973, pp. 251–5.

S. C. KLEENE, "Representations of Events in Nerve Sets and Finite Automata". AS, 1956.

D. E. KNUTH, "Von Neumann's First Computer Program". CS 2–4/1970: pp. 247–60.

M. KORMES, "Leibniz on his Calculating Machine", in D. E. SMITH, 1929: pp. 18–25.

G. A. KORN and T. M. KORN, *Electronic Analog Computers*. New York: McGraw-Hill, 1952.

F. C. KREILING, "Leibniz". SCAM 218–5/May 1968: pp. 94–100.

E. M. LANGLEY, *A Treatise on Computation*. London, 1895.

S. H. LAVINGTON, *A History of Manchester Computers*. Manchester: National Computing Centre Publications, 1975.

S. H. LAVINGTON, *Early British Computers*. Manchester: Manchester University Press, 1980.

H. LEGARD, F. P. McQUAID and A. SINGER, *From Baker Street to Binary*. New York: McGraw-Hill, 1983.

L. L. LOCKE, "The History of Modern Calculating Machines, an American Contribution". AMM, 3/1924: pp. 422–9.

A. A. LOVELACE, "Sketch of the Analytical Engine invented by Charles Babbage". TSM 3/1843, notes to article 29 by L. F. Menabrea: pp. 666–731.

P. E. LUDGATE, "On a Proposed Analytical Engine". SPRDS 12–9/1909: pp. 77–91.

P. E. LUDGATE, "Automatic Calculating Engines", in E. M. HORSBURGH, 1914, pp. 124–7.

R. C. LYNDON, "The Zuse Computer". MTAC, 2–20/1947: pp. 355–9.

N. MacLELLAN EDWARDS, *Office Automation*. White Plains: Knowledge Industry Publ., 1982.

K. MARSH, *The Way the New Technology Works*. New York: Simon and Schuster, 1982.

T. C. MARTIN, "Counting a Nation by Electricity". EEng, 12–184/11 November 1891: pp. 521–30.

S. T. McCLELLAN, *The Coming Computer Industry Shakeout. Winners, Losers, and Survivors*. New York: John Wiley and Sons, 1984.

P. McCORDUCK, *Machines who Think*. San Francisco, CA: Freeman, 1979.

W. S. McCULLOCH and W. PITTS, "A Logical Calculus of the Ideas Immanent in Nervous Activity". BMB, 5/1943.

J. C. McPHERSON, F. E. HAMILTON and R. R. SEEBER, "A Large-Scale General-Purpose Electronic Digital Calculator The IBM SSEC". AHC, 4–2/October 1982.

B. MELTZER and D. MICHIE (editors), *Machine Intelligence*. Edinburgh: Edinburgh University Press, 1972.

U. C. MERZBACH, *Georg Scheutz and the First Printing Calculator*. Washington, D.C.: Smithsonian Institution Press, 1977.

N. METROPOLIS, J. HOWLETT and G. C. ROTA (editors), *A History of Computing in the Twentieth Century*. New York/London: Academic Press Inc., 1980.

D. MICHIE, "Machines that Play and Plan". SJ, October 1968: pp. 83–8.

R. MOREAU (editor), *Biological Computers or Electronic Brains*, Les Entretiens de Lyon. Paris: Springer-Verlag France, 1990.

P. MORRISSON and E. MORRISSON, "Strange life of Charles Babbage". SCAM 186/April 1952: pp. 66–73.

P. MORRISSON and E. MORRISSON (editors), *Charles Babbage and his Calculating Engines: Selected Writings by Charles Babbage and Others*. New York: Dover Publications, 1961.

M. MOSELEY, *Irascible Genius Charles Babbage Inventor*. London: Hutchinson, 1964.

P. E. MOUNIER-KUHN, "Bull A Worldwide Company Born in Europe". AHC 11–4/1989.

F. J. MURRAY, *The Theory of Mathematical Machines*. New York: King's Crown Press, 1948.

F. J. MURRAY, *Mathematical Machines, vol. 1 (Digital Computers)*. New York: Columbia University Press, 1961.

J. VON NEUMANN, *First Draft of a Report on the* EDVAC, Contract no. W-670-ORD-4926, Philadelphia, PA: Moore School of Electrical Engineering, University of Pennsylvania, 30 June 1945. In B. RANDELL, 1973: pp. 355–64.

M. H. A. NEWMAN, "General Principles of the Design of All-purpose Computing Machines". PRS, A 195/1948: pp. 271–4.

M. H. A. NEWMAN, "Alan Mathison Turing, 1912–1954". BMFRS 1/1955: pp. 253–63.

M. D'OCAGNE (1893), *Simplified Calculation*. Cambridge, MA: MIT Press, 1986.

O. ØRE, *Number Theory and its History*. New York: McGraw-Hill, 1948.

S. T. PARKER, *Encyclopaedia of Electronic Computers*. New York: McGraw-Hill, 1984.

B. E. PHELPS, "Early Electronic Computer Developments at IBM". AHC, 2–3/1980.

E. W. PHILLIPS, "Binary Calculation". JIA 67/1936: pp. 187–221.

E. W. PUGH, L. R. JOHNSON and J. H. PALMER, *IBM's 360 and Early 370 Systems*. Cambridge, MA: MIT Press, 1991.

Z. W. PYLYSHYN (editor), *Perspectives on the Computer Revolution*. Englewood Cliffs, N.J.: Prentice-Hall, 1970.

B. RANDELL, *The Origins of Digital Computers, Selected Papers*. Berlin/Heidelberg/New York: Springer-Verlag, 1973.

B. RANDELL, "Ludgate's Analytical Engine". CJ 14–3/1971: pp. 317–26.

B. RANDELL, "On Alan Turing and the Origins of Digital Computers", in B. MELTZER and D. MICHIE, 7: pp. 3–20.

B. RANDELL, "From Analytical Engine to Electronic Digital Computer: the Contributions of Ludgate, Torres, and Bush". AHC 4–4/October 1982.

W. REITMAN (editor), *Artificial Intelligence. Applications for Business*. Norwood, N. J.: Albex, 1984.

M. REJEWSKI, "How Polish Mathematicians Deciphered the ENIGMA". AHC 3–3/July 1981.

F. J. REX, "Herman Hollerith, the first 'Statistical Engineer'". CA 10–8/1961: pp. 10–13.

R. K. RICHARDS, *Digital Computer Components and Circuits*. Princeton, N. J.: Van Nostrand, 1957.

R. K. RICHARDS, *Electronic Digital Systems*. New York: J. Wiley, 1966.

D. RITCHIE, *The Computer Pioneers*. New York: Simon and Schuster, 1986.

P. RODWELL, *Personal Computers*. London: W. H. Allen, 1982.

J. M. ROSENBERG, *The Computer Prophets*. New York: Macmillan, 1969.

BERTRAND RUSSELL (1919), *Introduction to Mathematical Philosophy*. London: Routledge, 1996.

C. E. SHANNON, "A Symbolic Analysis of Relay and Switching Circuits". TAIEE 723/1938.

C. E. SHANNON and W. WEAVER, *The Mathematical Theory of Communication*. Urbana: University of Illinois Press, 1949.

C. SINGER, *A History of Technology*. Oxford: Clarendon Press, 1955–1957.

D. E. SMITH (editor), *A Source Book in Mathematics*. New York: McGraw-Hill, 1929.

T. M. SMITH, "Some Perspectives on the Early History of Computers". See Z.W. PYLYSHYN, pp. 7–15.

D. DE SOLLA PRICE, "Portable Sundials in Antiquity. Including an Account of a New Example from Aphrodisias". CENT, 14/1969: pp. 242–66.

D. DE SOLLA PRICE, "Gears from the Greeks. The Antikythera Mechanism – a Calendar Computer from c.80BC". TAPS, November 1974.

D. DE SOLLA PRICE , "Calculating Machines". IEEE Micro, February 1984.

N. STERN, "The BINAC: A Case Study in the History of Technology". AHC, 1–1/July 1979.

N. Stern, *From ENIAC to UNIVAC, an Appraisal of the Eckert-Mauchly Computers.* Bedford, MA: Digital Press, 1981.

R. J. Sternberg (editor), *Handbook of Human Intelligence.* Cambridge: Cambridge University Press, 1982.

G. R. Stibitz, "The Relay Computers at Bell Laboratories". *Datamation,* 13–4/April 1967, pp. 35–44; 5/May 1967, pp. 45–9.

G. R. Stibitz and J. A. Larrivée, *Mathematics and Computers.* New York: McGraw-Hill, 1957.

S. Strandh, *A History of the Machine.* New York: A. and W. Publishers Inc., 1979.

N. I. Styazhkin, *History of Mathematical Logic from Leibniz to Peano, etc.* Cambridge, MA: The MIT Press, 1969.

M. Sullivan-Trainor, "The Precomputer Age. A time line, 500 B.C. to 1946 A.D.". *Computerworld,* 160/3 November 1986.

M. Sullivan-Trainor, "The Computer Age. A time line, 1946 A.D. to 2000 A.D.". *Computerworld* 160/3 November 1986.

D. Swade, *The Cogwheel Brain. Charles Babbage and the Quest to Build the First Computer.* London: Little, Brown, 2000.

A. H. Taub (editor), *John von Neumann: Collected Works.* Oxford: Pergamon, 1963.

S. Thomas, *Computers: their History, Present Applications and Future.* New York: Holt, Rinehart and Winston, 1964.

G. Tilghman Richards, *The History and Developments of Typewriters.* London: Science Museum, 1964.

K. Tilton, *International Diffusion of Technology: The Case of Semi-Conductors.* Washington, D.C.: Brookings Institution, 1971.

L. Torres y Quevedo, "Ensayos sobre Automatica. Su Definicion. Extension teorica de sus aplicaciones". For English version see B. Randell, 1973, pp. 87–105.

L. Torres y Quevedo, "Arithmomètre électromécanique". For English version see B. Randell, 1973, pp. 107–118.

M. Trask, *The Story of Cybernetics.* London: Studio Vista, 1971.

L. E. Truesdell, *The Development of Punch Card Tabulation in the Bureau of the Census 1890–1940.* Washington D.C.: Department of Commerce, Bureau of the Census, 1965.

J. A. V. Turck (1921), *The Origin of Modern Calculating Machines.* New York: Arno Press, 1972.

A. M. Turing, "On Computable Numbers, with an Application to the *Entscheidungsproblem*". PLMS, 2nd series, 42/1936: pp. 230–67.

A. M. Turing, *Proposals for Development in the Mathematics Division of an Automatic Computing Engine (A.C.E.).* Report E. 882, Executive Committee, National Physical Laboratory, Teddington, Middlesex, 1945.

S. Turing, *Alan M. Turing.* Cambridge: W. Heffer, 1959.

S. Ulam, "John von Neumann 1903–1957". BAMS 64/May 1958: pp. 1–49.

B. L. Van Der Waerden, *Science Awakening,* vol. 1, pp. 47–9. Groningen: Nordhoof, 1954.

M. H. Weik, "The ENIAC Story". JAOA, January–February 1961: pp. 3–7.

N. Wiener, *Cybernetics, or Control and Communication in the Animal and the Machine.* New York: J. Wiley, 1948.

M. V. Wilkes, "The Design of a Practical High-Speed Computing Machine". PRS, A 195/1948: pp. 274–9.

M. V. WILKES, "Programme Design for a High-Speed Automatic Calculating Machine". JSI 26/June 1949: pp. 217–20.

M. V. WILKES, "Automatic Calculating Machines". JRSA 4862/1951, vol. C: pp. 56–90.

M. V. WILKES, "The EDSAC Computer". REDC, Joint AIEE-IRE Computer Conference, 10–12 December 1951, pp. 79–83. New York: American Institute of Electrical Engineers, 1952.

M. V. WILKES, *Automatic Digital Computers*. London: Methuen, 1956.

M. V. WILKES, "Babbage as a Computer Pioneer". *Report of Proceedings, Babbage Memorial Meeting*. British Computer Society, London, 18 October 1971.

M. V. WILKES, *Memoirs of a Computer Pioneer*. Cambridge, MA: MIT Press 1985.

M. V. WILKES and W. RENWICK, "The EDSAC – an Electronic Calculating Machine". JSI 26/December 1949: pp. 385–91.

M. V. WILKES and W. RENWICK, "The EDSAC (Electronic Delay Storage Automatic Calculator)". MTAC 4/1950: pp. 61–5.

J. H. WILKINSON, "The Automatic Computing Engine at the National Physical Laboratory". PRS, A 195/1948: pp. 285–6.

W. F. WILLCOX, "John Shaw Billings and Federal Vital Statistics". JASA, 21/1926: pp. 257–66.

W. F. WILLCOX, "Herman Hollerith". MTAC, 3/1948: pp. 62–3.

F. C. WILLIAMS and T. KILBURN, "Electronic Digital Computers". NA, 162/September 1948: p. 487.

M. R. WILLIAMS, *A History of Computing Technology*. London: Prentice-Hall, 1985.

T. I. WILLIAMS, *A Short History of Twentieth Century Technology, 1900–1950*. Oxford: Clarendon Press, 1982.

P. WINSTON, *Artificial Intelligence*. Reading, MA: Addison-Wesley, 1982.

A. WOLF, *Calculating Machines. A History of Science, Technology and Philosophy in the Eighteenth Century*. London: George Allen and Unwin, 1938.

K. ZUSE, "German Computer Activities". CTF 13-6/17.

K. ZUSE, "Installation of the German Computer Z 4 in Zurich in 1950". AHC, 2–3/July 1980.

NO AUTHOR LISTING:

"A Manual of Operation for the Automatic Sequence Controlled Calculator (ASCC)". ACLHU, 1/1946.

"Description of a Relay Calculator". ACLHU, 14/1949.

"Description of a Magnetic Drum Calculator". ACLHU, 25/1952.

"New Calculating Machine". ILN, 26/ 30 June 1855: p. 661.

"The Swedish Tabulating Machine of Georg and Edvard Scheutz". ADO, 1/1866: pp. 116–26.

"Charles Babbage". NA, 5-106/9 November 1871: pp. 28–29.

"The Electrical Tabulating Machine Applied to Cost Accounting". AM, 25/ 16 August 1902: pp. 1073–5.

"The Burroughs Adding and Listing Machine". *Engineering*, 83/ 3 May 1907: pp. 580–1.

"IBM Selective Sequence Electronic Calculator". MTAC, 3/1948: pp. 216–17.

"The Thirsk Totalisator". ER, 106, no. 2724: pp. 268–9.

2. WORKS IN OTHER LANGUAGES

H. ADLER, J. BORMANN, W. KAMMERER, I. O. KERNER and N. J. LEHMANN, "Mathematische Maschinen". EMDDR, Berlin, 1974.

P. AIGRAIN, "Deux mille ans d'automates". SA, special issue, 49. Paris, September 1984.

P. AIGRAIN, "De l'organisation au calcul". IAF 27. Paris, January 1985.

D. ANDLER, "La Machine universelle". SA, special issue, 49. Paris, September 1984.

A. APOKIN and L. E. MAISTROV, *Rasvitie Vyichislenyih Mashin*. Moscow: Nauka, 1974.

L. APOSTEL, B. MANDELBROT and A. MORF, *Logique, Langage et Théorie de l'information*. Paris: PUF, 1957.

ARNAULD and NICOLE (1662), *Logique de Port-Royal, ou Art de penser*. Paris: Hachette, 1854.

J. ARSAC, *La Science informatique*. Paris: Dunod, 1970.

J. ARSAC, *Les Machines à penser*. Paris: Le Seuil, 1987.

H. ATLAN, *L'Organisation biologique et la Théorie de l'information*. Paris: Hermann, 1972.

CHARLES BABBAGE, "Note sur la machine suédoise de MM. Scheutz pour calculer les tables mathématiques par la méthode des différences et en imprimer les résultats sur les planches stéréotypes". CRAS 41/ 8 October 1855, pp. 557–60; 41/ 28 April 1856: pp. 798–800.

H. BAKIS, *IBM: une multinationale régionale*. Grenoble: Presses universitaires, 1977.

W. DE BEAUCLAIR, *Rechnen mit Maschinen. Eine Bildgeschichte der Rechentechnik*. Brunswick: Vieweg und Sohn, 1968.

C. BERTHO, *Télégraphes et Téléphones, De Valmy au microprocesseur*. Paris: Hachette, 1981.

C. BERTHO, *Histoire des télécommunications*. Paris: Ères, 1984.

R. BLANCHÉ, *La Science physique et la réalité*. Paris: PUF, 1948.

R. BLANCHÉ (1957), *Introduction à la logique contemporaine*. Paris: A. Colin, 1970.

R. BLANCHÉ, *L'Axiomatique*. Paris: PUF, 1965.

BLIN, "Un imbécile calculateur". BSMM, January 1910.

J. BOCCARDI, *Guide du calculateur*. Paris: Catane, 1902.

V. G. VON BOHL, *Pribory i mashiny dlya mehanicheskovo proizvodstva arifmeticheskih deistvii*. Moscow, 1896.

L. BOLLÉE, "Sur une nouvelle machine à calculer". CRAS 109/1889: pp. 737–9.

H. BOUCHER, *Organisation et fonctionnement des machines arithmétiques*. Paris: Masson, 1960.

BOULENGER, *Informatique et administration de l'entreprise*. Paris: Sirey, 1968.

P. BOUTROUX, *L'Idéal scientifique des mathématiciens*. Paris: Alcan, 1922.

J. BOUVERESSE, J. ITARD and E. SALLÉ, *Histoire des mathématiques*. Paris: Larousse, 1977

A. BOUVIER, *La Théorie des ensembles*. Paris: PUF, 1972.

V. BRANDEBOURG, *Méthodes simples de calcul extrarapide et de calcul mental*. Paris-Grenoble: Hachette, 1926.

V. BRANDEBOURG, "Les calculateurs prodiges et leurs expériences". BSA 65–7/1916: pp. 218–45.

P. BRETON, *Histoire de l'informatique*. Paris: Le Seuil, 1990.

L. BRILLOUIN, *La Science et la Théorie de l'information*. Paris: Masson, 1959.

L. DE BROGLIE, *Sur les sentiers de la science*. Paris: Albin Michel, 1960.

L. BRUNSCHVICG, *La Physique du vingtième siècle et la philosophie*. Paris: Hermann, 1936.

J. BUREAU, *Dictionnaire de l'Informatique*. Paris: Larousse, 1972.

A. BURLOUD, *La Pensée conceptuelle*. Paris: Alcan, 1927.

F. P. CAMPOS, *Applications mécano-comptables des machines à totalisateurs multiples*. Paris: Étienne Chiron, 1944.

P. CARRÉ, *Du Tam-Tam au Satellite*. Paris: Explora, Presses Pocket, 1991.

O. CARRERA, D. LOISEAU and O. ROUX, *Androïdes, les automates de Jaquet-Droz*. Lausanne: Scriptar, F. M. Ricci, 1979.

G. CASANOVA, *L'Algèbre de Boole*. Paris: PUF, 1972.

J. P. CHANGEUX, *L'Homme neuronal*. Paris: Fayard, 1983.

J. CHEVALIER and A. GHEERBRANT, *Dictionnaire des symboles*. Paris: Robert Laffont, 1982.

M. CHIVA, *Débiles normaux, débiles pathologiques*. Neuchâtel: Delachaux and Niestlé, 1973.

E. CLAPARÈDE, *L'Invention*, Centre international de Synthèse, 9e semaine. Paris: Alcan, 1937.

P. CONSO and P. POULAIN, *Informatique et gestion de l'entreprise*. Paris: Dunod, 1969.

O. COSTA DE BEAUREGARD, *Le Second Principe de la Science du temps, entropie, information, irréversibilité*. Paris: Le Seuil, 1963.

L. COUFFIGNAL, *Les Machines à calculer, leur principe, leur évolution*. Paris: Gauthier Villars, 1933.

L. COUFFIGNAL, "Calcul mécanique: Sur l'emploi de la numération binaire dans les machines à calculer et les instruments nomomécaniques". CRAS 202/1936, pp. 1970–2.

L. COUFFIGNAL, "Sur un problème d'analyse mécanique abstraite". CRAS 206/1938: pp. 1336–8.

L. COUFFIGNAL, *Sur l'analyse mécanique: application aux machines à calculer et aux calculs de la mécanique céleste*, Thèse présentée à la faculté des sciences de Paris, série A 1772. Paris: Gauthier Villars, 1938.

L. COUFFIGNAL, "Traits caractéristiques de la machine à calculer universelle de l'Institut Blaise Pascal". ACLHU 26/September 1949: pp. 374–86.

L. COUFFIGNAL, *Information et Cybernétique*. Paris: Gauthier Villars, 1958.

L. COUFFIGNAL, *La Cybernétique*. Paris: PUF, 1963.

L. COUFFIGNAL, *Les Machines à penser*. Paris: Les Éditions de Minuit, 1964.

L. COUFFIGNAL, and M. P. SCHUTZENBERGER, "La Cybernétique", in *Encyclopédie française*.

A. CROIX and J. QUÉNIART, *La Culture programme*. Roanne, 1993.

A. CUVILLIER (1954), *Cours de philosophie*. Paris: A. Colin, 1987.

M. DANLOUX-DUMESNILS, *Le Calcul analogique par courants continus*. Paris: Dunod, 1958.

M. DAUMAS (editor), *Histoire générale des techniques*, vol. 5. Paris: PUF, 1979.

E. DAVALO and P. NAÏM, *Des réseaux de neurones*. Paris: Eyrolles, 1989.

H. DELACROIX, *Le Langage et la Pensée*. Paris: Alcan, 1922.

C. E. DELAUNAY, "Rapport sur la machine à calculer présentée par M. Wiberg". CRAS 56/1863: pp. 330–9.

P. DEMARNE and M. ROUQUEROL, *Les Ordinateurs électroniques*. Paris: PUF, 1961.

L. DEPECKER, "Cinq notions de télédétection aérospatiale: un exemple de structuration d'un champ terminologique". *Meta*, 34–2/1989.

L. DEPECKER, *Introduction à l'étude de la télédétection aérospatiale et de son vocabulaire.* Paris: La Documentation française, 1991.

P. DEVAUX, *Automates et Automatisme.* Paris: PUF, 1941.

P. DEVAUX, *Automates, Automatisme, Automation.* Paris: PUF, 1967.

J. DHOMBRES, "La fin du siècle des Lumières: un grand élan scientifique", in *Deux Siècles de France*, (G. Ifrah, editor). *Total Information*, 111/June 1989): pp. 5–12.

J. DHOMBRES, *Nombre, Mesure et Continu: épistémologie et histoire.* Paris: Fernand Nathan, 1978.

M. DOROLLE, *Le Raisonnement par analogie.* Paris: PUF, 1949.

A. DOYON and L. LIAIGRE, *Jacques de Vaucanson, mécanicien de génie.* Paris: PUF, 1966.

O. DUCROT and T. TODOROV, *Dictionnaire encyclopédique des sciences du langage.* Paris: Le Seuil, 1972.

D. DURAND, *La Systémique.* Paris: PUF, 1987.

J. ERRARD DE BAR-LE-DUC (1584), *Premier Livre des Instruments mathématiques méchaniques.* Nancy: Berger-Levrault, 1979

R. ESCARPIT, *L'Écrit et la communication.* Paris: PUF, 1978.

J. FAVIER and R. THOMELIN, *La Mécanographie. Machines à calculer. Machines comptables. Machines à cartes perforées. Calculatrices.* La Chapelle-Montligeon: Éditions de Montligeon, 1965.

J. P. FLAD, "L'horloge à calcul de l'astronome W. Schickard semble avoir été la première machine à calculer à engrenages propre aux quatre opérations". *Chiffres* 1/1958: pp. 143–8.

J. P. FLAD, *Les Trois Premières Machines à calculer: Schickard, Pascal et Leibniz.* Lecture at the Palais de la Découverte, Paris, 8 June 1963.

J. M. FONT, J. C. QUINIOU and G. VERROUST, *Les Cerveaux non humains. Introduction à l'informatique.* Paris: Denoël, 1970.

P. FOULQUIÉ, *Nouveau précis de philosophie.* Paris: Éd. de l'École, 1960.

P. FOULQUIÉ, *Dictionnaire de la langue philosophique.* Paris: PUF, 1982.

S. FOURSIN, *La Mécanographie comptable.* Paris, 1950.

B. VON FREYTAG LÖRINGHOFF, "Wiederentdeckung und Rekonstruktion der Ältesten Neuzeitlichen Rechenmaschinen". *VDI Nachrichten* 14, no. 39.4 (December 1960).

B. VON FREYTAG LÖRINGHOFF, "Wilhelm Schickard und seine Rechenmaschinen von 1623" in M. GRAEF, pp. 11–20.

K. FRIEDRICHS, I. FISCHER-SCHREIBER, F. K. ERHARD and M. S. DIENER, *Dictionnaire de la sagesse orientale.* Paris: Robert Laffont, 1989.

T. DE GALIANA, *Dictionnaire des découvertes scientifiques.* Paris: Larousse, 1968.

A. GALLE, *Mathematische Instrumente.* Leipzig: B. G. Teubner, 1912.

M. GALLON, *Recueil des machines et inventions approuvées par l'Académie des Sciences,* vol. IV. Paris, 1730.

M. GARDNER, "Jeux mathématiques". PLS 10/1978: pp. 96–100.

C. J. GIESING, *Neuer Unterricht in der Schnellrechenkunst.* Döbeln, 1884.

B. GILLE, *Les Ingénieurs de la Renaissance.* Paris: Le Seuil, 1978.

B. GILLE, "Histoire des techniques". In *Encyclopédie de la Pléiade.* Paris: Gallimard, 1978.

B. GILLE, *Les Mécaniciens grecs, la naissance de la technologie.* Paris: Le Seuil, 1980.

F. GILLOT, *Algèbre et logique, d'après les textes originaux de G. Boole et W. S. Jevons, avec les plans de la machine logique.* Paris: A. Blanchard, 1962.

V. A. GISCARD D'ESTAING, *Le Livre mondial des inventions.* Paris: Fixot/ Compagnie 12, 1993.

E. GOBLOT, *Traité de logique.* Paris: A. Colin, 1918.

E. GOBLOT, *Le Système des sciences.* Paris: A. Colin, 1922.

F. GONSETH, *Qu'est-ce que la logique?* Paris: Hermann, 1937.

F. GONSETH, *Philosophie mathématique.* Paris: Hermann, 1939.

M. GRAEF, *350 Jahre Rechenmaschinen.* Munich: Carl Hanser Verlag, 1973.

M. GRANET, *La Pensée chinoise.* Paris: Albin Michel, 1988.

G. T. GUILBAUD, *La Cybernétique.* Paris: PUF, 1954.

P. GUILLAUME, *La Psychologie de la forme.* Paris: Flammarion, 1939.

S. GÜNTHER, "Die quadratischen Irrationalitäten der Alten". AGM 4/1882.

E. GUYOU. NAM 3–8/1889.

E. GUYOU, *Note sur les approximations numériques.* Paris, 1891.

C. HAMANN, *Über Elektrische Rechenmaschinen.* Neu Babelsberg, c.1932.

A. HENNEMAN, *Die Technische Entwicklung der Rechenmaschine.* Aix-la-Chapelle: Verlag Peter Basten, 1953.

D. HILBERT and W. ACKERMANN, *Grundzüge der theoretischen Logik.* Berlin, 1931.

L. A. D. HOYAU (1822), "Description d'une machine à calculer nommée arithmomètre de l'invention de M. le chevalier Thomas de Colmar". BSEIN 132/1920: pp. 662–70.

G. IFRAH, *Le Calcul: de l'abaque à la calculatrice programmable.* Paris, 1987.

F. JACOB, *La Logique du vivant.* Paris: Gallimard, 1970.

L. JACOB, *Le Calcul mécanique.* Paris: Octave Douin et Fils, 1911.

E. JACOBY, *Biographie de Henri Mondeux.* Paris, (no date).

R. JAKOBSON, *Essais de linguistique générale.* Paris: Éditions de Minuit, 1963.

W. KÖHLER, *L'Intelligence des signes supérieurs.* Paris: Alcan, 1927.

A. KOYRÉ, in: R. Taton, *Histoire générale des sciences: 2*, p. 51. Paris: PUF, 1957–1964.

J. LABADIÉ, "Les calculateurs prodiges ont plus de mémoire que de méthode". SV, 376/1949: pp. 19–22.

J. LARMAT, *La Génétique de l'intelligence.* Paris: PUF, 1973.

M. H. LAURENT, "Calcul mental", in *Grande Encyclopédie Larousse du XIXe siècle,* Paris.

A. LE GARFF, *Dictionnaire de l'informatique.* Paris: PUF, 1975.

F. LE LIONNAIS, *Les Grands courants de la pensée mathématique.* Paris: A. Blanchard,1962.

L. LEPRINCE-RINGUET, *L'Aventure de l'électricité.* Paris: Flammarion, 1983.

G. LHOSTE and P. PEPE, *Gestion automatisée des entreprises par les machines à cartes perforées.* Paris: Dunod, 1958.

L. J. LIBOIS, *Genèse et croissance des Télécommunications.* Masson, 1983.

R. LIGONNIÈRE, *Préhistoire et histoire des ordinateurs.* Paris: Robert Laffont, 1987.

R. LIGONNIÈRE, "Les origines de l'informatique anglaise". InPr 24/June–July 1984; 26/October 1984; 27/November 1984. (And see articles by R. Ligonnière in *Ordinateurs.*)

J. L. LIONS (editor), "De l'intelligence artificielle aux bio-sciences". *CESTA-AFCET, Colloque Cognitiva,* 4–7 June. Paris, 1985.

M. V. Locquin (editor), *Aux Origines de la vie*. Paris: Fayard, 1987.

A. Loewy, AMP, 33/1902.

Mackintosh, "L'ordinateur de John Atanasoff". PLS, Mensuel, 132/October 1988: pp.74–80.

M. Margenstern, *Langage Pascal et logique du premier ordre*. Paris: Masson, 1989.

E. Martin, *Die Rechenmaschinen und ihre Entwicklungsgeschicht*. Pappenheim, 1925.

L. Massignon and R. Arnaldez, "La Science arabe". In: R. Taton, *Histoire générale des sciences:* pp. 431–71. Paris: PUF, 1957–1964.

L. F. Menabrea, "Sur la Machine Analytique de Charles Babbage". CRAS, 28 July 1884: pp. 179–82.

R. Moreau, "Les langages de l'informatique". SA, special issue, 1969: pp. 41–7.

R. Moreau, *Introduction à la théorie des langages*. Paris: Hachette, 1975.

R. Moreau, "L'ordinateur électronique dans les sciences humaines". *Sciences* 43–44/May–August 1966: pp. 82–91.

R. Moreau, *Ainsi naquit l'informatique*. Paris: Bordas Informatique, 1987.

R. Moreau, PLS, Mensuel 132/October 1988: pp. 80–3.

E. Morin, *La Méthode*. Paris: Le Seuil, 1977.

H. de Morin, *Les Appareils d'intégration*. Paris, 1913.

P. Morvan, *Dictionnaire de l'informatique*. Paris: Larousse, 1986.

J. Mosconi, *Constitution de la théorie des automates*, thesis, University of Paris I, 1989.

P. E. Mounier-Kuhn, *L'Informatique en France depuis 1945: histoire d'une politique scientifique et industrielle*, thesis, Paris: Conservatoire nationale des Arts et Métiers, 1995.

P. E. Mounier-Kuhn, "Les premiers ordinateurs en France". HIF.2.

J. W. Nagler, "In memoriam Gustav Tauschek". BFT, 1966: pp. 1–14.

C. Nicoladze, "Arithmomètre à multiplication directe purement électrique". CRAS,186/1928: pp. 123–4.

M. d'Ocagne, "Histoire des machines à calculer". BSEIN 132/September 1920: pp. 554–69.

M. d'Ocagne, "Vue d'ensemble sur les machines à calculer". BDSM, 2nd series, 4/1922, pp. 102–44.

M. d'Ocagne, "Torres-Quevedo". *Larousse mensuel* 364/June 1937: pp. 727–8.

P. Oléron, *L'Intelligence*. Paris: PUF, 1986.

J. Payen, "Les exemplaires de la machine de Pascal". RHSA 16–2/April–June 1963: pp. 161–78.

M. Pelegrin, *Machines à calculer électroniques arithmétiques et analogiques*. Paris: Dunod, 1959.

J. Pérès, L. Brillouin and L. Couffignal, "Les grandes machines mathématiques". AT 2/1948, no. 11, pp. 329–46; no. 12, pp. 376–85.

J. Perriault, *Éléments pour un dialogue avec l'informaticien*. Paris: Mouton, 1971.

H. Piéron, *Vocabulaire de la psychologie*. Paris: PUF, 1979.

R. Pilorge, *Comprendre l'informatique*. Paris: J. Delmas, 1969.

Y. Ploton (with M. Barrois and the Bull Historical Archive), *Itinéraires*. Paris: Bull Communications, 1992.

H. Poincaré, *La Science et l'hypothèse*. Paris: Flammarion, 1902.

H. Poincaré, *Science et méthode*. Paris: Flammarion, 1909

J. Poyen and J. Poyen, *Le Langage électronique*. Paris: PUF, 1960.

M. Pradines, *Traité de psychologie générale*. Paris: PUF, 1943–1946.

D. DE PRAT, *Traité de tissage au Jacquard*. Paris: Ch. Béranger, 1921.

O. PRIEB, *Rechenmaschinen in Büro*. Baden-Baden, 1955.

I. PRIGOGINE and I. STENGERS, *La Nouvelle alliance*. Paris: Gallimard, 1979.

J. RAMUNNI, *Physique du calcul*. Paris: Hachette, 1989.

F. H. RAYMOND, *Calcul analogique. Principes et contributions et théorie générale*. Paris: Société de la Revue d'optique, 1952.

F. H. RAYMOND, "Les calculatrices numériques universelles". MAF, 3rd series, 1955; 4th series, 1956.

F. H. RAYMOND, *L'Automatique des informations*. Paris: Masson, 1957.

F. H. RAYMOND, "Deux calculatrices SEA". OE, December 1960.

F. H. RAYMOND, *Les Principes des ordinateurs*. Paris: PUF, 1969.

J. REGNAULT, *Les Calculateurs prodiges*. Paris: Payot, 1952.

B. RENARD, *Le Calcul électronique*. Paris: PUF, 1960.

T. RIBOT, *L' Évolution des idées générales*. Paris: Alcan, 1897.

F. ROSE, *L'Intelligence artificielle: histoire d'une science*. Paris: Payot, 1986.

R. RUYER, *La Cybernétique et l'origine de l'information*. Paris: Flammarion, 1954.

J. G. SANTESMASES, *Obra e inventos de Torres Quevedo*. Madrid: Coleccion Cultura y Ciencia, Instituto de España, 1980.

A. G. SCHRANZ, *Addiermaschinen einst und jetzt*. Aix-la-Chapelle, 1953.

P. M. SCHUHL, *Machinisme et philosophie*. Paris: PUF, 1938.

H. SEBERT, "Rapport fait par M. Sebert, au nom du Comité des arts économiques, sur la machine à calculer, dite arithmomètre, inventée par M. Thomas (de Colmar) et perfectionnée par M. Thomas de Bojano. . .". BSEIN 1879: pp. 393–425.

H. SEBERT, "Rapport fait par M. Sebert, au nom du Comité des arts économiques, sur les machines à calculer de M. Léon Bollée, du Mans". BSEIN 1895: pp. 977–96.

N. SILLAMY, *Dictionnaire de la psychologie*. Paris: Larousse, 1967.

J. C. SIMON, *L'Informatisation de la société*. Paris: La Documentation française, 1980.

J. C. SIMON, *La Reconnaissance des formes par algorithmes*. Paris: Masson, 1984.

C. SOTTOCORONA, *Come si diventa specialista di informatica*. Milan: Mondadori, 1986.

R. TATON (editor), *Histoire générale des sciences* (4 vol.), Paris: PUF, 1957–1964.

R. TATON, *Histoire du calcul*. Paris: PUF, 1969.

R. TATON, "Sur l'invention de la machine arithmétique". RHSA, 16/1963: pp. 139–60.

R. TATON and J. P. FLAD, *Le Calcul mécanique*. Paris: PUF, 1963.

G. TAUSCHEK, *Die Lochkarten-Bucchaltungsmaschinen ...* Vienna, November 1930.

G. TORRES-QUEVEDO, "Les travaux de l'école espagnole sur l'automatisme". *Les Machines à calculer et la Pensée humaine. Colloque international du CNRS.* Paris, 8–13 January 1951: Éditions du CNRS, Paris, 1953, pp. 361–81.

G. TORRES-QUEVEDO, "Présentation des appareils de Leonardo Torres y Quevedo". *Les Machines à calculer et la Pensée humaine. Colloque international du CNRS.* Paris, 8–13 January 1951: Éditions du CNRS, 1953, pp. 383–406.

L. TORRES Y QUEVEDO, "Ensayos sobre Automatica. Su Definicion. Extension teorica de sus aplicaciones". RACE, 12/1913: pp. 391–418.

L. TORRES Y QUEVEDO, "Arithmomètre électromécanique". BSEIN, 199/ September–October 1920: pp. 588–9.

B. A. TRAHTENBROT, *Algorithmes et machines à calculer*. Paris: Dunod, 1963.

J. TRICOT, "1940–1950: Les premiers ordinateurs à programme enregistré". SV, no. 741, June 1979: pp. 114ff.

H. J. UFFLER, E. HONORÉ and E. TORCHEUX, "Le calculateur universel 'ANALAC 101'". OE, 405/December 1960: pp. 979–94.

M. VAJOU, "Alan Turing, ou la mécanisation de l'intelligence". SA, special issue, no. 49, September–October 1984.

P. VALÉRY, *Variété*, IV. Paris: Gallimard, 1939.

R. L. A. VALTAT, "Calcul mécanique: machine à calculer fondée sur l'emploi de la numération binaire". CRAS, 202/1936: pp. 1745–8.

G. VERROUST, *Structure d'un automate algorithmique universel: l'ordinateur*. Séminaire du CNSNSM, CNRS, Orsay, 1965.

G. VERROUST, *De la maîtrise du feu à une révolution de l'intelligence*. Conférence du MURS, Paris, December 1984.

G. VERROUST, *La Révolution informationnelle du XXe siècle, une étape fondamentalement nouvelle de l'histoire des hommes*. European Philosophy Conference, 24–28 June, Athens, 1985.

G. VERROUST, "Vers une civilisation de robots qui prendront la place de l'homme sur terre?". SA, Autumn 1992.

G. VIEILLARD, *L'Affaire Bull*. Paris: Spag-Chaix, 1968.

H. VIGNERON, "Les automates". NAT, 13 June 1914: pp. 56–61.

H. VIGNERON, "L'arithmomètre de M. Torres y Quevedo". NAT, 7 August 1920: pp. 89–93.

R. WILHELM, *Le Yi-King, le livre des transformations*. Paris: Librairie de Médicis, 1973.

K. ZUSE, *Der Computer – Mein Lebenswerk*. Munich: Verlag Moderne Industrie, 1970.

NO AUTHOR LISTING:

Instruments et machines à calculer. Catalogue du musée du Conservatoire national des Arts et Métiers. Section A. Paris, 1942.

Le Centenaire de NCR. Vol. 1 (*1884–1922, L'Ère de la caisse enregistreuse*). NCR Corporation, 1984.

Le Centenaire de NCR. Vol. 2 (*1923–1951, L'Ère de la machine comptable*). NCR Corporation, 1984.

Le Centenaire de NCR. Vol. 3 (*1952–1984, L'Ère de l'ordinateur*). NCR Corporation, 1984.

Terminologie du traitement de l'information. Paris: IBM France, 1987.

De la machine de Pascal à l'ordinateur: 350 ans d'informatique. Paris: Musée national des Techniques, 1990.

Informatique: nouveaux concepts scientifiques. Colloque en l'honneur de JeanClaude Simon. Paris, 3–4 October 1990.

INDEX